EARTH PRESSURE and EARTH-RETAINING STRUCTURES

Third Edition

EARTH PRESSURE and EARTH-RETAINING STRUCTURES

Third Edition

Chris R.I. Clayton
Rick I. Woods
Andrew J. Bond
Jarbas Mililitsky

CRC Press
Taylor & Francis Group
Boca Raton London New York

CRC Press is an imprint of the
Taylor & Francis Group, an **informa** business

A SPON PRESS BOOK

CRC Press
Taylor & Francis Group
6000 Broken Sound Parkway NW, Suite 300
Boca Raton, FL 33487-2742

First issued in paperback 2018

© 2013 by Taylor & Francis Group, LLC
CRC Press is an imprint of Taylor & Francis Group, an Informa business

No claim to original U.S. Government works

ISBN-13: 978-1-138-42729-7 (hbk)
ISBN-13: 978-1-4665-5211-1 (pbk)

Visit the Taylor & Francis Web site at
http://www.taylorandfrancis.com

and the CRC Press Web site at
http://www.crcpress.com

Contents

Preface

This book is about the calculation or estimation of the pressures that soil—or more often, soil and water—can apply to retaining structures. Retaining structures are used on almost all construction projects, so an understanding of their interaction with the ground is essential for structural and geotechnical engineers.

The first edition of this book was published in 1986, almost 30 years ago. The IBM PC had been launched in 1981, but affordable computing remained a little-known world, even for many engineers. So we covered the derivations of classical earth pressure equations and included sets of tables giving earth pressure coefficients.

In the second edition, published in 1993, we included additional material on compaction pressures and notes on the effects of wall installation methods on the lateral pressures to be supported by retaining structures. Gabion and crib walls had become widely used, as had soil reinforcement, so we included details on their design. Charts were provided to allow the preliminary sizing of cantilever and anchored sheet-pile walls and to estimate the effects of seepage on passive resistance. We introduced basic material on the numerical modelling of retaining structures. Tables of earth pressure coefficients were moved to an appendix, and the code for a slope stability program was included.

For this third edition, the text has been re-organised once more. However, we have attempted to retain as much material from the second edition as possible. The two parts of the book now deal with 'Fundamentals' and 'Design'. Previous chapters on the development of earth pressure theory and on graphical techniques have been moved to an appendix. Chapter 3 brings together and describes the wide range of possible interactions between the ground and a retaining wall. There are now a number of reliable and easy-to-use software packages dealing with seepage and slope instability, so in Chapters 4 and 5, we have assumed these to be accessible to the reader and have included material to allow the design issues to be understood, and computer output checked.

In the second part of the book, we have, as before, included descriptions of different types of wall. In the UK and elsewhere, Eurocode 7 is being implemented for most types of earth-retaining structure. In Chapter 7, we describe the background, but we have not revised the book in the expectation that all readers will wish to design in this way. However, the final three chapters now follow the Eurocode 7 structure in that they deal with gravity walls, embedded walls and composite walls. More recent material on propped and braced excavations has been included, as has work on soil nailing, anchored walls and cofferdams.

Chris R. I. Clayton
Rick I. Woods
Andrew J. Bond
Jarbas Mililitsky

Authors

Chris Clayton (BSc, MSc, DIC, PhD, CEng, FICE, CGeol, FGS) is Professor of Infrastructure Engineering at the University of Southampton, UK. After graduation, his early career was spent in industry working as a specialist geotechnical engineer in site investigation and civil engineering contracting. In 1972, he took a year off to study for his master's degree at Imperial College, which he completed with distinction. On his return to industry, he combined commercial and academic research and was awarded a collaborative PhD by the University of Surrey in 1978. On completion of his PhD, he joined the University of Surrey as a lecturer in geotechnical engineering, becoming Professor of Geotechnical Engineering in 1992. He joined the University of Southampton in 1999.

Chris has published over 200 papers, books, national reports and guidance documents, including graduate textbooks on 'site investigation' and 'earth pressure and earth-retaining structures', and national guidance (CIRIA/DETR) reports on procurement of ground investigation, on engineering in chalk, on the SPT and on 'managing geotechnical risk'. He has been the editor for the *Proceedings of Institution of Civil Engineers— Geotechnical Engineering, the Quarterly Journal of Engineering Geology and Hydrogeology, Géotechnique,* and the *Journal of the South African Institution of Civil Engineering.*

As a result of his geotechnical research, on retaining structures, site characterisation and fundamental soil and weak-rock behaviour, he has been invited to deliver a number of prestige lectures, including the British Geotechnical Association Géotechnique Lecture (1999), the Institution of Civil Engineers Unwin Memorial Lecture (2001), the South African Institution of Civil Engineering Jennings Memorial Lecture (2006), the British Geotechnical Association 50th Rankine Lecture (2010), and the 2nd Korean Geotechnical Society Lecture (2012).

Rick Woods (BSc, PhD, CEng, MICE, FHEA) is a senior lecturer in geotechnical engineering and an associate dean of faculty at the University of Surrey, UK. After graduating with a first-class honours degree (for which he was awarded the Institution of Civil Engineers Prize), he joined Binnie & Partners (now Black & Veatch) in 1981 and worked on a variety of civil and geotechnical engineering projects. He joined the City University (London) in 1984 as a research assistant before being appointed lecturer in geotechnical engineering. He moved to the University of Surrey in 1988, obtaining his PhD in 2004 for research into applications of finite-element analysis in retaining wall design.

Rick has taught, researched, consulted and published on a wide range of geotechnical topics—principally in the field of soil–structure interaction. He has served on the UK committee of the International Geosynthetics Society and the editorial panel for *Proceedings of the Institution of Civil Engineers—Ground Improvement*.

Andrew Bond (MA, MSc, PhD, DIC, MICE, CEng) gained first-class honours from Cambridge University in 1981 before working for WS Atkins and Partners on a variety of civil, structural and geotechnical engineering projects. He obtained his MSc with distinction from Imperial College in 1984 and his PhD in 1989, for pioneering research into the behaviour of driven piles and design of the Imperial College Pile. He won the London University's Unwin Prize and ASTM's Hogentogler Award for his research. Andrew joined the Geotechnical Consulting Group in 1989, becoming a director in 1995. While there, he developed the computer programs ReWaRD® (for embedded retaining wall design) and ReActiv® (for reinforced slope design).

In 1999, Andrew formed his own company, Geocentrix, for which he developed the pile design program Repute® in 2002. He has delivered a wide range of Eurocode training courses and lectures, both publically and privately, under the title 'Decoding Eurocode 7'. He co-authored the best-selling book of the same title in 2008.

Andrew was appointed chairman of TC250/SC7, the Eurocode 7 committee, in 2010, having served as a UK delegate since 1997. Andrew is a former editor of *Geotechnical Engineering*, part of the *Proceedings of the Institution of Civil Engineers*.

Jarbas Milititsky graduated with a diploma in civil engineering from the School of Engineering of the Federal University of Rio Grande do Sul (UFRGS–Brazil) in 1968. He received his MSc in geotechnics from COPPE at the Federal University of Rio de Janeiro (UFRJ–Brazil) and a PhD degree in foundation engineering from the University of Surrey, Guildford, UK, in 1983.

As a member of the academic community, he has been the dean of the Engineering School of the UFRGS and the chairman of the State Research Council (FAPERGS) of Rio Grande do Sul, Brazil. He was the president of the Brazilian National Geotechnical Society (ABMS) and a member of a number of scientific and community boards and councils. During his term as the head of the geotechnical group of the School of Engineering at UFRGS, he supervised MSc and PhD theses and published a number of journal and conference papers and books.

As a specialist geotechnical consultant, Professor Milititsky has participated in the design of foundations and retaining structures for refineries, bridges, industrial areas and wind turbines, and in the rehabilitation of historical monuments. He has served on a number of standardisation committees in Brazil, for example, the foundation design code of practice and the standard for pile testing.

Part I

Fundamentals

Chapter I

Soil behaviour

1.1 INTRODUCTION

An understanding of soil and groundwater is essential for the safe design of earth-retaining structures. Soil strength acts to reduce the load that must be carried by a retaining structure, whilst its stiffness significantly affects the ground movements that occur. This chapter considers soil behaviour. Chapter 2 discusses ground investigation and the determination of soil properties. Chapter 4 examines groundwater issues.

All engineering designs rely on the successful identification of situations and mechanisms which may make a structure, or a part of it, unfit for the purpose it is intended to serve. For example, a reinforced concrete beam might perform unsatisfactorily because the steel within it failed in tension, or the concrete failed in compression, but equally, it might prove to be too flexible, giving unacceptably large deflections under the loads applied to it.

The recognition of these so-called limit states is also of fundamental importance in soil mechanics, although few soil engineers notice that they are applying the same principles as their structural colleagues. In soil mechanics, the two common limit states occur due to the following:

1. Shear failure of the soil, leading to excessive distortion of a structure or disruption of highways and services
2. Excessive displacement of the soil, inducing unacceptably high stresses in a structure as a result of differential movements

Thus, routine soil mechanics problems divide into those where a prediction of displacement is required and those which attempt to calculate the reserve of strength left in the soil after the application of shear stress during construction. In this second case, the traditional lumped 'factor of safety' obtained from calculations is the ratio of the available shear strength divided by the applied shear stress. The 'partial factor' approach now used in Europe and elsewhere divides the factor of safety into components that can be applied both to applied loads (actions) and resistances (reactions) (see Chapter 7).

When ensuring that a 'structure' (whether composed of soil, reinforced concrete, steel or some combination of these) does not come to any limit states, the designer must envisage all the mechanisms which may lead to such unsatisfactory performance. In soil mechanics, this may require considerable imagination, partly because no two construction sites are the same and because the exact geometry and properties of the subsoil are never precisely known.

In structural design, statistics and probability theory are used to define the properties of materials, such as concrete. For example, the characteristic strength used in reinforced concrete design is based on the assumption that if a large number of concrete cube tests were carried out on a mix, their results would have a normal Gaussian distribution. The characteristic strength corresponds to the lower 95% confidence limit, i.e. the strength below which only 5% of test results should fall.

In soil mechanics, this type of approach is considered by many to be over-simplified and, in most cases, impractical. Soil can fail because the designer has not appreciated that the weakest 5% of the soil occurs in one location, rather than being spread evenly throughout, and is in precisely the most unfavourable place from the point of view of the structure. Thus, while the limit state concept is potentially useful in soil mechanics, an over-complex statistical approach to material properties (which often accompanies it in other areas of civil engineering design) may not be. A feel for the variability of soils and their likely properties requires some basic understanding of their origins, which are discussed in the next section.

1.2 ORIGIN, COMPOSITION AND STRUCTURE OF SOILS AND ROCKS

From an engineering geological point of view, materials can be categorised according to their origin as follows:

- Fresh rock
- Weathered rock and residual soil
- Sedimentary weak rock and soil
- Pedogenic soil

Soils and weak rocks are often particulate (i.e. formed of clay, quartz or calcite particles, for example). As a result of time and range of geological processes, these particles can become interlocked and cemented, causing significant increases in strength and stiffness.

1.2.1 Rock

Even when fractured, rock can generally stand unsupported at steep angles, for example, slopes of 70° or more. Some type of light structure or coating may be necessary to prevent weathering of rock (e.g. the use of '*chunam*'— a mixture of decomposed granite, cement and hydrated lime—in Hong Kong), but the design of these facings is not considered here. This book is aimed at the design of weaker near-surface materials, which are termed 'soils' by geotechnical engineers.

1.2.2 Weathered rock and residual soil

Weathering is the means by which rocks, soils and their constituent minerals are broken down as a result of near-surface processes. Weathering occurs *in situ*. It is often associated with erosion, which involves the movement of rocks and minerals by water, wind, ice and gravity. Weathered rocks can retain much of the strength and stiffness of the materials from which it is derived. There are two main types of weathering processes— mechanical and chemical.

Mechanical (also termed 'physical') weathering involves the breakdown of rocks and soils through stress relief, and contact with atmospheric conditions such as heat, water, ice and salt crystallization. The rock is broken down into blocks, with joints and fissures forming as planes of weakness. Mechanical effects in general dominate the breakdown of rock in temperate climates, whilst chemical effects dominate in tropical climates. Mechanical weathering effects tend to be relatively shallow (of the order of 10 m) whilst chemical weathering effects can be much deeper (of the order of 50 m).

Chemical weathering results from the effects of atmospheric or ground-water chemicals, or biologically produced chemicals (also known as biological weathering), and typically involves either dissolution or alteration of the rock. Chemical weathering attacks the rock itself, causing a general 'rotting' of the material and changes in its composition. 'Saprolite' is the term used for a chemically weathered rock. It is weaker than the unweathered material but commonly retains the structure of the parent rock since it is not transported but formed in place. Besides resistant relic minerals of the parent rock, saprolites contain large amounts of quartz and a high percentage of kaolinite, along with other clay minerals which are formed by chemical decomposition of primary minerals, mainly feldspars. More intense weathering conditions produce laterite and residual soils.

The term 'soil' is commonly used in engineering to refer to any kind of loose, unconsolidated natural near-surface material that can be easily separated into its constituent particles. Residual soils remain at the location at which they are formed by weathering. Transported soils (see Section 1.2.3) accumulate elsewhere, after weathering and erosion. Residual soils are therefore the final *in situ* product of chemical weathering and are common in tropical climates. They are derived from rock, but the original rock texture has been completely destroyed.

Dissolution is a quite different process, which occurs widely in carbonate rocks, both in temperate and tropical climates. Limestone and chalk are dissolved, along joints, bedding planes and other discontinuities, as a result of the slight acidity of rain water. The rock structure is loosened as a result of this action, large voids may form, and the bedrock surface typically becomes extremely irregular.

1.2.3 Sedimentary weak rock and soil

In common engineering parlance, 'soil' is any geotechnical material of low strength. The precise undrained strength limit between soil and weak rock is not universally agreed but, for practical purposes, can be taken as about 500 kPa. As a guide, at this strength, it becomes difficult to make an impression on a flat surface of the material using a thumbnail. The materials described in this section will have been transported and deposited during their formation, in contrast to those discussed in the previous section.

Because of this division on the basis of strength, geotechnical engineers tend to characterise, test and analyse alluvium, inter-glacial and glacial deposits and sedimentary weak rocks in much the same way, even though some soils may be hundreds of millions of years old and, as a result of diagenesis, have undergone significant cementing, whilst others are very young. Examples of sedimentary weak rock and soil found in the UK include

- Lower Cretaceous sands (heavily overconsolidated, these sands normally have deformed and interlocking grains, as a result of sustained loading, and may be cemented by iron or calcium carbonate)
- Chalk (found over much of northern Europe, in the Middle East and Texas)
- London clay (strictly part of the 'London Clay Formation', found in the Anglo-Paris basin, and consisting of very stiff clays and dense sands)
- Terrace gravel (formed during warmer, interglacial periods, on the sides of valleys in southern England and elsewhere)
- Glacial till (usually predominantly stiff clay, deposited during the ice age)
- River alluvium and lake deposits (consisting of compressible peat, loose sand and soft clays)

1.2.4 Pedogenic soil

A pedogenic soil is formed from sedimentary material that has been exposed near the ground surface in an arid environment for a sufficiently long period to develop structure, texture and mineralogy (leading to cementing and increased strength) through one or more physical, chemical or biological processes. Even though pedogenic processes take place at normal temperatures and pressures and in the physical/chemical environment at today's Earth's surface, the results can be similar to low-grade metamorphism in that the material becomes stronger and stiffer and eventually has properties similar to weak rock.

There are three types of pedogenic soil: ferricrete (bonded by iron oxide, hematite), calcrete (bonded by calcium carbonate) and silcrete (bonded by silica). The first sign of this rock forming is the presence of small nodules (about the size of a pea) distributed at approximately 10- to 100-mm centres in a zone about 1–3 m thick below ground surface. This zone is where the dissolved minerals concentrate due to evaporation in the upper zone of the capillary fringe. The nodules then become larger and later form clusters of poorly cemented material (e.g. ferricrete). This process continues until the final stage when hardpan is formed, i.e. large boulders or 'slabs' of rock, a few metres in diameter in a zone of approximately 1–3 m thick. Pedogenic soil can be strongly cemented, despite its recent origin.

1.2.5 Variability

The materials that a geotechnical designer may need to consider during the design of an earth-retaining structure vary from very soft clayey sediment, which will barely support the weight of a person, through to hard rock, with strength properties similar to those of concrete. Within each type of material on a given site, there will be further variability, both in composition and properties. This variability results not only from the nature of the depositional environment, which controls the type of material (e.g. gravel, sand, clay, peat) being deposited, but also from diagenetic processes (depth of burial, age, etc.) and weathering. This variability needs to be recognised and assessed during design.

In addition, the designer needs to recognise the possibility that the upper layers of the ground to be supported may consist of man-made ground. This material is likely to be highly variable, can be very weak and compressible in places and may contain contaminants (including toxic substances).

1.3 SOIL STRENGTH AND EFFECTIVE STRESS

Much of the earth pressure theory, and the behaviour of earth-retaining structures, is dominated by considerations of shear strength. Most structural engineers think in terms of strength characteristics (such as the compressive strength of concrete or the ultimate tensile strength of steel) that are constant and unaffected by ambient compressive stress levels. Geotechnical engineers, on the other hand, are concerned with materials that are composed of at least two and sometimes three phases (soil particles, water and air). Soils generally have relatively high compressibility and low strength when compared with other construction materials. Such strength as they have is highly stress dependent and frictional in its nature but is also a function of density.

Consider a block placed on a flat frictional surface and subjected to a normal force N (Figure 1.1). The maximum shear force T that can be applied horizontally to the block before it slides can be related to the coefficient of friction, μ, between the block and the surface, and to the normal force. In soil mechanics terms, we would write

$$T = N \cdot \tan \phi \tag{1.1}$$

or by dividing by the contact area to obtain stresses,

$$\tau = \sigma.\tan \phi \tag{1.2}$$

where $\tan \phi$ is equivalent to the coefficient of friction, μ.

Soils behave in a similar if more complex way. First, they are usually a mixture of soil particles and water (and possibly air—see succeeding text). An element of saturated soil under external pressure ('total stress') will contain water which is also under pressure ('pore water pressure'). Consider a sealed rubber balloon full of soil and water (Figure 1.2). The total stress applied to the outside of the balloon is carried partly by the pore water pressure, and only the difference between the pore water pressure and the total stress is applied to the soil structure, i.e. to increase the forces between individual particles. Because, at normal rates of shear, water has negligible shear strength and its properties are unaffected by pressure increases, the pressure taken by the pore water does not contribute to the overall strength of the soil.

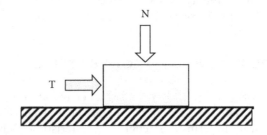

Figure 1.1 Sliding block analogy.

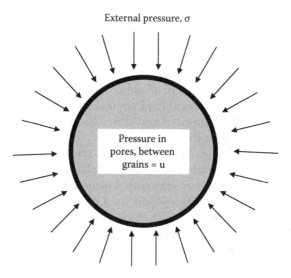

Figure 1.2 Total and effective stress.

For saturated soil, the numerical difference between total stress and pore water pressure is termed as the 'effective stress', σ', where

$$\sigma' = \sigma - u \tag{1.3}$$

In much the same way as with the block sliding on a surface, an increase in inter-particle force leads to an increase in shear resistance:

Increase in total stress and constant pore pressure
or
Constant total stress and decreasing pore pressure

$\Rightarrow$ strength increase

and conversely

Constant total stress and increasing pore pressure
or
Decreasing total stress and constant pore pressure

$\Rightarrow$ strength decrease

Effective stress has an additional effect, particularly in soils with platy or clayey particles. It causes the soil particles to pack more closely together. A decreased porosity (and therefore water content) resulting from an increase in effective stress produces a further increase in strength. It is observed that if a soil exists at the same effective stress level, but different water contents, the lower the water content and the higher the density, the higher the strength.

The strength of soil is routinely measured in the triaxial apparatus (Figure 1.3). The soil specimen is sealed in a rubber membrane, pressurised by cell water, and has additional vertical load applied to it by a ram, until failure occurs. The axial load is measured and this, divided by the cross-sectional area of the specimen, gives the 'deviator stress', $(\sigma_1 - \sigma_3)$, which is plotted as a function of axial strain. Figure 1.4 shows typical stress/strain curves. Drained tests on dense sands and undrained tests on carefully sampled natural clays tend to produce results with a pronounced peak, whilst loose sands and remoulded normally consolidated clays do not. The results of such tests are plotted as Mohr circles of effective stress at failure, as shown in Figure 1.5. Each Mohr circle represents the stresses on a single specimen at failure. The position of the circle is defined by the minor effective principal stress at failure

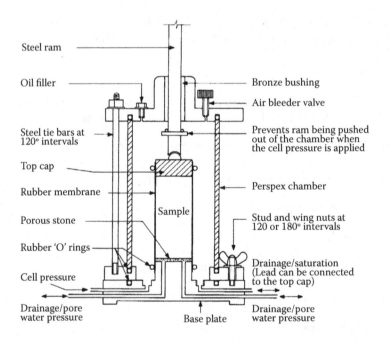

Figure 1.3 Triaxial apparatus. (From Clayton, C.R.I. et al., Site Investigation, 2nd ed. Blackwell Scientific, Oxford, 1995. Downloadable from www.geotechnique. info.)

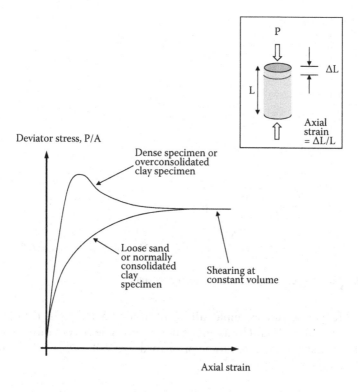

Figure 1.4 Generalised stress–strain behaviour of granular soil in a drained triaxial test.

$\sigma_3' = $ (cell pressure – pore water pressure)

and the major effective principal stress at failure

$\sigma_1' = \sigma_3' + $ (deviator stress)

which as an example are labelled on Figure 1.5 for one of the circles. The failure envelope for five circles is shown as a dashed line in Figure 1.5. Typically, it is curved. As a result, triaxial tests should be carried out at approximately the normal effective stresses in the field, which are often low, of the order of 20–100 kPa. In conventional interpretation, the results from three triaxial test specimens (for example, shown by the full circles in Figure 1.5) are interpreted using a best fit straight line envelope (shown by the full line in Figure 1.5) to determine values of effective cohesion intercept, c', and effective angle of friction, ϕ'. Testing at unrealistically high effective stresses leads to high values of effective cohesion intercept, c', which is unsafe.

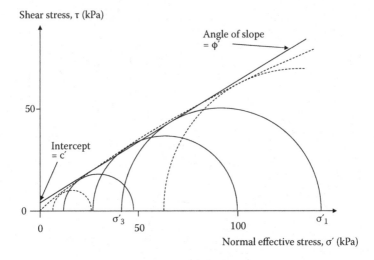

Figure 1.5 Example effective stress triaxial test results.

For soft, young (for example, alluvial) or compacted soils, the effective cohesion intercept should be assumed to be zero. The effective angle of friction of clay may be expected to be of the order of 20°–30°, whilst that of a sand or gravel will exceed 30°.

1.4 DILATANCY AND THE CRITICAL STATE

In soils composed of more bulky particles, such as most sands and gravels, the packing of the material can make a significant contribution to its strength, as a result of dilatancy. Figure 1.6 shows a schematic diagram of spherical soil particles under shear. Initially, the particles are densely packed (Figure 1.6a). As the shear force is applied, the particles must either ride over each other or must break. Under low effective stresses (relative to their intact strength), the particles tend to ride over each other, doing work against the confining stress, and producing a higher (peak) strength than if they were in a loose packing (Figure 1.4). Once they achieve their loosest packing (Figure 1.6b), they can continue to shear at constant volume, without the additional effects of dilatancy. The combination of effective stress level and void ratio at which shearing at constant volume takes place is known as a 'critical state'. Because it does not allow for the effects of packing, interlocking or bonding, a critical state effective angle of friction, ϕ'_{crit}, will give a conservative (low) estimate of soil strength. Critical state theory is widely used for clays as well as sands.

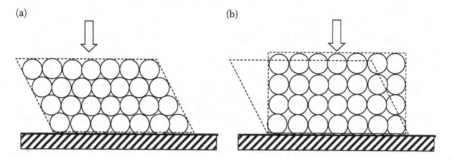

Figure 1.6 Dilatancy. (a) Before shearing. (b) After shearing to constant volume.

1.5 STRENGTH ON PREEXISTING FAILURE PLANES

If clays are subjected to very large displacements on confined rupture surfaces, as happens when landslides move slowly down slope for many hundreds or thousands of years, then their platy particles align with the failure surface. The two sides of the failure surface become polished, and the shear strength becomes much reduced, tending toward what is termed as the 'residual' strength. The residual effective stress strength envelope may have zero effective cohesion, but it is often curved. For plastic clay, the residual effective angle of friction may be as low as 8°–10°.

Since retaining structures are often constructed to retain soil in sidelong ground, it is likely that from time to time the preexisting failure planes of ancient landslides will be encountered. The residual effective angle of friction can be determined using a 'ring shear' apparatus. As with the determination of peak effective strength parameters (see above), it is vital that tests are carried out at very low effective stress levels, similar to those in the field at the level of the failure planes.

1.6 SOIL STIFFNESS AND GROUND MOVEMENTS

If, as is common during design, a large factor of safety is applied to the available peak soil strength, then the shear stress may be restricted to 1/2 or less of the available peak strength. Soil strains will become small, and an equivalent Young's modulus (E) or shear modulus (G) may be used with elastic stress distributions or computer modelling, to predict the deformation that will occur if the soil is loaded or unloaded. This situation is common where embedded retaining walls are used in inner-city sites to prevent ground movements that would otherwise damage adjacent existing buildings and infrastructure such as tunnels.

Unfortunately, the stiffness of soils is more complex than that of steel. First, its stiffness is strongly dependent on effective stress, and second, its stress–strain behaviour is non-linear, so that stiffness relevant to the expected strain levels in the soil needs to be used in calculations. Finally, its stiffness is loading path and loading history dependent, and there is some evidence that it is also affected by the strain rate during testing. However, despite all this complexity, it has been found that non-linear stress–strain models can deliver useful predictions of ground movements around basements in stiff clays.

The measurement of small-strain stiffness requires high-quality sampling and advanced laboratory testing. Hall effect devices or submersible linear variable differential transformers (LVDTs) are mounted on the side of the specimen in order to avoid bedding and apparatus compliance effects. Figure 1.7 shows the results of an undrained triaxial compression test carried out on three soils using local small-strain instrumentation. Note

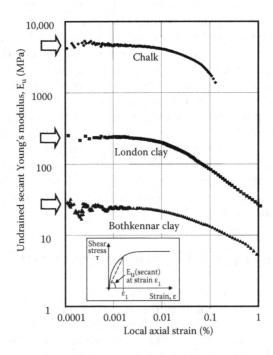

Figure 1.7 Triaxial small-strain stiffness measurements for three soils: soft/firm Bothkennar clay, London clay, and weak chalk. (Redrawn from Clayton, C.R.I. and Heymann, G., Géotechnique, 51, 3, 245–256, 2001.)

the logarithmic scales used both for local axial strain and stiffness. The materials all produce approximately linear stress–strain behaviour up until axial strains of the order of 0.02%–0.03%, after which their stiffnesses decreases until failure occurs. Numerical modelling (Jardine et al. 1986) has shown that the dominant strains behind a strutted retaining structure can be expected to be low, and typically between 0.01% and 0.1%. The stiffness of soil at 0.01% strain is typically between 0.8 and 0.5 times that at very small strains (shown by the arrows in Figure 1.7) and then decreases to about 40% of this value as strains increase from 0.01% to 0.1%.

Figure 1.8 shows the effect of loading path direction on the stiffness measured on a specimen of London clay. In the inset box, the loading paths are shown in MIT ($t = (\sigma_1 - \sigma_3)/2$, $s' = (\sigma'_1 + \sigma'_3)/2$) stress space. It can be seen that loading toward triaxial compression (labelled 1) produces much less reduction than loading toward extension (labelled 2). Results such as these show that the stiffness of soil is stress-path dependent, and that if better accuracy of movement prediction is required, it is necessary to mimic in the laboratory the loading paths to be applied in the field.

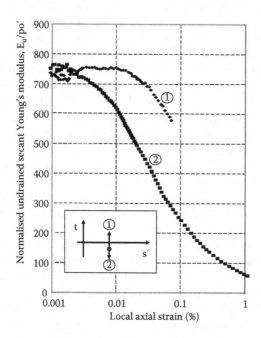

Figure 1.8 Triaxial small-strain stiffness measurements for London clay, loaded towards compression and towards extension. (Redrawn from Clayton, C.R.I. and Heymann, G., Géotechnique, 51, 3, 245–256, 2001.)

There are four principal issues that need to be addressed when assessing ground movements around retaining structures (see Chapter 11):

1. What will be the contribution of wall installation effects?
2. What strain level(s) should be considered when determining soil stiffness?
3. How non-linear will the stress–strain behaviour of the soil be within that range?
4. Are ground conditions sufficiently uniform and well understood to justify determining soil stiffness using representative loading (stress) paths?

1.7 CONSOLIDATION AND SWELLING—'SHORT-TERM' AND 'LONG-TERM' CONDITIONS

Construction on, or in, the soil normally involves stress changes at the boundary between the soil and the structure. For example, footings for an office block will normally apply an increased stress to the soil, while excavation to form a motorway cutting will (because soil is excavated) result in a stress decrease.

Total stress changes at the boundary of a sealed specimen of soil will result in pore pressure changes within the soil. If the soil is 'saturated' (i.e. it contains no free air), the pore fluid within the soil skeleton will be very rigid compared with the soil skeleton. Of course, the individual soil particles will be less compressible than the water, but since soil compression occurs as a result of expulsion of pore fluid due to a change in the arrangement of the soil particles, this is unimportant to the process.

We can use a spring and dashpot analogy (Figure 1.9) to describe what happens when saturated soil is loaded. With the valve closed, if a weight is placed on the piston, water cannot escape. Since the water is incompressible compared with the spring, and since the spring must compress if it is to carry additional load, all the weight is supported by an increase in water pressure. The piston does not move, indicating that no volume change occurs, and the load carried by the spring does not change, indicating that the effective stress and hence the strength of the soil remains unchanged.

If the valve is now opened, water will flow out of the container until the water pressure dissipates to its original value, in this case atmospheric. As water escapes through the valve, the piston will move downward, indicating that the soil is changing volume ('consolidating'). Thus, compression is associated, in this case, with a water pressure decrease. Since the weight has not been removed, and the water pressure no longer supports it, its load is thrown onto the spring. Thus, a volume decrease is associated with an

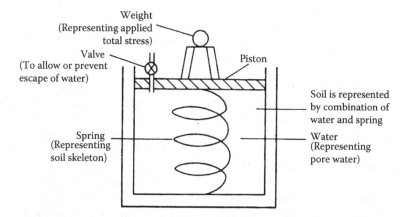

Figure 1.9 Spring and dashpot analogy of consolidation of soil. (From Clayton, C. et al., *Earth Pressure and Earth-Retaining Structures*, Second Edition, Taylor & Francis, New York, 1993.)

effective stress increase (an increase in the spring load) and therefore an increase in strength.

In practice, soil in the ground is not normally bounded by a totally impermeable barrier. Therefore, as soon as loads are applied, consolidation will start. Conversely, as soon as loads are removed, the soil will start to swell. It is the rate at which swelling or consolidation occurs which is significant. The permeability of soil to water can vary through 10 orders of magnitude, with water flowing out of the clay at perhaps 10^{-10} m/s and gravel at 10^{-1} m/s, for a hydraulic gradient of unity. It therefore takes a very long time for the water to be squeezed out of the clay beneath a large foundation, or to be sucked into the soil beneath a highway cutting or basement excavation. Whereas construction of a building might take 18 months, it could take 15 years for a substantial proportion of the final settlement due to consolidation to occur. Thus, for a clay, it is reasonable to assume that material in a zone of changing stress is unable to change volume, and remains 'undrained' at least in the short term until the 'end of construction'.

Since the strength of saturated soil is a function of effective stress and moisture content, and since neither can change if volume does not change, geotechnical engineers carry out analyse for two cases:

1. 'Short term', or 'end of construction', when the maximum shear stresses are applied to the soil, but there has been little time for consolidation or swelling. The soil strength is assumed not to have changed from the original value. Tests can be carried out before the start of construction, to measure the initial strength of the soil.

2. 'Long term', when the shear stresses and total stresses due to construction have been applied, all volume changes due to consolidation or swelling have occurred, and the groundwater in the soil is assumed to have come to an equilibrium level.

As an example, consider a shallow foundation for an office block. It applies a total stress increase to the clay subsoil, and positive pore pressures are induced (Figure 1.10). In the short term, the shear strength of the soil will remain unchanged, but shear stresses will be applied to the soil by the foundation. As consolidation occurs, water will be driven out of the soil, and its strength will increase. If we were (simplistically) to define a factor of safety against failure of the soil beneath the foundation as the ratio (available shear strength/applied shear stress), it can be seen that this 'factor of safety' decreases during construction (as more shear stress is applied) but increases after construction as consolidation occurs. The critical period, when the factor of safety is at a minimum, occurs at the end of construction (Figure 1.10), and it is not normally necessary to carry out an analysis for the long-term case.

A second group of problems exists which requires analyses in the long term, because this is the critical time for stability. Most earth pressure determinations fall into this category. Consider a form of construction such as a cutting for a motorway, where the total stress in the soil is reduced by the work carried out (Figure 1.11). As a result of a reduction in total stress, swelling will eventually occur, and the soil will lose strength in the period of pore pressure stabilisation between the end of construction and the long term. Clearly, the prudent engineer will design his structure for the long-term case, when the factor of safety is lowest.

There is, however, the problem of temporary works, such as cuttings or earth-retaining structures, which are only required to function during the construction period. Temporary works constitute a considerable part of the cost and risk associated with construction. They are normally designed by the contractor and might, for example, be cuttings or retaining walls required during excavation for a basement. It is tempting to analyse such cases in the short term, because they will not be used beyond the end of construction, but in reality, this is a risk. The simplified models in Figures 1.10 and 1.11 ignore the fact that some drainage of pore water will occur during the construction period. In cases where construction involves unloading (Figure 1.11), it is impossible to predict with certainty the rate at which pore pressures will rise, and there are very few case records to give guidance. It may be that the real situation approaches the dotted line in Figure 1.11, and the engineer should therefore be cautious, and carry out a long-term analysis.

The degree to which drainage and dissipation of excess pore water pressures (set up by loading or unloading) occur during the construction period

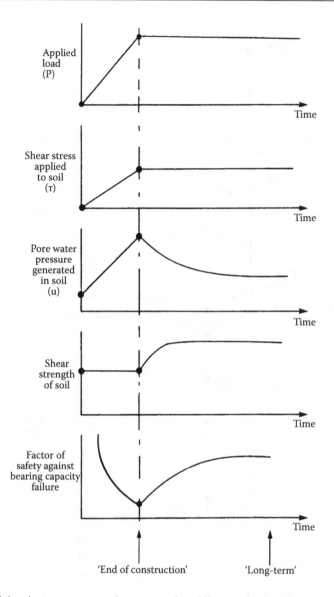

Figure 1.10 Load, pore pressure, shear strength and 'factor of safety' for a clay beneath an embankment. (From Clayton, C. et al., *Earth Pressure and Earth-Retaining Structures*, Second Edition, Taylor & Francis, New York, 1993.)

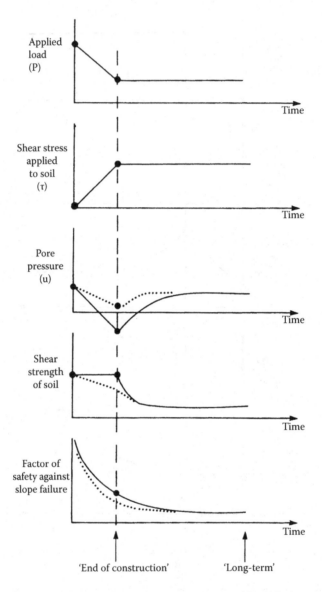

Figure 1.11 Load, pore pressure, shear strength and 'factor of safety' for a clay beneath a motorway cutting slope. (From Clayton, C. et al., *Earth Pressure and Earth-Retaining Structures*, Second Edition, Taylor & Francis, New York, 1993.)

is a function of a number of factors, such as soil particle size, fabric (i.e. fissuring, presence of silt or sand laminations in clays), the availability of free water (either from the ground surface as a result of rainfall, or because of the existence of a high water table), and the time taken for construction to be completed (which may be significantly increased if unforeseen problems occur during construction).

It is clear, however, that granular soils (clean silts, sands and gravels) have such a high permeability that full dissipation of pore pressures will occur during the construction period, and certainly in this case, it will be unrealistic to carry out a short-term, end-of-construction analysis which assumes that effective stresses and volumes remain unchanged.

1.8 CONSEQUENCES FOR ENGINEERING DESIGN

As we have just seen, soils can be divided into those that clearly will undergo rapid dissipation of excess pore pressures set up by loading or unloading (e.g. granular soils), and those that will not (i.e. cohesive soils). As far as engineering design is concerned, loading cases can be divided into those which increase the total stresses in the soil (which give a lower factor of safety at the end of construction than in the long term), and those which decrease total stresses, where the long-term factor of safety is the lowest. The limiting factors considered in the design may be deformation, or total failure of the soil (as, for example, the settlement and safe bearing pressure of a foundation). The principal assumptions, and their consequences for design, are shown in Table 1.1.

1.9 STRUCTURED SOILS

The previous sections of text have made the assumption that the soil loading a retaining wall is saturated and, therefore, that the effective stress (the stress controlling the strength and compressibility of the soil) can be assumed to be the numerical difference between the total stress and the pore water pressure. Moreover, it has assumed that the soil behaves as a granular, uncemented material. In reality, many natural and compacted soils are unsaturated, and many natural soils benefit from some kind of inter-particle bonding or cementing.

In reality, smaller retaining structures, and structures constructed in arid environments, support a significant amount of unsaturated soil. In such material, there is a pore air pressure as well as a pore water pressure, and the strength and compressibility of such soils are controlled by the net normal stress (the difference between total stress and pore air pressure) as well as the matric suction (the difference between the pore air pressure and the

Table 1.1 Short- and long-term behaviour, and consequences for design

Behaviour	Consequences for design
'Short term' or end of construction	
(1) Assumes insufficient time for drainage of water from soil during construction period.	Short-term design is inappropriate for granular soils. Always use long-term, effective stress analyses.
(2) No drainage implies no volume change and no significant effective stress change in the soil.	Shear strength remains the same as it was before the start of construction. Use total stresses and undrained shear strength for stability assessment.
(3) Displacements and soil boundaries occur as a result of change in soil shape, without change in volume.	Short-term settlements and displacements may be estimated from elastic stress distribution theory, coupled with Young's moduli from undrained compression tests.
(4) Loading increase produces positive pore pressures, which dissipate with time, giving a strength increase in the long term. For this case the factor of safety against shear failure is lowest in the short term.	Structures applying a load increase to clay are normally analysed for stability only in the short term.
(5) Some dissipation of excess pore water pressures will probably occur during the construction period, even in clay.	It is unwise to use total stress, short-term design methods for temporary works, in unloading situations.
'Long term'	
(1) Following a load increase or decrease, the soil will consolidate or swell, and its strength will also change.	Long-term analysis cannot be based on the available shear strength before construction. Effective stress must be used in connection with c', ϕ' and an equilibrium pore pressure to obtain a new value of shear strength.
(2) If there is a change in geometry or boundary conditions, the equilibrium groundwater conditions will probably change.	It will be necessary to predict long-term pore pressures in the soil, before its strength can be obtained. This will probably be difficult, and will lead to imponderables in the design calculations.
(3) If construction applies a load decrease, pore water pressure within the soil will drop. In the long term the soil will swell, and the factor of safety against shear failure will fall after the end of construction.	Structures applying a load decrease should be analysed for the long-term case, using effective stress analysis.
(4) Some dissipation of excess pore pressures will probably occur during construction.	It is unwise to use total stress, short-term design methods for temporary works.

pore water pressure). Menisci at the pore throats in the soil add an element of inter-particle force that strengthens and stiffens the material, helping the stability of a retaining structure.

Strength and compressibility are improved as natural material becomes cemented, and as a result retaining wall performance is likely to be better than expected on the assumption of uncemented material. However, both unsaturated and cemented behaviour are difficult to guarantee. Testing of unsaturated materials is highly complex, and the effects of lack of saturation may be removed as a result of flooding and heavy rainfall. Cementing is often spatially variable, requiring numerous complex effective stress tests to determine design parameters. For this reason, the effects of structure are not normally incorporated in retaining wall design.

Chapter 2

Soil properties

This chapter considers the soil data required for earth pressure and soil deformation analyses, and methods of obtaining them.

2.1 SOIL MODELS USED FOR EARTH PRESSURE AND RETAINING STRUCTURE ANALYSIS

In order to carry out geotechnical analysis of a retaining structure, the engineer needs to adopt one or more models of soil behaviour, and then determine the parameters for the model selected, and for the ground at the site of the proposed structure(s). This section describes commonly used soil models. In the following section, methods of determining the parameters for those models are discussed.

In principal, the simplest soil model that will deliver good predictions should be used for any analysis. The sophistication necessary or feasible will be controlled by a number of factors. Examples are as follows:

- The requirements of the analysis and of the design (e.g. whether calculations aim to avoid failure or, in addition, aim to predict displacements)
- Soil and groundwater complexity and spatial variability
- Availability of personnel, equipment and experience needed to determine the parameters

These issues will be discussed for each of the types of soil models commonly used, in the sections below.

The most frequently used soil models can be classified as follows:

- Rigid plastic models
- Winkler spring models
- Elastic (linear, non-linear, inhomogeneous and cross-anisotropic) models
- Elasto-plastic models

In addition, strain-softening behaviour and progressive failure need to be considered, as does the possibility of preexisting (much weaker, residual strength) failure surfaces existing on the site.

2.1.1 Rigid-plastic soil models

Figure 2.1 shows a simplified view of the deflection of a rigid cantilever embedded wall moving under ground loading. Where the wall moves toward the soil (below excavation level), the horizontal pressure (σ_h) is increased, whilst the vertical stress (σ_v), which is controlled by self-weight and wall friction, remains fairly constant. As the wall moves in on the soil, the shear stress (approximately ($\sigma_h - \sigma_v$)/2) changes and eventually, when the applied shear stress reaches the available shear strength, soil failure will occur. This condition (failure with $\sigma_v < \sigma_h$) is known as the passive earth pressure state.

Where the wall moves away from the soil (on the retained side of the wall), the vertical stress is again more or less controlled by the mass of soil being supported; the reduction of horizontal stress also leads to an increase in shear stress in the soil, and eventually to soil failure. Failure of the soil with $\sigma_v > \sigma_h$ is known as the active earth pressure state.

Analyses based on classical earth pressure theories for active and passive conditions consider that the strength of the ground is fully mobilised regardless of the amount of wall movement, once movement of the wall takes place either away from (active state) or toward (passive state) the

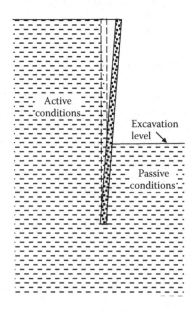

Figure 2.1 Active and passive states for a propped embedded wall.

soil. As soon as the wall moves toward the soil, for example, the full soil strength is mobilised to produce passive failure conditions. This is equivalent to considering the soil material to be 'rigid plastic', as shown in Figure 2.2a. Because the soil is modelled as rigid, no strains (or displacements) need to occur before failure is reached.

Rigid plastic models require only a definition of failure strength, and then only in two-dimensional (plane strain) conditions. In short-term (end-of-construction) conditions the strength of clay is assumed to remain essentially the same as before construction.

$$\tau_f = c_u \ (\phi_u = 0) \tag{2.1}$$

whereas for sands and gravels (non-cohesive soils), where drainage of water to dissipate any excess pore pressures set up during construction occurs immediately, the available strength of the soil depends upon the effective strength parameters (c′ and φ′), as well as the effective stress (σ′), which is the numerical difference between the total stress (σ) and the pore pressure (u):

$$\tau_f = c' + \sigma'.\tan \phi' = c' + (\sigma - u).\tan \phi' \tag{2.2}$$

The drained condition is also relevant for clays in the long term (when dissipation of excess pore pressures is complete).

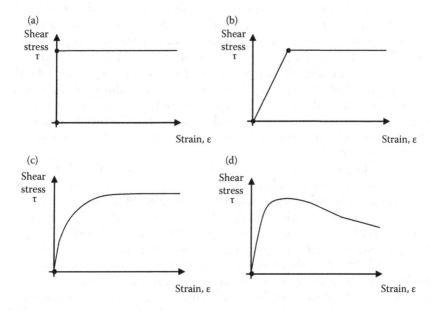

Figure 2.2 Stress–strain curves for some soil models used in retaining structure analyses. (a) Rigid-plastic behaviour; (b) linear elastic plastic behaviour; (c) nonlinear elastic plastic behaviour; (d) strain-softening behaviour.

Therefore the simplest models require knowledge of

 i. The direction of wall movement at all levels on either side of the wall
 ii. Variation of pore pressure (u) with depth, and either
 iii. Undrained shear strength (c_u) in clay, or (more usually)
 iv. The effective strength parameters (c' and ϕ') as a function of depth

Earth pressures calculated on the basis of a rigid plastic model can produce significant overestimates of passive pressures, as noted by Rowe and Peaker (1965). Strain softening (Figure 2.2d) is often observed when testing dense granular soils or natural clays. Compressible soil undergoes 'progressive failure'; different elements of soil around the wall will be at different strain levels, and therefore mobilising different proportions of the shear strength of the soil. Since at no time can all the soil be at the strain necessary to mobilise the soil's shear strength, a rigid-plastic model will overestimate the stability of a retaining structure. This concept is discussed in Chapter 3.

2.1.2 Winkler spring models

Winkler spring models are often used in situations when earth pressures are required in order to determine shear forces, bending moments and anchor or prop loads, but the ground geometry is too complex to justify the use of full continuum analysis. Winkler spring models (Figure 2.3) assume that the pressure on a retaining wall is a function of its horizontal displacement toward or away from the soil, up to the point that (active or passive) failure conditions are reached. The wall is modelled as a beam, either using a finite difference or a finite element approximation. Without calibration with local experience, such models are not considered reliable for predicting the movements in the ground and at the ground surface behind embedded retaining structures, which are required in order to estimate the effects on adjacent (existing) construction.

Winkler spring models are a major simplification of wall/soil behaviour. In reality, the size of a loaded area as well as the stiffness (and strength) of the material it bears upon will control displacement. For a retaining wall, differences in overall stress distribution (as well as the stress at a particular section of the wall) will have an effect on the wall displacements. Nonetheless, the evidence suggests that these relatively simple models can give good estimates of the deformed shape of embedded walls, their prop or anchor loads, and shear force and bending moment distributions. Winkler analyses have been successfully used for many major retaining structures, e.g. for multi-propped embedded retaining walls, such as those originally constructed for the World Trade Center in New York, where the depth to rockhead was highly variable.

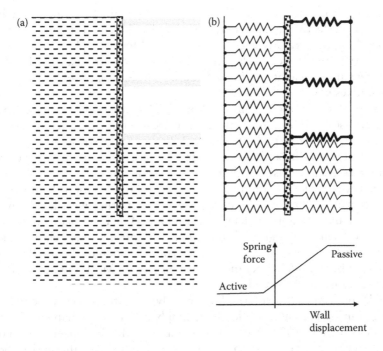

Figure 2.3 Winkler spring idealisation of a retaining wall. (a) Envisaged wall geometry; (b) analytical idealisation.

As can be seen from Figure 2.3b, the Winkler model requires an estimation of the spring stiffness at the different levels on both sides of the wall. These are usually based upon measured undrained shear strength (for clays) or density and stress level (for sands and gravels). Computer programs that use this form of analysis will normally give guidance based upon experience. In addition, the limiting (active or passive) forces must be estimated at each level. Computer codes generally do so by calculating active and passive pressures from either total or effective strength parameters, as for simple rigid-plastic models.

2.1.3 Elastic models

Elastic models are often used (typically in conjunction with plastic soil behaviour—see below) to make estimates of wall and ground movements for embedded walls in inner-city sites. Soil conditions may be modelled as

- Linear elastic, with or without
- Non-homogeneous stiffness (typically stiffness increasing with depth)
- Cross-anisotropic stiffness
- Non-linear elasticity

Figure 2.2b gives an example of the stress–strain behaviour of a linear elastic perfectly plastic material. Up until yield (equals failure for most soil models), the strain is linearly proportional to the applied shear stress. Beyond this, the soil can take no further load and strains, and therefore displacements accelerate as load shedding to other parts of the soil-wall system takes place.

Linear elastic models require estimates of stiffness parameters, which in natural soils are usually a function of location, depth (effective stress), strain level, and stress path. As Figure 2.2c shows, the stress–strain behaviour of soil is non-linear from small strain levels. Therefore if linear elastic soil models are to be used, a characteristic strain must be estimated, either for the entire structure, or varying with depth and location. Estimates of stiffness can be obtained by using stress path testing with local (small) strain instrumentation (see later in this chapter) or from back analysis of similar walls in similar ground conditions. An isotropic linear elastic model requires two stiffness parameters; either Young's modulus, E, and Poisson's ratio, v, or bulk modulus, K, and shear modulus, G. Most, but not all, computer codes use an E, v model.

Under undrained (short term, for clay only) conditions there is no volume change during loading or unloading, simply a change in shape of the soil as it deforms under the changing shear stresses. In the longer term, both E and v change in response to volumetric straining. The alternative shear modulus/bulk modulus (G, K) model has the advantage that the transition from undrained to drained behaviour can be modelled by keeping the shear modulus of the soil constant and reducing the bulk modulus from that of water to that of the soil skeleton.

Non-homogeneous stiffness is also normally required, even in apparently uniform soils, because soil stiffness is a function of effective stress, and effective stress increases with depth. The rate of increase in stiffness with depth can be estimated from tests on materials at different depths, or from back analysis of similar wall types in similar ground conditions. For example, back analysis of basement excavations in the London clay formation in central London has suggested (Hooper 1973)

$$E = 10 + 5.2\,z \tag{2.3}$$

where E is the Young's modulus in MPa and z is the depth below ground level in metres. Non-homogeneity can also arise because of soil layering. If reasonable estimates of the order of magnitude of ground movements are to be obtained, then non-homogeneity must be included in any analysis. Most soils are also layered, with different strength and stiffness properties in each layer.

Cross-anisotropic stiffness is important when analysing embedded walls in heavily overconsolidated soils because the *in situ* stresses before excavation are greater, near surface, in the horizontal direction than in the vertical. The horizontal stiffness is greater than the vertical, and therefore, for a given stress reduction as a result of excavation, horizontal strains will be lower than vertical. However, a cross-anisotropic stiffness model requires five stiffness parameters (for example, E_v, E_h, v_{vh}, G_v and G_h, but see Clayton [2011]). Young's modulus is normally measured in the vertical direction and estimated in the horizontal. A value of Poisson's ratio is required, plus two values of shear modulus. Opinions vary as to the importance of estimating these accurately, and in practice they are often assumed, without any attempt at measurement.

Non-linear elasticity is now widely used in numerical modelling in the UK and many other parts of the world, particularly in situations where ground movements adjacent to a retaining structure are considered important. A number of models exist. It is convenient to use the undrained Young's modulus, measured as the secant to the stress–strain curve (Figure 1.7 inset) and to normalise this by the initial effective stress in the specimen at the start of testing (for an example, see Figure 1.8). The stiffness decrease between 0.01% and 0.1% strain is also required. As a check, triaxial tests with small strain measurements frequently give $(E_u)_{sec}/p'_0$ at 0.01% axial strain in the range 500–1000.

2.1.4 Plasticity and failure

Most geotechnical modelling programs routinely incorporate a number of yield and failure functions. Because the analysis involves stresses in three directions, these must work in 3D space, even though the structure may be (and often is) simplified to a 2D (plane strain) problem. The most common failure conditions modelled are defined by the

- Tresca
- Modified Tresca
- Mohr–Coulomb criteria

In addition, Camclay uses work-hardening plasticity to model the effects of volumetric strains occurring under drained conditions.

For most retaining structures for which ground movements are an issue, there should be a sufficiently large margin of error that the inclusion of plasticity in the model will have relatively little effect, since only a small volume of ground will be yielding. The use of a Mohr–Coulomb criterion is convenient, since estimates of the effective strength parameters c' and ϕ' will normally be made as a routine for any retaining wall design.

2.2 SITE INVESTIGATION AND ACQUISITION
OF SOIL PARAMETERS

This section gives a brief introduction to the types of geotechnical parameters required for earth pressure calculation and retaining wall design, and
considers the equipment and techniques that are commonly used to obtain
these parameters.

Site investigation (also termed 'ground investigation') is the process by
which the ground conditions on a site and their characteristic parameters
are determined. A key component of ground investigation is the desk study
and walk-over survey, where all the available existing information on a
site is acquired and interpreted, including topographical and geological
maps, records of previous site investigations, and current and previous land
use (including archaeological heritage and the possibility of contaminated
land). These data are also used to plan the direct ground investigation,
using boreholes, etc. UK national guidance on site investigation is given in
the Site Investigation Steering Group documents (revised in 2013), and in
the Institution of Civil Engineers document on '*Managing Geotechnical
Risk*' (Clayton 2001). Guidance on desk studies and walk-over surveys, as
well as a comprehensive guide to site investigation processes, can be found
in Clayton et al. (1995).

Earth-retaining structures are often conceived of, or introduced into
a design scheme, late in the project design process. They are also often
relatively low-cost, near-surface structures, perhaps of considerable lateral
extent. Near-surface soils tend to be softer, weaker and more variable than
deep soils, and they therefore may require more extensive and more refined
methods of site investigation than for deep foundations, for example.
Because of the often low percentage of overall construction cost of retaining structures, however, there may be little or no money available for site
investigation. Regrettably, small earth-retaining structures are sometimes
designed in an almost total absence of site-specific soil parameters.

Even when soil information is available, however, the problem of the
design earth pressure may not be soluble, unless the construction method
can be precisely defined before the design takes place. If the contractor is
free to build a retaining structure by any method then, for example, he
might (i) place the structure with a minimum of disturbance to the existing soil, or (ii) excavate the soil adjacent to the structure, and replace it
after construction. These two techniques would lead to completely different pressure distributions and magnitudes between the soil and the structure. In the first case, the structure might have to sustain lateral earth
pressures close to those *in situ* (K_0) existing before construction, or more
likely, if the wall was allowed to yield away from the soil sufficiently, active
earth pressures which would be determined from *in situ* soil properties.
If the soil were excavated and replaced, then compaction methods would

largely control the final pressures. If the soil was excavated and replaced by imported backfill then data on the local ground conditions would have little value when calculating the earth pressure to be supported by the wall. When the construction method is not defined, it is difficult to argue that any design is either reasonable or possible.

2.2.1 Key parameters

The magnitude of earth pressure mobilised on a structure will be related to

- Wall/soil placement techniques
- Wall movement relative to the soil
- Shear strength developed between the wall and the soil
- Shear strength of the soil itself
- Wall, soil, groundwater geometry
- External loads

It is essential that the designer correctly anticipates the wall/soil placement technique to be used, the direction of wall movement relative to the soil, and the probable worst groundwater conditions. Compared with these factors, the choice of soil parameters is less important, provided that those parameters are relevant to the design method, are realistic, and are relatively conservative.

In general, the designer will require knowledge of

i. The geometry of the problem—height and inclination of supported soil face, geometry of ground surface, distribution of soil types in three dimensions, position of any external loads, and position of any existing structures to be protected.
ii. The bulk density of the soil—this will often be in the range 1.6–2.2 Mg/m^3, and may well increase with depth.
iii. Groundwater conditions—these are required for the worst conceivable condition in the life of the structure. It may be necessary to carry out seepage studies, perhaps using flow net sketching, in order to obtain an estimate of the final groundwater level, and pore water pressure variations on the shear surface and the back of the wall. Techniques for estimating water pressures can be found in Chapter 4.
iv. Soil strength parameters—these will vary according to the type of problem. In general, the designer will require a good knowledge of the peak effective strength parameters (c', ϕ') for the soil that the wall is to support, since active earth pressures are nearly always very much greater in the long term than the short term. When carrying out empirical design for strutted excavations supporting clay, the short-term 'undrained' shear strength (c_u) is required. Undrained shear

strength may also be used if short-term (undrained) conditions can be justified (e.g. for very rapid excavations in uniform plastic clays) for the life of a temporary structure.

v. Soil stiffness parameters—soil stiffness will be required for analyses of wall and ground displacements, typically carried out using finite difference or finite element computer code (see Chapter 8), and for foundation settlement prediction. Soil stiffness values can be obtained from *in situ* tests, from laboratory tests, or from back-analysis of similar structures in similar ground conditions.

vi. *In situ* horizontal stress profile—when computer analysis involves constitutive soil models based upon both elasticity and plasticity, it will be necessary to have a detailed knowledge of soil stiffness, effective strength parameters, and the initial *in situ* horizontal stress. *In situ* horizontal stress is both difficult and expensive to measure, and can usually only be derived from either self-boring pressuremeter tests *in situ*, or from 'suction' tests on laboratory specimens.

The development of earth pressure behind a retaining structure is generally extremely complex. Therefore, even sophisticated analytical methods cannot approach the real situation, and all design is, in effect, semi-empirical or empirical. In the absence of good field observations of existing structures, to allow objective judgements on the validity of available predictive methods, the design methods that are commonly in use are applied because experience shows that structures designed by these methods neither fall down nor generally behave in an undesirable way. Therefore, the designer should remember that any calculations used are semi-empirical in nature, and may well not be an accurate reflection of the real behaviour of the soil and the structure supporting it. It should also be borne in mind that in purely empirical methods of calculating pressure distributions (such as for braced excavations—see Chapter 10), the soil parameters must be obtained in the same way as they were obtained when the original database for the empiricism was prepared.

2.2.2 Methods of investigation

Site investigation should (at least) consist of

- A desk study and site walk-over survey, to gather existing information on the site
- Probing, boring or drilling, to determine the soil profile and variability of the ground
- *In situ* testing to obtain data on soil conditions
- Sampling and laboratory testing, for direct determination of the properties of clay soils

A common dilemma in general geotechnical design is that while the soil conditions beneath a site will control the type of foundation adopted for a structure, it is necessary to have some idea of the type of foundation to be used before the site investigation can be designed. This dilemma also applies in the case of earth-retaining structures, but the problem is further aggravated because the structure may have to support imported fill. It is, therefore, essential that a good desk study of existing soils information is carried out as early as possible in the design. Once this has been completed, the approximate ground conditions on the site will be known, the likely type and depth of retaining structure can be predicted, and the details of an effective site investigation can be put together.

Depending upon local ground conditions, and therefore local practice, investigation may be carried out using probing (typically the cone penetration test, or CPT, see below), boring (in soft ground), or drilling (in rock). During boring, samples will be taken and *in situ* tests will be conducted. In non-cohesive (sand and gravel) soils, sampling is of little use, since it is impossible to obtain high quality samples for laboratory testing.

The common methods of sampling and *in situ* testing used during boring and drilling are described below (see further sections and Clayton et al. 1995) for more detailed descriptions).

i. Firm to very stiff clays. Thick-walled driven 'undisturbed' 100-mm-diameter open drive tube samples are typically taken at 1.0- to 1.5-m centres down the hole. Small disturbed samples (1–2 kg) are taken between undisturbed samples.

ii. Very soft to firm clays or peats. Thin-walled piston drive samples may be taken at 1.00-m centres, if laboratory tests are required, or alternatively *in situ* vane tests can be carried out to obtain undrained shear strength.

iii. Sands and gravels. Standard penetration tests (SPT) are carried out at 1.0- to 1.5-m centres. Small disturbed samples are recovered from the SPT 'split spoon' in sand. In sands and gravels large disturbed samples (25–50 kg) are taken at 1.0-m centres between SPT tests.

iv. Rock. Continuous rotary core is taken, with the aim of determining whether rockhead or boulders have been encountered during soft-ground boring.

v. All soils. Small disturbed samples should be taken immediately when a new soil type is encountered. Water samples (0.5–1.0 l) should be taken every time water is struck, and boring should be suspended for as long as possible to allow the water level to equalise in the borehole, for groundwater level determination.

The objectives of site investigation drilling and testing are three-fold: to divide the subsoil into simplified soil types or groups (determining the extent of each soil type), to determine the variability of each soil type or group, and to obtain parameters representative of each group for design purposes. All of these activities are normally carried out at the same time.

Soil grouping and the assessment of soil variability within each group is normally carried out on the basis of visual sample description, moisture-content tests, Atterberg limit (liquid and plastic limit) tests, and particle size distribution tests. These last three types of test are often referred to as 'classification tests', because they are used to classify the soil into groups of materials which are expected to have similar behaviour. Visual sample description is an extremely good technique for soil grouping, but it must be carried out by a trained and experienced geotechnical engineer.

A large number of classification tests should be carried out if the true variability of the soil is to be determined. In addition, the results in *in situ* tests such as the SPT or *in situ* vane test are very useful for assessing variability. Even better, however, are tests which give a more or less continuous record with depth, such as the static cone test (CPT) or the continuous dynamic penetration test.

2.2.3 Desk studies and walk-over surveys

Desk studies should include examination of records of the previous use of the site, current and old topographical maps, geological maps and memoirs, and air photographs, when these are available. Previous site investigation reports may be available for nearby sites. All this information will help in giving an idea of the probable ground conditions and their variability beneath the site.

Desk studies are important because the type of retaining structure to be used will often depend on soil conditions. They are also important because the type of work to be carried out during site investigation will depend on the soil conditions encountered. It is not practical to take undisturbed soil samples in sands and gravels, and so *in situ* tests must be used to obtain the required data for design. In clays, however, samples can normally be recovered from the boreholes and sent to a soil-testing laboratory for testing and analysis.

Once all the available records have been collected and assimilated, a thorough examination of the location proposed for retaining-wall construction should be made. During the visit to the proposed site, the available records should be compared with what can be seen on the ground. Exposures of soils or rocks may be seen, and used to supplement information gained from geological maps and records. Information may be gained on ground-water levels (e.g. boggy ground, wells), on access for drilling rigs, on local construction techniques, and most particularly on the presence of preexisting slope instability. In this last respect, clay slopes standing at more than about 7° to the horizontal will always need serious investigation.

Visual evidence of slope movement may be obtained from the presence of hummocky ground, boggy ground on slopes, and from trees with kinked trunks.

2.2.4 Depths and spacing of boreholes or probe holes

Ground investigation is normally carried out by boring (in soft ground), or rotary core drilling (in rock). General rules for the depth of ground investigation for structural foundations were given by Hvorslev (1949), and these remain valid for retaining structures.

The borings should be extended to strata of adequate bearing capacity and should penetrate all deposits which are unsuitable for foundation purposes—such as unconsolidated fill, peat, organic silt and very soft and compressible clay. The soft strata should be penetrated even when they are covered with a surface layer of high bearing capacity.

When structures are to be founded on clay and other materials with adequate strength to support the structure but subject to consolidation by an increase in the load, the borings should penetrate the compressible strata or be extended to such a depth that the stress increase for still deeper strata is reduced to values so small that the corresponding consolidation of these strata will not materially influence the settlement of the proposed structure.

Except in the case of very heavy loads or when seepage or other considerations are governing, the borings may be stopped when rock is encountered or after a short penetration into strata of exceptional bearing capacity and stiffness, provided it is known from explorations in the vicinity or the general stratigraphy of the area that these strata have adequate thickness or are underlain by still stronger formations. When these conditions are not fulfilled, some of the borings must be extended until it has been established that the strong strata have adequate thickness irrespective of the character of the underlying material.

When the structure is to be founded on rock, it must be verified that bedrock and not boulders have been encountered, and it is advisable to extend one or more borings from 10 to 20 ft [3–6 m] into solid rock in order to determine the extent and character of the weathered zone of the rock.

In regions where rock or strata of exceptional bearing capacity are found at relatively shallow depths—say from 100 to 150 ft [30–45 m]— it is advisable to extend at least one of the borings to such strata, even when other considerations may indicate that a smaller depth would be sufficient. The additional information thereby obtained is valuable insurance against unexpected developments and against overlooking foundation methods and types which may be more economical than those first considered.

The depth requirements should be reconsidered, when results of the first borings are available, and it is often possible to reduce the depth of subsequent borings or to confine detailed and special explorations to particular strata.

Hvorslev has suggested that the preliminary depth of exploration for retaining walls should be 3/4–1 times the wall height below the bottom of the wall, or the bottom of its supporting piles if these are to be used. Since it is rare to carry out more than one ground investigation, it is generally better to err on the safe side, perhaps by boring or drilling to two times the wall height below the proposed level of the bottom of the wall.

The Recommendations of the Committee for Waterfront Structures (1975) give much more specific guidance for site investigation. It is assumed that, since deeper soils tend to be more uniform than shallow soils, the evaluation of passive pressure will require fewer boreholes than the evaluation of active pressure. Three groups of boreholes may be used (Figure 2.4):

a. Main borings, at 50-m centres along the centreline of the wall, are taken to twice the proposed wall height below the upper level, or to a known geological stratum.
b. First-phase intermediate borings are carried out after the completion of the main borings, on the active and passive sides, again at 50-m centres along the proposed wall position.
c. If the ground proves to be very variable, a second phase of intermediate boring is carried out, with a boring depth dictated by the findings of the previous boreholes.

This amount of investigation would be rare in the UK. Rather than carry out large numbers of boreholes, it may be better to extend the information gained from the main borings by carrying out static cone penetration tests (CPT) or continuous dynamic probing or either side of the centreline, since these will be cheaper and faster to carry out. Both these types of test are carried out without a borehole, and so it is necessary to have some borings to identify the soil types on the site. Details for the CPT can be found below, and in Clayton et al. (1995). They are described, together with continuous dynamic penetration tests, in BS EN 1997-2 (2007).

BS EN 1997-1 (2004) and BS 6349 (2000) also give recommendations with respect to the depth and distribution of borings. In particular, BS 6349-1 (2000) suggests that borings should go to a depth of twice the retained height of the soil for sheet-pile walls, or 1.5 times the width of the base below the level of the bore for monoliths or gravity structures.

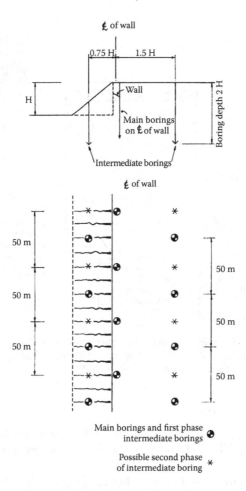

Figure 2.4 Layout of site investigation borings according to German recommendations. (From Clayton, C. et al., *Earth Pressure and Earth-Retaining Structures*, Second Edition, Taylor & Francis, New York, 1993.)

The layout of boreholes should be made in such a way that the full longitudinal and lateral extent of the works and the soil influenced by the works is investigated. Thus, for the investigation of a quay, boreholes should be spaced at intervals along the waterside frontage of the wall itself, but in addition should extend in a direction normal to the wall to investigate areas of passive resistance, soil in which anchors are to be placed, and possible risks due to slope instability on the landward side, and overall stability. Figure 2.5 gives an example from BS 6349 for a piled wharf.

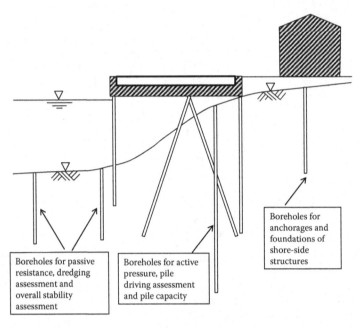

Boreholes for anchorages and foundations of shore-side structures

Boreholes for passive resistance, dredging assessment and overall stability assessment

Boreholes for active pressure, pile driving assessment and pile capacity

Figure 2.5 Layout and depths of investigation boreholes for a piled wharf. (Based on BS 6349, Codes of Practice for Maritime Structures, Part 1. General Criteria. British Standards Institution, London, 2000.)

2.2.5 *In situ* testing

The two main types of *in situ* test used around the world are the standard penetration test (SPT) (Clayton 1995) and the cone penetration test (CPT) (Lunne et al. 1997).

The SPT equipment is shown in Figure 2.6, which is based upon UK practice. This test is carried out in a borehole, and essentially involves counting the number of hammer blows (N) that are necessary to drive the thick-walled tube (outside diameter 51 mm) 300 mm into the ground at the bottom of the hole. Points to note are

1. Borehole disturbance can change (and generally reduces) the N value (by up to five times in extreme circumstances, whether in sands or sandy gravels). If there is a choice, small diameter boreholes, or boring methods that do not apply suction to the base of the hole (e.g. wash-boring and rotary open-holing) are to be preferred, since the extent and amount of loosening (respectively) are reduced. In UK soils, this is not practical because of the frequently high coarse gravel content of the soil.
2. The hammer used for the SPT varies around the world. Different hammers deliver different energies, and since the N value is inversely

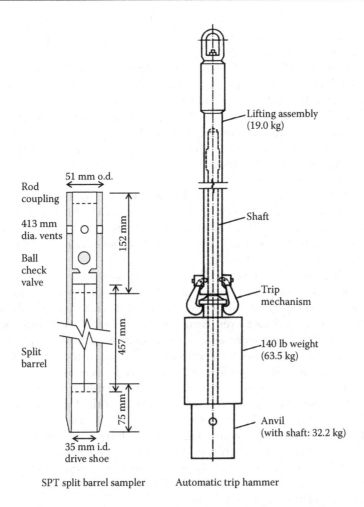

Figure 2.6 The standard penetration test (SPT). (From Clayton, C.R.I. et al., *Site Investigation*, Second Edition, Blackwell Science, Oxford, UK, 584 pp., 1995.)

proportional to the input energy, it is necessary (if possible) to correct the measured N value to a standard energy level. It has been agreed internationally that this is 60% of the free fall energy of the weight (i.e. 60% of the free-fall energy delivered by a 63.5 kg weight dropping through 760 mm). The N value corrected for hammer energy is denoted 'N_{60}'.

3. In sands and gravels (non-cohesive soils), the effective stress at the test level has a significant effect on penetration resistance. The same sand at the same density will have a higher strength and stiffness at depth than it has near the surface, and this will be reflected in the measured N value. It is therefore necessary to correct N values for sands and

gravels to a standard level of vertical stress, namely 100 kPa (or 1 tonne/m²). The N value corrected for both overburden pressure and hammer energy is denoted '$(N_1)_{60}$'.

4. Therefore in clays, the SPT N value obtained in the field must be corrected for energy (to give N_{60}), whilst in sands and gravels it must be corrected for both energy and vertical effective stress (to give $(N_1)_{60}$).

5. The SPT N value is increased by soils containing coarse particles (greater than about 1 mm diameter). Since most correlations have been obtained for sands, tests in gravels may give over-optimistic results (but see item 1 above).

The SPT test is far from perfect, but it can be used in almost all soil types, and a small sample is obtained (except in gravels) for visual description. It is therefore widely adopted in practice, except in soft soils where very low (unusable) blow counts result. In softer and looser soils, the use of the CPT (see below) is preferable.

The CPT is shown in Figure 2.7. A 60° cone with a 10-cm² face area is pushed into the ground at a rate of 2 ± 0.5 cm/s. The force necessary to advance the cone is measured. A 150 cm² friction sleeve is typically fitted behind the cone, and measures shear stress between the soil and cone at this location. A pore pressure sensor may also be used, and is typically fitted between the cone and the friction sleeve. The forces applied to the cone as it is pushed into the ground may be measured mechanically or electrically. Wherever possible, an electric cone should be used.

The cone resistance (q_c) is calculated by dividing the force necessary to advance the cone by its (10-cm²) cross-sectional area, and the sleeve friction (f_s) is calculated by dividing the vertical shear force on the friction sleeve by its 150 cm² surface area. Friction ratio (R_f) is obtained by dividing sleeve friction by cone resistance. Typically, near-continuous plots of q_c and R_f are produced as a function of depth (Figure 2.7).

The CPT has many advantages over the SPT.

- No borehole is required, so borehole disturbance is avoided.
- The process is more or less automated, so operator independence is achieved.
- Because data are taken electronically, using strain-gauged sensors, very frequent readings (i.e. at very small increments of depth) can be obtained, allowing the ground profile to be examined in detail.

There are one or two important disadvantages, however:

- The equipment cannot always penetrate the ground to the depth envisaged, as obstructions such as dense gravel, cobbles and boulders can cause refusal.
- No sample is obtained for visual description.

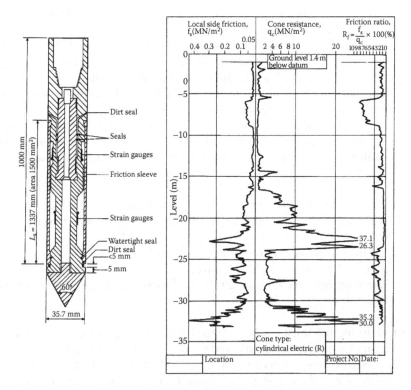

Figure 2.7 The cone penetration test. (From Clayton, C.R.I. et al., *Site Investigation*, Second Edition, Blackwell Science, Oxford, UK, 584 pp., 1995; Meigh, A.C., 'Cone penetration testing—methods and interpretation'. CIRIA Report. Butterworths, London, 141 pp., 1987; te Kamp, W.C., Sondern en funderingen op palen in zand. Fugro Sounding Symp., Utrecht, 1977.)

2.2.6 Sampling and laboratory testing

In clays, it is possible to take samples that are reasonably undisturbed, and can therefore be used in laboratory tests to determine parameters for retaining wall design. However, it is important to bear in mind that the boring and sampling processes produce soil disturbance, which changes the properties of the material to be tested.

A full description of the causes and effects of soil disturbance is beyond the scope of this book. It is sufficient to note that

- The amount of soil displaced by a sampling tube affects the degree of sample disturbance (Clayton and Siddique 1999). Thin-walled tubes (displacing less than 10% of the sample area) are better than thick-walled tubes for example, displacing 30%–45% of the sample area. It often proves impossible to recover samples of soft clay when using a thick-walled sampler.

- The detailed cutting shoe geometry also affects disturbance; sharp cutting shoes (for example with a cutting edge of 5°) are better than blunt ones (e.g. with a 15°–30° cutting edge taper angle).
- The method of advancing a sample tube has an effect on disturbance. Pushed tubes are better than driven tubes.
- Block samples, hand cut from pits, tend to be less disturbed, because disturbance caused by the boring rig is avoided. However, it is difficult to take block samples at depth in soft clay, or below the water table.

Samples and tests also need to be representative of the ground from which they are taken. This means that they must be large enough to contain

- All the particle sizes present in the ground
- Planes of weakness such as fissures
- Drainage features such as silt-covered fissures, laminations or lenses of sand

In addition, they must not be damaged during transportation to the laboratory, and they must not be allowed to dry out during storage.

Soft alluvial soils are very easily disturbed, and their behaviour modified, by boring and sampling. Such material is typically destructured (for example, any inter-particle cementing may be destroyed) and its effective stress is reduced. To minimise this, samples should be taken with a pushed 100-mm thin-walled piston sampler (the piston helps reduce disturbance, but also prevents loss of the soil as it is pulled up the borehole), with a 5° cutting edge taper angle. Stiff fissured soils are routinely sampled in some countries using 100-mm-diameter open tube samplers, driven (hammered) into the bottom of the borehole. Better samples can be obtained with thin-walled pushed sampling. All samples should be carefully sealed as soon as they are brought to the surface, to prevent moisture loss.

Laboratory testing for strength properties is routinely carried out in the triaxial test (Figure 1.3). Tests can be carried out either

1. 'Drained', where a drainage vent is left open to a fixed water pressure during shear, which must be carried out slowly enough to allow excess pore pressures set up by deviatoric stress increase to drain.
2. 'Undrained with pore water pressure measurement', where specimen drainage is prevented, and the pore pressure in the specimen is measured during shear using a pore pressure transducer.
3. 'Undrained', where the vent is closed during shear, pore pressure is not measured, and the test is interpreted in terms of total stress.

Drained tests and undrained tests with pore water pressure are known as 'effective stress' triaxial tests and are routinely used to measure the effective strength parameters (c' and ϕ') of clays. The third type of test is used to determine the undrained shear strength of the soil, which is useful for the design of structures for short-term (end of construction) conditions, e.g. for the estimation of bearing capacity, and sometimes when a structure is only required to stand for a very short time, and excess pore pressures caused by construction are unlikely to dissipate.

Effective stress tests are used to obtain the effective strength parameters of soils. The cell pressure ($\sigma_c = \sigma_h$, see Figure 1.3) provides the horizontal total stress on the soil, and remains constant during the test. The ram load (P) is measured and is increased until failure occurs, and the cross-sectional area of the specimen (A_s) is calculated, so that at failure,

$$\sigma_v = \sigma_h + P_f/A_s \tag{2.4}$$

At failure, therefore, $\sigma_v > \sigma_h$, so that the vertical total stress, $\sigma_v = \sigma_1$, and the horizontal total stress, $\sigma_h = \sigma_3$. The horizontal and vertical effective stresses are obtained by subtracting the measured pore water pressure at failure, so that

$$\sigma'_1 = \sigma_h + P_f/A_s - u_f \tag{2.5}$$

$$\sigma'_3 = \sigma_h - u_f \tag{2.6}$$

The results of a number of specimens tested at different initial effective stress levels are plotted as Mohr circles to determine the effective strength parameters. Figure 1.5 gives an example result. The major and minor effective principal stresses are labelled for illustrative purposes on the largest full circle. The test specimens are normally trimmed, placed in rubber membranes, and consolidated at effective stresses equivalent to one-half, one, and two times the calculated vertical effective stress at the depth in the ground from which the soil sample was extracted. In an undrained test with pore pressure measurement, the specimens must be sheared slowly to allow pore pressure equalization throughout the specimen, so that the pore pressure developed in the centre of the specimen can be measured by a transducer connected to its base. Three different shear strengths result, giving three Mohr circles (shown in full in Figure 1.5), and an inclined failure envelope in terms of effective stress.

It is normal practice to test at least three specimens (shown in this example as full lines) and to draw the best fit straight line tangential to the circles to determine the effective cohesion intercept, c', and the effective angle of friction, ϕ'. However, most soils do not have straight line failure envelopes, as shown by the additional, dashed, Mohr circles. The use of a significant cohesion intercept in earth pressure calculations can lead to a large overestimate of wall stability, and therefore it is important to

1. Carry out effective stress triaxial tests at effective stress levels that simulate the low effective stress levels that will exist in the ground
2. Ignore any effective cohesion intercept that is measured, unless sufficient high-quality testing can be carried out to establish its existence beyond doubt

In the undrained triaxial test, pore pressure is not measured, and so only total stresses are known. Tests are normally carried out on sets of three specimens prepared from the same sample but tested at different cell pressures, or on single specimens. For saturated clay, differences in applied cell pressure lead to equal differences in the pore water pressure in the specimen, so that the effective stresses in the specimens remain constant. As a result (Figure 2.8), the failure deviator stress should not vary between different specimens tested at different cell pressures, and if this is so, the undrained shear strength, c_u, can be obtained as shown in the figure.

In the UK and USA, it is common to prepare either sets of smaller diameter (35–38 mm) specimens taken from the same position in a sample, or single specimens (70–100 mm diameter) using the full sample cross section. Larger samples are required in order to obtain undrained shear strengths that reflect the weakening effects of natural fissures in the soil, which are common in stiff clays. Fissuring and other causes of desaturation may give an apparent increase in undrained shear strength with increasing cell pressure. It is therefore normal to calculate the undrained shear strength from multiple specimen tests by averaging the undrained shear strengths from the individual specimens, rather than by attempting to fit an inclined failure envelope. In undrained, end of construction calculations, ϕ_u is taken as zero.

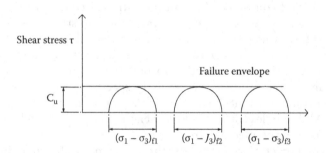

Figure 2.8 Expected result from unconsolidated undrained triaxial test on saturated clay. (From Clayton, C. et al., *Earth Pressure and Earth-Retaining Structures*, Second Edition, Taylor & Francis, New York, 1993.)

Another form of laboratory test that may be useful is carried out in the oedometer. Here, a disc of soil (typically about 75 mm in diameter and 20 mm high) is confined laterally by a metal ring whilst being compressed under increasing vertical stress (Figure 2.9a). The results are plotted either as vertical strain ($\Delta H/H_o$), as a function of logarithm of applied vertical stress, or void ratio (e), or specific volume (= 1 + e) as a function of

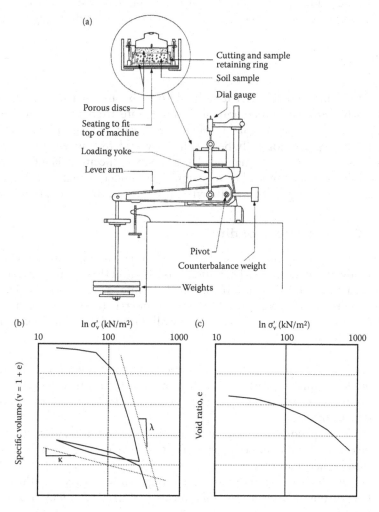

Figure 2.9 The oedometer test. (a) Apparatus; (b) soft clay; (c) stiff clay. (From Clayton, C.R.I. et al., *Site Investigation*, Second Edition, Blackwell Science, Oxford, UK, 584 pp., 1995.)

logarithm of applied vertical stress. The vertical stress is typically doubled at each loading stage, and there may be unloading as well as loading stages (Figure 2.9b).

The rate of compression under each load is interpreted in terms of the coefficient of consolidation, c_v, which can be used to estimate the speed at which consolidation of foundations, for example, might occur. For soft clays, the test is routinely used to estimate pre-consolidation pressure and the compressibility index, c_c. When specific volume is plotted against the natural logarithm of vertical stress, the test may also be used to produce the critical state parameters λ and κ. For overconsolidated clays, the test is typically used to calculate values of 'coefficient of compressibility', m_v, where

$$m_v = \frac{\Delta e}{(1 + e_0)} \qquad (2.7)$$

Results from the oedometer test should only be used when preliminary, conservative, estimates of compression and its rate are required. The small specimen size, and the fact that drainage occurs vertically, mean that in soft alluvial clays c_v will normally be significantly underestimated (in extreme cases by several orders of magnitude), and small specimen height means that bedding effects will lead to very significant underestimates of the stiffness of stiff and very stiff clays (in extreme cases by a factor of about 10).

2.2.7 Groundwater conditions

The loads on earth-retaining structures are very significantly influenced by groundwater conditions. Therefore, it is essential that ground investigation for earth-retaining structures should look into, as fully as possible, the regimes of the groundwater, and in waterfront structures the seasons and tidal variations of water in front of the wall. Groundwater may exist in a number of forms.

i. *Hydrostatic conditions*. The pressure of pore water within the ground increases linearly with depth, such that at any distance (z) below the groundwater table, $u = \gamma_w.z$. This means that there is no vertical component of seepage.

ii. *Artesian groundwater*. Groundwater in a permeable aquifer, confined below a relatively impermeable material such as clay, exists at a pressure higher than that due to its distance below the top of the aquifer. When a hole is drilled into the aquifer, groundwater rises up the borehole through the relatively impermeable material, and may rise above ground level.

iii. *Perched groundwater*. Rainfall seeping from ground surface through relatively permeable soil rests on top of a less permeable layer, below which a more permeable soil exists. When boring through this ground, a water strike will occur at or above the level of the impermeable soil, but the groundwater will drain down the hole once the lower, more permeable material is reached. In developing countries, this situation will sometimes be found in cities, where the water may be untreated effluent.

iv. *Under drainage.* This is the reverse of artesian conditions, and has occurred in many areas of heavy population (e.g. London) as a result of exploitation of the aquifer (in this case the chalk) for domestic and industrial water supply. The deeper a cased borehole extends in the clay, the lower will be the equilibrium water level within it.

The form of groundwater is important not only for the design of the permanent works; it has very serious implications for construction costs because of its influence on the difficulty of carrying out groundworks. In this respect, it is important to know in some detail not only the geometry of the subsoil and the forms of groundwater regime, but also the permeability of each soil type (in both the horizontal and vertical directions if possible). British Standard BS 5930 (1999) describes *in situ* permeability tests in some detail. Suitable piezometers should be installed (see Clayton et al. 1995) in sufficient positions to give a clear idea of the groundwater regime. Some types of piezometer can be used to conduct *in situ* permeability tests. Once installed, piezometers may need to be read regularly for several months, until water levels are seen to stabilise. Groundwater is considered in more detail in Chapter 5.

2.3 OBTAINING REQUIRED SOIL PARAMETERS FROM SITE INVESTIGATION DATA

Probing, borehole sample descriptions and the results of classification tests are used to divide the soil samples into groups with expected similar engineering behaviour. The subsoil geometry (layering, etc.) can then be idealised.

It is then necessary to determine the relevant soil parameters required for the specific design, for each group of soils. The following parameters are frequently required:

- Bulk unit weight
- Effective strength parameters c' and ϕ'
- Undrained shear strength of clay, c_u
- Stiffness (m_v, E, etc.) for prediction of movements
- Coefficient of consolidation, c_v
- Coefficient of permeability, k

2.3.1 Bulk unit weight

The bulk unit weight of soil is normally expressed in kN/m³. It can be obtained from bulk density by multiplying it by the gravitational constant (*g*). Bulk density is expressed as Mg/m³. Bulk unit weight should not be confused with buoyant unit weight, which is the difference between bulk unit weight and the unit weight of water. It is recommended that buoyant weights should not be used in earth pressure calculations, since this can lead to confusion.

When a coarse granular material is above the groundwater table, its pore space will contain air. When below the groundwater table, the pore space will be full of water, and as a consequence the bulk unit weight of the soil will be higher. Clays will not normally have a significantly different bulk unit weight above and below the groundwater table, since capillary suction can maintain their pore spaces full of water.

Some typical values of bulk unit weight are given in Table 2.1.

2.3.2 Peak effective strength parameters, c′ and φ′

A number of different laboratory and *in situ* tests are available for determining the peak strength of soils. Except in the case of very small structures, the strength properties of cohesive soils are normally obtained by testing good-quality undisturbed samples, in the laboratory.

The stability of earth-retaining structures will normally be at its minimum in the long term, so that for most designs, it is the peak effective

Table 2.1 Bulk unit weight of soils

Soil type	Relative density or state of consolidation	Bulk unit weight (kN/m³)	
		Drained, above ground water level	Submerged, below ground water level
Gravel	Loose	16	20
	Dense	18	21
Sand	Loose	16.5	20
	Dense	18.5	21.5
Silt	Loose	16	16
	Dense	18	18
Clayey silt		17	17
Silty clay	Soft, normally consolidated	15–19	15–19
	Stiff, overconsolidated	19–22	19–22
Peat		11	11

Note: Values partly taken from BS 6349.

strength parameters that are required for earth pressure calculations. These can be obtained as follows:

- *Clays:* Consolidated undrained triaxial compression tests (described earlier in this chapter) with pore water pressure measurement, to find c' and ϕ', on undisturbed tube samples obtained from boreholes, or block samples obtained from trial pits.
- *Sands and gravels:* The effective strength parameters are required for the *in situ* soil, but it is not possible to obtain undisturbed samples. Generally, although the failure envelope for the soil may be curved, c' is assumed to be zero. The peak effective angle of friction, ϕ', is determined from correlations with *in situ* penetration test results.

Clays. The peak effective angles of friction (ϕ') of cohesive soils do not vary widely, but the effective cohesion intercept (c') (which has a relatively large effect on the calculated long-term earth pressures) may be very different from one soil type to another and from one test result to another. It is difficult to determine with good accuracy. For normally consolidated (soft) clays $c' = 0$, and as a soil becomes more heavily overconsolidated or bonded its effective cohesion intercept is found to increase. Typically,

$$0 < c' < 10 \text{ kN/m}^2$$

$$18° < \phi' < 30°$$

It is unwise to assume an effective cohesion intercept of more than about 3 kN/m² , even if the soil is heavily overconsolidated, and it may be sensible to assume $c' = 0$ in most cases. For example, Chandler and Skempton (1974) found by back analysis of highway cutting slopes that for the stiff London and Lias clays in the UK effective cohesion intercepts of the order of only 1.5–2 kN/m² were mobilised.

In the absence of any triaxial test data, the angle of friction can be estimated from the plasticity index (liquid limit–plastic limit) from results by Kenney (1959) (Figure 2.10). The methods described above give effective strength parameters for failure in triaxial compression conditions (i.e. $\sigma_1 > \sigma_2$, $\sigma_2 = \sigma_3$). In reality, the earth pressure problem is one of plane strain (i.e. $e_2 = 0$), and research testing indicates that a slightly higher value of ϕ' may be obtained for granular soils in plane strain. For clays, the difference is thought to be slight (about 10%), however, so that triaxial values can be used with confidence since they will be on the safe side.

Granular soils. The effective strength parameters of a granular soil are a function of particle size distribution, soil density, imposed stress level, angularity and cementation. They could be obtained from a series of laboratory tests of representative disturbed samples obtained during *in situ* investigations, and re-compacted to a range of densities, but in practice, c'

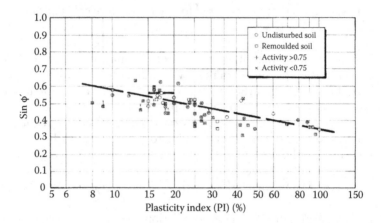

Figure 2.10 Sine of effective angle of friction of clays as a function of plasticity index (PI). (From Clayton, C. et al., *Earth Pressure and Earth-Retaining Structures*, Second Edition, Taylor & Francis, New York, 1993.)

is assumed zero and ϕ' is commonly estimated from SPT (N) or CPT (q_c) values, such as those given in Figure 2.11 and Figure 2.12. Note that these charts were derived for sands, and as noted above, coarser-grained soils such as gravels will have an effect on penetration resistance which may not be completely a reflection of an increased angle of friction.

Alternatively, Lundgren and Brinch Hansen (1958) suggest the following method of estimating ϕ' for granular soils:

$$\phi' = 36° + \phi_1 + \phi_2 + \phi_3 + \phi_4 \tag{2.8}$$

where

ϕ_1 is the correction for particle shape

(+1° for angular grains, 0° for 'average' grains, −3° for slightly rounded grains, −5° for well-rounded grains)

ϕ_2 is the correction for particle size

(0° for sands, +1° for fine gravels, +2° for medium and coarse gravels)

ϕ_3 is the correction for grading

(−3° for uniformly graded soils, 0° for 'average' grading, +3° for well-graded materials)

ϕ_4 is the correction for relative density

(−6° for the loosest packing, 0° for 'average' packing, +6° for the densest packing).

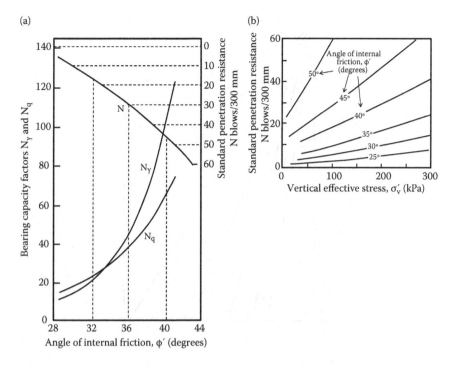

Figure 2.11 (a) Chart to obtain φ′ and bearing capacity factors for sands from SPT results. (Peck, R.B., Hanson, W.E., and Thornburn, T.H.: *Foundation Engineering.* 1974. Copyright Wiley-VCH Verlag GmbH & Co. KGaA. Reproduced with permission.) (b) Chart to obtain φ′ and bearing capacity factors for sands from SPT results. (After Mitchell, J.K. et al., The measurement of soil properties in situ, present methods—their applicability and potential. US Dept. of Energy report, Dept. of Civil Engineering, University of California, Berkeley, 1978.)

For preliminary design, Table 2.2 gives a further estimate of effective angles of shearing resistance (φ′) for a number of soil types. These parameters seem suitably conservative when compared to the values cited by other sources. Density must be estimated from *in situ* penetration test results. For example, the density descriptor used on borehole records in the UK is derived from the SPT value on the following basis:

$(N_1)_{60} < 4$ 'Very loose'
$4 < (N_1)_{60} < 10$ 'Loose'
$10 < (N_1)_{60} < 30$ 'Medium dense'
$30 < (N_1)_{60} < 50$ 'Dense'
$(N_1)_{60} > 50$ 'Very dense'

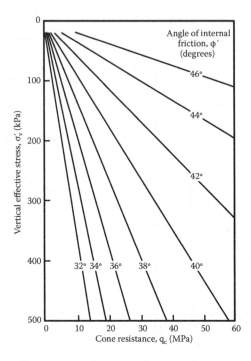

Figure 2.12 Determination of effective angle of friction of uncemented normally consolidated quartz sands from CPT cone resistance. (From Durgunoglu, H.T. and Mitchell, J.K., Static penetration resistance of soils: I Analysis, II Evaluation of theory and implications for practice. Proc. A.S.C.E. Symp. on In-situ Measurement of Soil Properties, Rayleigh, I, 151–171, 1975; Meigh, A.C., Cone penetration testing—methods and interpretation. CIRIA Report. Butterworths, London, 141 pp., 1987.)

2.3.3 Effective wall friction and adhesion

The amount of mobilised wall friction and wall adhesion (the components of shear strength between the back of the wall and the soil it supports) are a function of

    i. The strength parameters of the soil (since the shear strength between the soil and the wall cannot exceed the shear strength along a parallel plane in the soil a short distance away)
    ii. The frictional properties of the back of the wall
    iii. The direction of movement of the wall with respect to the soil
    iv. The amount of relative wall/soil movement
    v. The ability of the wall to support the vertical force implied by the wall friction and adhesion

Table 2.2 Effective angles of shearing resistance, ϕ', for preliminary design

Soil type	Compacted state	Effective angle of friction, ϕ' (°)	
		Active state	Passive state[a]
Gravel	Loose	35	35
	Medium dense	38	37
	Dense	41	39
	Very dense	44	41
Sand	Loose	30	30
	Medium dense	33	32
	Dense	36	33
	Very dense	39	34
Silts	Loose	24–27	
Clayey silts		21	
Silty clays		15–18	

Source: Data from BS 6349, *Codes of Practice for Maritime Structures, Part 1*. General Criteria. British Standards Institution, London, 2000.

[a] Passive values of ϕ' have been reduced to allow for strain softening and progressive failure, following Rowe and Peaker (1965).

Thus, the mobilised effective angle of wall friction, δ', must be less than the effective angle of friction of the soil.

$$0 \le \delta' \le \phi' \tag{2.9}$$

and the effective wall adhesion (c_w') must be less than any effective soil cohesion that is assumed:

$$0 \le c_w' \le c' \tag{2.10}$$

Wall friction has the effect of decreasing active pressures and increasing passive pressures. Therefore, it is necessary to include a realistic value if an economical design is to be obtained, but if the value is overestimated then the design will be unsafe. British Standard Code of Practice No. 2 recommended the values in Table 2.3 for sand, in the active state.

For bitumen-coated steel piles, a value of $\delta' = 0°$ would appear wise, since bitumen coating has been used to reduce negative skin friction on foundation piles. Based on model wall tests, Rowe and Peaker (1965) have made the recommendations in Table 2.4 for the passive case for the ratio between the maximum possible wall/soil friction and the amount actually

Table 2.3 Effective angles of wall friction in sand, for the active state

Wall material	Effective angle of wall friction, δ' (°)
Concrete or brick	20°
Uncoated steel	15°
Walls subjected to vibration	0°
Walls unable to support a vertical force (e.g. sheet-piles in soft clay)	0°

Source: BS Code of Practice No. 2, *Earth Retaining Structures*. Institution of Structural Engineers, London, 1951.

Table 2.4 Effect of direction of wall movement and restraint on mobilised wall friction

Problem	Soil density	δ' mob/δ' max
Masonry walls—horizontal movement only	Loose	0
	Dense	1/2
Light walls for anchors	Loose	0
	Dense	0
Sheet-pile walls—passive side, free embedment in sand	Loose	1
	Dense	1
Sheet-pile walls bedded on rock	Loose	0
	Dense	1/2

Source: Rowe, P.W. and Peaker, K., Passive earth pressure measurements. *Géotechnique* 15, 57–78, 1965.

mobilised, taking into account the relative vertical movement between the soil and the wall. These values were adopted in BS 6349 (2000).

For anchored sheet-pile walls, Terzaghi (1954) has recommended

$\delta' = 1/2 \ \phi'$ for the active case
$\delta' = 2/3 \ \phi'$ for the passive case

whilst British Standard CP2 (1951) recommended that in the passive case, the values for the active case should be halved.

The German Committee for Waterfront Structures (EAU 1978) has recommended

$\delta_a = 2/3 \ \phi'$ active case, planar surface of sliding
$\delta_p = 2/3 \ \phi'$ with a planar sliding surface, but only up to $\phi' = 35°$
$\delta_p = \phi'$ for curved sliding surfaces

The ability of the structure to give the restraint necessary for the vertical equilibrium of the mass of failing soil must be verified, and the use of a factor of safety on vertical sliding is recommended.

It is quite clear from the data given above that there is considerable inconsistency in the codes of practice regarding the choice of an angle of wall friction. This results from a lack of good data in this area. It is suggested that the following procedure is adopted:

i. Determine the effective angle of friction of the soil (ϕ'): ($\delta' \leq \phi'$).
ii. Where possible, determine the maximum effective angle of wall friction based on soil and wall material (e.g. CP2, sand on uncoated steel, $\delta'_{max} = 15°, \delta'_{mob} \leq \delta'_{max}$).
iii. Consider the influence of direction of wall movement, and vertical restraint (Table 2.4) ($\delta'_{mob} < \delta'_{max}$).
iv. Check that generally accepted good practice is not violated (active, $\delta' < 1/2\ \phi'$; passive, $\delta' < 2/3\ \phi'$).

Since there are virtually no records of measured shear stresses on retaining structures to be found in the literature, it is wise to adopt conservative values of c'_w and δ'. In the active case, it may be reasonable to assume

$$c'_w = 0$$

$$\delta' = 1/2\ \phi'$$

provided that the wall is free from vibration, and has reasonable resistance to vertical movement.

Much of the previous discussion has related specifically to the effective angle of friction on near-vertical surfaces (i.e. the front or the back of a wall). For structures such as gravity walls, or cantilever reinforced concrete walls, there is a need to assess the effective angle of friction between the base of the wall and the soil upon which it is founded, since this resists the tendency of the wall to slide under the influence of active pressures. If the wall were cast directly on to a rough surface of soil, it might be reasonable to assume $\delta' = \phi'$, but in practice it is common to assume $\delta' = 2/3\ \phi'$.

2.3.4 Critical state and CamClay parameters

At the critical state the soil has been remoulded and sheared to constant volume. There is no effective cohesion, and the relationship between shear stress and normal effective stress at failure can be expressed either in terms of the critical state effective angle of friction, ϕ'_{crit}, of the critical state parameter M. M is equivalent to ϕ'_{crit}, and can be converted to it using the equation (e.g. Powrie 2004)

$$M = \frac{6.\sin\phi'_{crit}}{(3 - \sin\phi'_{crit})} \tag{2.11}$$

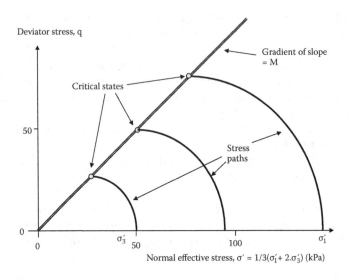

Figure 2.13 Determination of critical state parameter M for normally consolidated reconstituted clay.

The critical state angle of friction of granular soil is thought to depend primarily on mineralogy and pore water composition, although particle shape characteristics may well be important. In sands, Bolton (1986) has suggested that for quartz $\phi'_{cs} = 33°$, whilst for feldspar $\phi'_{cs} = 40°$.

The critical state angle of friction of soft to firm clay can be obtained by remoulding the material at high water content (say 1.5 times the liquid limit), and carrying out a series of consolidated undrained triaxial tests with pore pressure measurement, continuing to strain the material until changes in deviator stress and pore water pressure cease. Plotting the final values of deviator stress, $(\sigma_1 - \sigma_3)$ $(= q)$ as a function of mean effective stress $(\sigma'_1 + 2\sigma'_3)/3(= p')$, the critical state parameter M can be obtained. The stress paths and critical states for such a test are shown in Figure 2.13.

An estimate of the critical state angle of friction can be obtained for stiff overconsolidated clay by determining the peak effective strength parameters, c'_{peak} and ϕ'_{peak}, and adopting $\phi'_{cs} = \phi'_{peak}$ (with $c' = 0$).

2.3.5 Residual effective strength parameters

Residual strength parameters are relevant only for analyses on preexisting shear surfaces (i.e. in ancient unstable slopes), in materials such as clays, where large displacements may have led to alignment of particles and polishing of the shear surface, producing very low effective angles of friction. Although a value of residual effective cohesion intercept may be produced

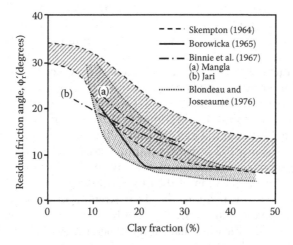

Figure 2.14 Relationship between residual friction angle and clay fraction. (From Lupini, J.F. et al., *Géotechnique* 31, 2, 181–213, 1981.)

from a best fit to laboratory test data at different effective stress levels, it is normal to assume that $c'_r = 0$, and values of $\phi'_r = \tan^{-1}(\tau/\sigma'_n)$ determined for the (low) effective stress levels on the residual shear surface in the field. Data from Lupini et al. (1981) suggest that residual conditions are unlikely for natural soils with clay contents less than 30%–40% by weight, and plasticity indices (liquid limit – plastic limit) less than 20%.

For materials that are affected by large displacements, the residual shear strength can be determined on remoulded soil using the ring shear apparatus. A simplified version of the apparatus, suitable for commercial testing, is described by Bromhead (1979). Alternatively, a cut-plane shear box test may be used. Details of the procedures for these tests are given by Clayton et al. (1995). Figure 2.14 (Lupini et al. 1981) gives a summary of values, from which it can be seen that the residual angle of friction, ϕ'_r, of plastic clay soil can be as low as 7°–8°. Residual shear envelopes are often curved, and it is important, therefore, that the soil is tested at normal effective stress levels similar to those on the preexisting shear surface in the field. Most will not be deeper than 5–10 m, and since the water table is generally high, suitable normal effective stress levels will be in the range of 25–100 kN/m².

2.3.6 Undrained shear strength, c_u

Values of undrained shear strength may be required for

- Determination of earth pressures exerted by clays under undrained (short-term) conditions

- Calculation of earth pressures exerted by clays on braced excavations
- Determination of bearing capacity, for example for clays supporting the foundations of gravity walls, and for reinforced earth

The undrained shear strength, c_u, of a saturated soil will normally be determined by an unconsolidated undrained triaxial test, which is specified in most national soil testing standards. The specimen is compressed at a constant rate of strain (often 2%/min in the UK) and failure should take place in about 10 minutes. Three 38-mm-diameter specimens are prepared from a single level of a 102-mm-diameter undisturbed sample, and are tested at confining pressures equivalent to one half, one and two times the vertical total stress at the position of the sample in the ground. If the specimen is saturated and is not fissured, the shear strength of the three specimens should be the same (Figure 2.8). Commonly, however, the failure envelope is slightly inclined, indicating that the specimen is either unsaturated or fissured. In this case, the average undrained shear strength $\left(\Sigma(\sigma_1 - \sigma_3)_f/6\right)$ of the three specimens should be used in calculations. On no account should an undrained angle of friction (ϕ_u) be used. ϕ_u should be assumed zero.

Other methods of obtaining the undrained shear strength of cohesive soils are

i. The *in situ* vane test (for soft and saturated soils)
ii. Plate bearing tests (rarely used in the UK because of expense)
iii. Via correlations with the SPT 'N' value (Stroud 1975); approximately, $c_{u(U100)} = 5. N_{60}$ for overconsolidated cohesive soils
iv. Via regional or preferably site-specific correlations with static cone resistance; $c_u = q_c/N_k$, where for most soils $10 < N_k < 20$, and $N_k = 15$ is commonly adopted

It should be noted that undrained shear strength is dependent on test technique, and different values will therefore result from each type of test in the same soil.

In the absence of test data, and for small structures only, an approximate value of undrained shear strength can be obtained from the sample descriptions given on borehole records (Table 2.5). These descriptions, which are carried out in a standardised way, relate strictly to 'consistency' rather than undrained shear strength because they ignore the weakening effects of fissures, etc., but if used conservatively can at least provide a starting point for design.

2.3.7 Soil stiffness

For small walls, where design can be over-conservative without unacceptable financial penalty, it is acceptable to use the results of oedometer testing to determine a value of Young's modulus of clay (e.g. to determine

Table 2.5 Standard clay soil strength descriptors used on borehole records

State	Descriptor	Approximate undrained shear strength (kPa)
Cannot indent with thumb nail	Hard	>300
Can be indented by thumb nail, but not thumb	Very stiff	150–300
Cannot be moulded by fingers; can be indented by thumb	Stiff	80–150
Moulded by strong finger pressure	Firm	40–80
Moulded by light finger pressure	Soft	20–40
Exudes between fingers	Very soft	<20

Source: BS 5930, *Code of Practice for Site Investigations (formerly CP 2001).* British Standards Institution, London, 1999.

settlement of a gravity wall). Strictly, and ignoring bedding, the constrained effective modulus, E_c' (rather than Young's modulus), is obtained from the coefficient of compressibility determined from an oedometer test, i.e.

$$E_c' = \frac{1}{m_v} \tag{2.12}$$

In stiff clays, bedding between the apparatus and the specimen has a significant effect on measured strains, reducing the modulus obtained by up to an order of magnitude, so that any settlements will be over-estimated. In soft clays the values obtained will be more reasonable. For embedded walls, especially in inner city sites, more accurate estimates of compressibility are required. Stiffness should be determined using the triaxial test, with local strain measurement, as discussed earlier in this chapter. As noted there, a typical value of normalised undrained secant Young's modulus $\left(E_{usec}/p_o'\right)$ will lie between 500 and 1000. Stress path testing may be necessary on critical projects, because of the loading path effects shown in Figure 1.8.

The stiffness of granular soils will normally be determined from the results of penetration testing. Back analysis of Burland and Burbidge's data (Burland and Burbidge 1985) suggests the relationship between E and N (for immediate settlement) given in Table 2.6. Stroud's relationship, which depends upon the mobilised bearing capacity q_{net}/q_{ult} is shown in Figure 2.15.

Stiffness can also be estimated from the CPT test. Correlations between effective constrained modulus (M) (i.e. effective Young's modulus measured

Table 2.6 Young's Modulus E/SPT N

SPT N (blows/300 m)	E'/N (MPa) at		
	Mean	Lower limit	Upper limit
4	1.6–2.4	0.4–0.6	3.5–5.3
10	2.2–3.4	0.7–1.1	4.6–7.0
30	3.7–5.6	1.5–2.2	6.6–10.0
60	4.6–7.0	2.3–3.5	8.9–13.5

Source: Burland, J.B. and Burbidge, M.C., Settlement of foundations on sand and gravel. *Proc. I.C.E.*, Part 1, 78, 1325–1371, 1985.

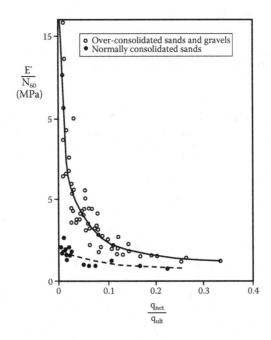

Figure 2.15 Stroud's relationship.

under conditions of no lateral strain) and cone resistance have been expressed in the following form:

$$M = \alpha_M \cdot q_c \qquad (2.13)$$

where α_M is 'often stated to be in the range 1.5–4' (Meigh 1987).

Chapter 3

Factors affecting earth pressure

Retaining structures may be built for a number of reasons, for example,

- To provide a platform (e.g. for housing or highway construction) by excavating into sloping ground (Figure 3.1)
- To provide a quay in a dock (Figure 3.2)
- To support the sides of temporary excavations, for example, around basement construction for buildings (Figure 3.3)
- As part of the permanent structure of a building, to provide a basement
- To provide abutments for a bridge deck

Whatever the reason for construction, a retaining wall is typically needed to support any vertical or near-vertical face of soil or rock. Only in exceptional circumstances (for example, unfractured or horizontally bedded rock) will an excavation stand for any significant length of time without support.

3.1 WALL CONSTRUCTION

Retaining walls may be needed in order to provide support for both natural and made ground. The wall may be constructed in natural ground, or it may be built before backfill is placed (see Figure 3.4). Because of the many situations and ground conditions in which they are used, retaining walls are built from a variety of materials (principally concrete and steel) and use different methods to support the retained ground. The type of retaining structure used will depend upon the type of ground it is to support, and the original and desired ground surface profile. Some example wall types are given below, to illustrate this point.

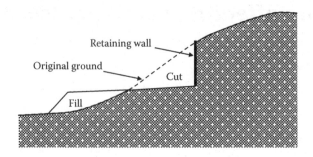

Figure 3.1 Cut and fill on sloping ground.

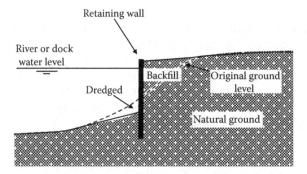

Figure 3.2 Retaining structure for a quay wall.

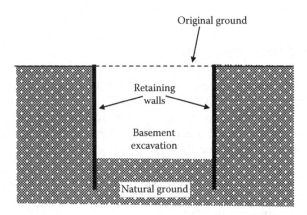

Figure 3.3 Retaining structures for a basement excavation.

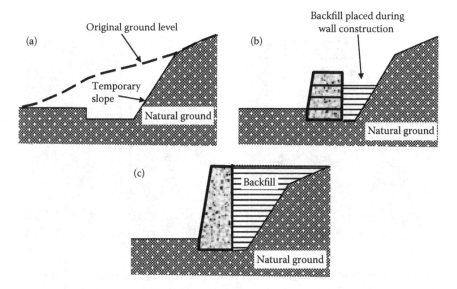

Figure 3.4 Gravity retaining wall construction. (a) Excavation, (b) wall construction and (c) after completion.

3.1.1 Types of wall and their support

Retaining structures come in many shapes and sizes. In this book, we have classified them into the following groups:

- Gravity walls
- Embedded walls
- Composite/hybrid walls

A *gravity wall* (Figure 3.4) uses its self-weight and the strength of the ground itself to maintain equilibrium. Lateral sliding of the wall is prevented largely by friction between its base and the founding soil. Gravity walls have been used for centuries, because they are simple to construct, typically being made of masonry or mass concrete. Because of their large mass, they need reasonable foundation soil and are not generally efficient for retaining great heights of material.

Embedded walls (Figure 3.5) prevent lateral movement partly or wholly by embedding the base of the wall in the ground, normally to significant depths below the excavation level. Additional support can be provided to the upper part of the wall by propping, or by anchoring into the natural ground on the retained side of the wall. Embedded walls are commonly used in two forms:

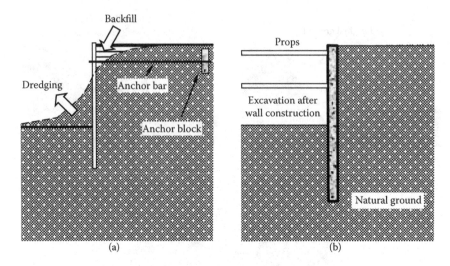

Figure 3.5 Embedded retaining wall. (a) Pre-formed wall construction. (b) *In situ* wall construction.

- Pre-formed walls (Figure 3.5a). 'Sheet-pile' walls are made by driving thin steel, timber or concrete 'sheets' into the ground. There is no excavation during sheet-pile construction, although some minor ground displacement occurs as the sheets are driven. After the sheeting and any anchors have been constructed, the ground is re-profiled (Figure 3.5a).
- *In situ* walls (Figure 3.5b). Diaphragm or bored-pile walls are made by excavating deep trenches or auger holes, placing reinforcement, and then filling them with concrete. Diaphragm and bored pile walls are popular forms of basement construction.

Composite/hybrid walls use a range of components to support the ground, for example,

- Walls working on the gravity principle (see above) can be constructed of selected granular material or rock (rather than concrete or masonry), reinforced by metal, polymer grids or fabrics, for example, in bottom-up gabion or reinforced earth wall construction (Figure 3.6).
- Ground anchors can be used in conjunction with pre-cast or *in situ* facing units, and sometimes vertical 'king piles', to support top down excavation (Figure 3.7).
- In coastal or dock construction, multiple sheet-pile retaining walls can be used to provide cells that are backfilled with granular fill or rock, for example, to provide a quay (Figure 3.8).

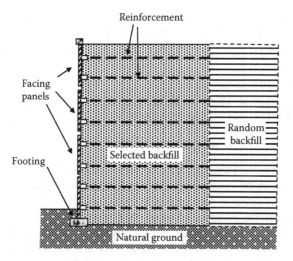

Figure 3.6 Reinforced soil (mechanically stabilised earth) wall.

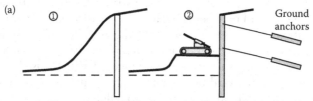

1. Excavate vertical holes from original ground level and insert H-section steel column (king piles).
2. Cast *in situ* reinforced concrete panels between king piles as soil is excavated from front of wall, then anchor.

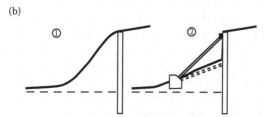

1. Drive H-section steel king piles from original ground level.
2. As excavation proceeds, install timber planking between king piles, and place steel struts.

Figure 3.7 Examples of king pile and planking construction. (a) Anchoring an embedded wall. (b) Propping an embedded wall.

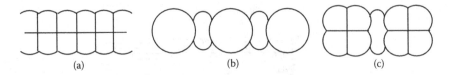

Figure 3.8 Plan views of cellular cofferdam geometries. (a) Arc and diaphragm cells. (b) Circular cells. (c) Clover leaf cells.

3.1.2 Construction of gravity walls

The construction of a gravity wall requires considerable site preparation. The ground must be excavated down to a suitable founding level (Figure 3.4a), where the ground has sufficient bearing capacity to take safely the loads that must be eventually supported, due to the self-weight of the wall and the vertical, horizontal and overturning components of the force between the retained soil and the wall. In order to make a safe excavation for the wall, the ground on the retained side must be cut back at a safe slope, typically of 45° or so. The ground in front of the wall is often removed, at least down to the required final dig level, at this stage.

Once the temporary excavation has been made, then the foundation can be constructed, typically using blinding concrete to trim and protect the foundation formation level. The wall is then constructed, often in lifts, and backfill placed. The backfill may be excavated soil, or selected granular material, or a combination of the two. If (unusually) the excavated material is suitable (for example, a free-draining granular material), it may be reinstated behind the wall. More usually an imported granular fill will be used in a wedge immediately behind the wall in order to ensure that significant water pressures do not develop behind the wall, and to reduce earth pressures. In order to prevent unacceptable settlement of the backfill during the life of the wall, it is generally placed and compacted in thin layers. Although the traditional wall was constructed of masonry or mass concrete, crib walls, gabion walls and interlocking block retaining walls are more commonly used today.

Because of their breadth, mass and method of construction, gravity walls are generally used for low retained heights, above water, and where founding conditions are reasonable. The method of wall construction used for gravity walls means that the type of imported fill that is used, and the way that it is compacted, will determine the earth pressures on the wall. However, the *in situ* ground will typically determine its foundation behaviour, since this will be left in place below the wall.

3.1.3 Construction of embedded walls

Unlike gravity walls, embedded walls do not require a good founding soil stratum, and are therefore favoured (in the form of steel sheet piling) for waterfront structures. They are widely used for temporary support, for example, during basement construction or for service trenches, because steel sheeting can be extracted and re-used. Embedded walls are also used as permanent support for basement excavations, where they are generally constructed from contiguous (touching or more normally with a small gap) or secant (interlocking) individual foundation piles to form a continuous structure. Diaphragm wall panels may also be used for this purpose, particularly on larger projects.

As noted above, there are two basic types of *in situ* retaining wall. The pre-formed walls typically used in river protection works and quay construction are formed using interlocking steel sheet piling (although in the past, timber and concrete sheeting have been used) (Figure 3.5a). Depending upon the starting geometry, steel sheeting may be pitched (i.e. aligned) and driven from a barge, or from land. Anchors (blocks or sheet piling) are placed on shore, and the anchors themselves are then connected to the wall. Finally, if necessary, the rear of the wall is backfilled and the front is dredged. In this type of wall most of the soil is *in situ*, with the important exception of fill behind the top of the wall. Ground conditions are often quite poor (loose sands, soft clays, organic material and man-made ground) in the majority of the profile, and a satisfactory design relies on the toe of the wall penetrating good ground. Little attempt is made to control ground movements, and because ground conditions are poor, these can be quite large.

In situ walls are typically used to support basement excavations on inner city sites, and to construct highway underpasses in urban settings, where land take must be minimised. The ground is often relatively level, at least between the front and back of the wall (Figure 3.5b). The wall is constructed from ground level, by excavating panels or boreholes (typically under bentonite, to provide some support for the open hole), then placing reinforcement cages or steel column sections, and finally concreting using a tremie (to avoid the high-slump concrete mixing with the bentonite; see Part 2 for more description). After the wall is complete, excavation starts. Props (or ground anchors) are placed as each support level is reached. The ground around a cast *in situ* wall remains largely undisturbed. Because these types of wall are often constructed in urban settings, there may be considerable effort put into designing and constructing a wall that produces minimal movement in the surrounding ground. Under these conditions (and depending on the construction method), *in situ* soil pressures may be high, and similar to those in the ground before construction.

3.1.4 Construction of composite/hybrid walls

The form of this type of wall is very varied. Retaining structures working on the gravity principle, but constructed of selected granular material or rock, include gabion walls (wire baskets), crib walls (constructed of inter-locking pre-cast concrete members) and reinforced soil construction. The general construction methods are essentially as discussed above for gravity walls, except that the wall itself must be developed. Gabion and crib walls are discussed in later chapters.

Reinforced soil or 'mechanically stabilised earth' (MSE) walls improve the strength of selected soil by using metallic (strip, bar, or mat) or geosyn-thetic (geogrid or geotextile) reinforcement that is connected to pre-cast concrete or prefabricated metal facing panels to create a reinforced soil mass (Figure 3.6). In effect, the strength of the backfill is increased to the level that the soil is capable of maintaining a vertical face, with the fac-ing panels providing only local support. Without the reinforcement, the backfill would only stand at its angle of repose, typically about 30° to the horizontal for a good granular material.

Figure 3.6 shows the typical components for a mechanically stabilised or reinforced earth soil wall. After preparation of the site by general excavation, a small foundation is made. The lowest level of facing units is placed, and given temporary support, and selected fill is then compacted to the first level of reinforcement. The reinforcement is placed, and attached to the facing unit. This process is repeated to the top of the wall, where a small capping beam is cast in order to knit the top level of facing units together. This type of wall is useful where founding conditions are moderate to good. Because of its flexibility, it can survive modest amounts of ground movement, for example, from settlement under load on soft soil, or from mining subsidence. They are also very good at absorbing energy without failing, making them well suited to use in seismically active areas. These walls behave much as a gravity wall, except that the wall itself is composed of reinforced soil, and the random backfill is placed as the wall increases in height.

Ground anchoring can also be used in conjunction with vertical 'king piles' and timber or cast *in situ* concrete planking to provide temporary support for basement excavation (Figure 3.7a). In this application, steel col-umns are either driven to the required depth or concreted in auger holes drilled from the ground surface before the start of basement excavation. Excavation takes place in stages, and at each stage, the zone between the king piles is cut back and either supported by timber planking, or shut-tered and filled with reinforced *in situ* concrete. Anchors, bracing struts or inclined props are then installed to support the king piles at that level, and following this, the excavation is taken down to the next level. This form of construction is similar to that of an *in situ* embedded wall, but is generally used as temporary rather than permanent support. Because of the tempo-rary removal of support when panels are excavated and cast, and the sys-tem's greater flexibility, earth pressures carried on the wall are likely to be more significantly reduced. For narrow service trenches, steel sheet piling, supported by horizontal bracing across the excavation, is commonly used.

Ground anchoring is also used in conjunction with sprayed concrete (shotcrete) to create low-cost retaining structures for near-vertical excavations in good ground conditions, such as weathered rock or residual soil, for example in highway cuttings. A top-down approach can be taken, with panels being sprayed over a limited height and width. The ground must be strong enough to support itself over small depths for sufficient time to allow the shotcrete panel to gain strength, and anchors to be drilled to support it. When the ground can be excavated to a relatively smooth profile, pre-cast or cast *in situ* panels can be used in conjunction with anchoring. This technique is also used with a form of passive anchor termed a 'soil nail'.

In coastal or dock construction, multiple sheet-pile retaining walls can be used to provide cells that are backfilled with granular fill or rock, for example, to provide a quay. Essentially, the construction process is the same as for a quay wall, except that the sheets will generally be driven into the river or sea bed under conditions where the bed profile is relatively flat, thus requiring considerable amounts of fill to be placed on the retained side of the wall, and the walls are anchored to each other, rather than to anchor blocks. Figure 3.8 shows some plan views of different cellular cofferdam arrangements.

3.2 WALL AND GROUND MOVEMENTS

Because the ground surface geometry is altered during retaining wall construction, there is redistribution of forces, and displacements occur. The following sections examine the causes of wall movements, the basic patterns of movement that occur, and the effect of these on the movement of the surrounding ground.

3.2.1 Causes of wall and ground movements

Wall movements occur because of the changing forces that are applied to a wall. For example, compaction of fill behind a gravity wall pushes it forward. Stressing the anchors in a wall pulls it toward the retained soil. Wall movements inevitably occur. The forces applied by the wall to support the soil are matched (because of equilibrium) by equal and opposite forces on the wall, and these result in deformations in the wall/soil system. Figure 3.9 shows this for a gravity wall constructed above the water table. Note that additional (and often important) forces can occur as the result of water pressures.

The magnitude of wall movements depends upon

- The type of retaining wall used
- The method of construction

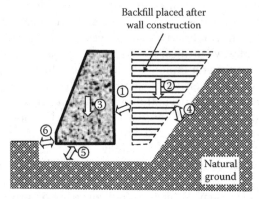

Backfill placed after wall construction

Natural ground

1. Resultant of earth pressure between backfill and wall
2. Self-weight of backfill
3. Self-weight of retaining wall
4. Resultant of forces between backfill and natural ground
5. Resultant of forces between soil and wall foundation
6. Toe force

Figure 3.9 Forces between soil and a gravity wall.

- Design details
- Stiffness of supports
- Wall flexibility
- Ground stiffness
- Loading due to water pressures

3.2.2 Rigid body movements

For a well-designed wall, movements will occur primarily because of the relatively low stiffness of soil under the loads the wall imposes on it (Figure 3.9). For a stiff mass-concrete gravity wall such as that shown in Figure 3.9, the wall will move as a rigid body, undergoing translation, settlement, and/or rotation. If the geotechnical design is not adequate, then the wall may fail, resulting in large and unacceptable movements occurring.

For example, a gravity wall placed on a weak rock foundation may slide horizontally as the backfill is placed and compacted behind it (Figure 3.10a). If the toe restraint is good, then sliding may be prevented, and ultimately the wall will try to rotate about the toe (Figure 3.10b). A cantilever sheet-pile wall driven through soft alluvium into good ground will also rotate and translate as a result of the loads applied by the soil it supports.

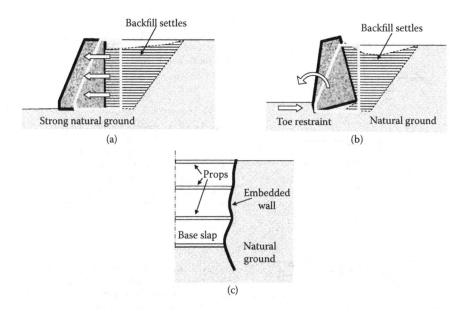

Figure 3.10 Examples of wall movements. (a) Horizontal translation (sliding) of a gravity wall. (b) Rotation about toe of a gravity wall. (c) Flexing (exaggerated) of an embedded wall.

3.2.3 Wall flexibility

Most walls are not completely rigid, so they undergo both rigid body movement and flexing. Steel sheet piling is particularly flexible, but even bored piles will flex significantly when used to support a deep excavation. A braced, strutted or anchored embedded wall constructed to retain the soil below adjacent buildings will undergo relatively small amounts of lateral translation at the support levels, once the supports are placed, but will flex between them (Figure 3.10c). However, it is inevitable that some horizontal movement of the wall will occur as the construction of the wall is carried out, resulting from

- The reduction in support for the soil when excavating for bored piles or diaphragm walls
- Wall movements as top-down excavation occurs in each stage of excavation, before support can be installed
- Wall flexure
- Movement in the support system as load comes onto it, for example, due to compression of struts and packing between struts and the wall, or extension of ground anchor tendons

Figure 3.11 gives sketches of some basic patterns of movement for different wall types.

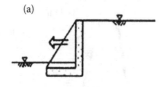

(a)

Reinforced concrete buttressed wall: structure does not bend, due to stiffening effect of buttresses. The wall may translate rather than rotate about its base, when loads are applied.

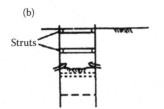

(b)

Struts

Strutted steel sheet piling: interlocking steel sheet piling is driven from ground level before excavation starts. Steel wales and struts are placed in position at the bottom of the excavation as it is dug. Between strut levels the sheets tend to deflect inward, effectively giving rotation about ground level.

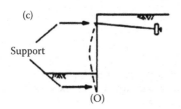

(c)

Support

(O)

Anchored sheet-pile wall: for pressure calculation is assumed to rigidly rotate about its toe (O). In reality, the sheets deflect as shown by the dotted line, causing a redistribution of pressures.

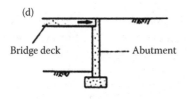

(d)

Bridge deck   Abutment

Bridge abutment: thermal expansion of the bridge deck (perhaps after casting *in situ* concrete) forces abutment against soil, and earth pressures rise. This effect also applies to strutted excavations under large ambient temperature changes.

(e)

Diaphragm wall: carefully constructed and restrained to restrict settlements beneath adjacent structures. Lateral movements are restricted, and full shear strength is not mobilized. Forces probably lie between 'at rest' (c) and 'active' (a), but rotation is not about bottom of wall.

Figure 3.11 Examples of retaining wall movements.

3.2.4 Associated ground movements

Ground movements are caused partly by the loads resulting from changes made to the ground surface geometry, and partly by wall movements. For example, the excavation of a basement causes unloading, leading to heave, whilst the inward movement of the wall toward the excavation leads to settlement behind the wall. In built-up areas, ground movements can damage adjacent structures. The design must not only ensure equilibrium of the wall-soil system but also keep movements of the wall and adjacent ground acceptably small.

3.3 EARTH PRESSURE PRINCIPLES

This section considers the basic principles controlling the earth pressures applied to walls. Classical earth pressure derivations are given in Appendix A. These provide earth pressure coefficients (Appendix B) for some detailed but simple methods of analyses for a range of wall and soil conditions, as will be discussed in Part II.

In principle, an excavation can be made in strong, intact rock without the need for a retaining structure. The rock has sufficient strength to support its own weight, and therefore does not apply earth pressure. However, weaker ground must be supported.

In the case of a very inflexible braced diaphragm or bored pile wall, it might be possible to prevent any significant ground movement by placing bracing at close centres as excavation proceeded, but this would mean that high pressures would develop behind the wall. Because soil is not a rigid material, some movement of the wall toward the excavation is required to mobilise the component of soil shear strength which helps to support the retained soil. Thus, movement of a wall away from the soil progressively reduces the effective stress applied to it, but only up to the point that the full shear strength of the soil is mobilised (i.e. the soil—but not necessarily the structure—is yielding).

For example, in Figure 3.7, excavation in front of the wall inevitably allows a little horizontal movement before the anchors (Figure 3.7a) or struts (Figure 3.7b) are placed, and during this time, horizontal pressures decrease behind the wall, in the supported soil, as its strength is mobilised. In an embedded wall, the load coming onto the back of the wall in the retained height produces movement toward the excavation (as in Figure 3.10c) and this causes an increase in the earth pressure between the wall and the soil below the excavation, on the excavated side of the wall.

Many small walls are constructed in a temporary excavation, as shown in Figure 3.4, and backfill is then placed behind the wall. It is normal, good practice to compact the backfill, to limit settlements of the ground surface behind the wall. Compaction can lead to much higher horizontal pressures behind a structure than would normally be expected from an *in situ* soil.

Four different situations can therefore be identified:

1. The ground is sufficiently strong that it can be cut vertically, or at a sufficiently steep angle for project needs, without falling into the excavation. In practice, it is possible to dig relatively shallow vertical faces in stiff clay, but they do not remain stable for long. As water gets into the retained soil, it may weaken, weather or fail. Unsupported vertical or near-vertical cuts are only possible in weak or strong rocks, and even here it is normally necessary to provide some degree of protection to the face in order to prevent the material weathering, and creeping down slope.
2. An unyielding, rigid wall is used, in order to protect adjacent structures from ground movements. The horizontal stresses to be supported by the retaining structure may be close to those in the soil before the start of construction, although installation effects, wall flexibility, support movements, etc. will generally mean that the stresses finally applied to the retained side of the wall are less than these.
3. In a further situation, where there are no adjacent structures, a wall may be allowed to flex and yield. The horizontal pressures in the supported soil are reduced, because some of the soil strength is mobilised, and helps to support the self-weight of the retained soil.
4. A temporary excavation is made, after which the wall is built, and backfill is compacted behind the wall. The type of backfill and the method of placing and compacting the soil will control the horizontal pressures applied to the wall.

3.3.1 Components of earth pressure

The loads applied to a retaining structure are produced by earth and water pressures that are the sum of the

- Effective stress normal to the face of the wall
- The pore water pressure at the wall face
- Shear stress between the wall and the soil

It is important to remember that most retaining structures must support any groundwater that is present, and that this can provide the largest component of load on the wall. A relatively strong soil may impose little effective stress, but the full water pressure will still have to be supported.

The magnitude and pattern of earth pressure is also significantly affected by

- The way in which a wall is placed
- Whether the wall moves way from, or toward, the soil
- The amount of movement it undergoes
- The pattern of wall movement (translation, rotation, flexure, etc.)

Simple (e.g. limit equilibrium) analysis cannot take account of all these factors. More complex continuum analysis (using finite element or finite difference methods) must be used.

3.3.2 Earth pressure at rest

Consider a deposit of soil formed by sedimentation in thin layers over a wide area. No lateral yield occurs as a result of the imposition of load upon it by the deposition of successive layers above. The *in situ* horizontal effective earth pressure (σ'_h) in such a soil is known as the '*earth pressure at rest*'.

Terzaghi used the concept of an earth pressure coefficient, K,

$$K = \sigma'_h / \sigma'_v \tag{3.1}$$

where

σ'_h is the horizontal effective stress at any depth below the soil surface, and

σ'_v is the vertical effective stress at any depth below the soil surface, which for the simple case of a uniform dry soil equals the product of the depth below the soil surface (m) and the bulk unit weight (kN/m^3) of the soil.

The effective horizontal and vertical pressures in the at-rest state are related by K_0.

$$K_0 = \left[\sigma'_h / \sigma'_v \right]_{\text{at rest}} \tag{3.2}$$

In principle, if it were possible to insert an embedded wall into the ground without disturbance, the earth pressures on either side of the wall would remain the same as in the ground before wall construction. Theoretical and experimental determinations have shown that the value of the coefficient of earth pressure at rest (K_0) lies between the active and passive earth pressure coefficients, K_a and K_p (see below).

Jaky (1944) developed the following theoretical equation for a granular material under first loading, i.e. in the normally consolidated state where the vertical effective stress has never been higher than at present:

$$K_0 = (1-\sin\phi').\frac{\left(1+\frac{2}{3}\sin\phi'\right)}{(1+\sin\phi')} \tag{3.3}$$

$$\approx (1-\sin\phi')$$

For example, if $\phi' = 30°$, $K_0 = 0.5$. Another analytical solution was found by Hendron (1963) who showed that, for an assembly of uniform frictional spheres,

$$K_0 = \frac{1}{2}\left[\frac{1+\sqrt{\frac{6}{8}}-3.\sqrt{\frac{6}{8}}\sin\phi'}{1-\sqrt{\frac{6}{8}}+3.\sqrt{\frac{6}{8}}\sin\phi'}\right] \tag{3.4}$$

In reality, however, many near-surface soils have at some time during their geological history been buried under considerable thicknesses of overburden, and have become 'over-consolidated'. The effect of over-consolidation is to increase the undisturbed lateral stress in the ground, and therefore to increase the *in situ* value of K_0. Mayne and Kulhawy (1982) give an empirical equation for K_0 which takes into account over-consolidation

$$K_0 = (1 - \sin\phi').OCR^{\sin\phi'} \tag{3.5}$$

where OCR = over-consolidation ratio, the ratio of the maximum effective vertical stress that has ever been imposed on the soil (the 'pre-consolidation pressure') to its current vertical effective stress level. Methods of determining pre-consolidation pressure can be found in most standard soil mechanics texts. Values of K_0 for typical values of ϕ' and OCR are given in Table 3.1.

Eurocode 7 combines Meyerhof's formula (Meyerhof 1976) with Kezdi's modification for sloping ground (Kezdi 1972) to give

$$K_0 = (1-\sin\phi').\sqrt{OCR}.(1+\sin\beta) \tag{3.6}$$

where β is the slope angle of the ground surface.

Cementation (e.g. during diagenesis) and weathering will alter the effects of over-consolidation. As a result, the *in situ* stress regime is rarely simple, and the horizontal *in situ* stress (σ_h) may need to be measured *in situ*. In heavily over-consolidated clays such as the London clay, K_0 can rise to 2 or 3 at shallow depth (Burland, Simpson and St. John 1979). It can be seen from Table 3.1 that at high over-consolidation ratios, the effective stress ratio can approach the value at passive failure.

Table 3.1 Coefficient of earth pressure at rest, K_0 (compared with failure values)

Over-consolidation ratio	Coefficient of earth pressure at rest (K_0)	
	$\phi' = 20°$	$\phi' = 30°$
1	0.66	0.50
2	0.83	0.71
4	1.06	1.00
10	1.45	1.58
20	1.83	2.24
	At failure (Rankine active and passive values)	
Active	0.49	0.33
Passive	2.04	3.00

Source: Mayne, P.W. and Kulhawy, F.H., *J. Geot. Eng. Div. ASCE*, 108, GT6, pp. 851–872, 1982.

3.3.3 Active and passive states

As noted above, and as shown in Figure 3.11, wall movement can occur in a number of different ways. The precise way in which the wall moves has an important effect on the distribution of pressures between the structure and the soil. In the late 1920s and early 1930s, Karl Terzaghi reported the results of a series of experiments on large model retaining walls. The apparatus used basically consisted of a box of dry sand, on one side of which was a wall, hinged at its base. When the wall was rotated about its base (as in the inset to Figure 3.12) the horizontal pressure between the soil and wall was found to increase linearly with depth below the surface of the soil, and thus, the moment applied to the wall could be measured and converted to a triangular horizontal pressure distribution.

Simplified sketched results of such a test are shown in the main part of Figure 3.12. It can be seen that when the sand is poured behind the propped wall, it exerts a pressure on it, even before the wall is moved. This is the earth pressure at rest discussed above. If the wall is allowed to move under this pressure, then it moves away from the sand, and the pressure (and hence the earth pressure coefficient, $K = \sigma'_h/\sigma'_v$) decreases. After a relatively small displacement, the minimum value of earth pressure coefficient is reached. Further displacement gives no further decrease in pressure.

If, on the other hand, the wall is moved toward the sand from its original position, the coefficient of earth pressure rises and continues to rise for much larger displacements. Eventually, however, a constant value is reached once again. The minimum value of earth pressure coefficient, when the wall yields away from the soil, is termed the coefficient of active earth pressure, K_a, while the maximum value (when the wall is pushed toward the soil) is termed the coefficient of passive earth pressure, K_p.

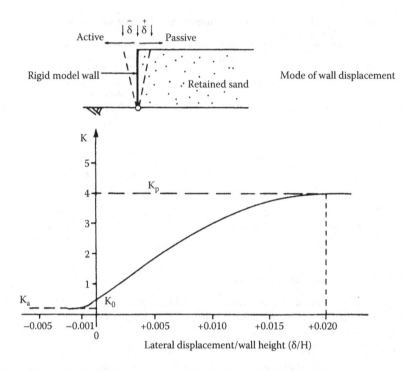

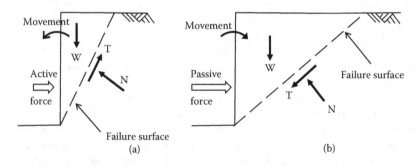

Figure 3.12 Model retaining wall tests, based on Terzaghi's experiments.

The reason for change in the earth pressure coefficient due to wall movement is that, as the wall moves away from its initial 'at rest' condition,
increasing shear stresses are applied to the soil. Eventually, these shear
stresses mobilise the full shear strength of the soil, and the soil then fails.
Figure 3.13 shows how failure surfaces might develop in the simple case

Figure 3.13 Failure states for soil supported by a smooth wall. (a) Active case. (b) Passive
case.

for sand supported by a smooth wall. In the active case (Figure 3.13a), the shear stresses on the failure surface in the soil, τ, produces a shear force, T, that helps to support the weight of the soil W. If the supported material were sufficiently strong, the shear stress would not reach its shear strength, and the face would stand unsupported. When active failure does occur in the soil, the angle of the shear surface will be steeper than 45°. In the passive case (Figure 3.13b), the shear stress, τ, and resultant shear force, T, act against the force pushing the wall into the soil. In this case, the normal stresses on the failure plane (σ_n) combines with the shear force and a relatively large force must be applied to the wall in order to bring the soil to failure.

3.3.4 Earth pressure coefficients

In the simplest case of a smooth rigid vertical wall retaining horizontal granular backfill, Rankine theory (see Appendix A for derivation) predicts

$$K_a = \frac{(1 - \sin\phi')}{(1 + \sin\phi')}$$

and (3.7)

$$K_p = \frac{(1 + \sin\phi')}{(1 - \sin\phi')} = \frac{1}{K_a}$$

For this case, when $\phi' = 30°$, $K_a = 0.33$ and $K_p = 3.0$. However, more sophisticated earth pressure coefficients are required in most situations, because there is friction between the wall and the soil, the soil may have cohesion, and the back of the wall and the surface of the retained ground may not be vertical and horizontal respectively. Table 3.2 summarises the development of analytical expressions for earth pressure, aimed at providing for increasingly complex geometries and soil/wall characteristics (Figure 3.14). Appendix A gives the derivations of methods that have been used to cater to the many different situations that arise, whilst Appendix B gives tables of the earth pressure coefficients obtained from these equations.

Recently Eurocode 7 (BS EN 1997-1: [2004] Annex C) has provided charts that allow active and passive earth pressures to be calculated for vertical and inclined rough walls, supporting both horizontal and sloping ground surfaces, as well as frictional and cohesive backfill. The active and passive (*total*) earth pressures acting normal to the back of the wall can be calculated using (Bond and Harris 2008):

Table 3.2 Development of analytical and graphical earth pressure coefficients

Source	Capability						
	α (°)	β (°)	c'	ϕ'	c_w'	δ'	A/P
Rankine (1857)	90	β		ϕ'		$= \beta$	A/P
Mayniel (1808)	90	0		ϕ'		δ'	A
Müller-Breslau (1906)	α	β		ϕ'		δ'	A
Bell (1915)	90	0	c'	ϕ'	0	0	A/P
Caquot and Kerisel (1948)	α	β		ϕ'		δ'	P
BS CP2 (1951) based on Packshaw (1946)	90	0	c'	ϕ'	c_w'	δ'	A/P
BS 8002 (1994) based on Kerisel and Absi (1990) and Bell (1915)	90	β	c'	ϕ'	$c_w' \, c_w'$	δ'	A/P
BS EN 1997 (2004)	α	β	c'	ϕ'	c_w'	δ'	A/P

Note: A, active; P, passive.

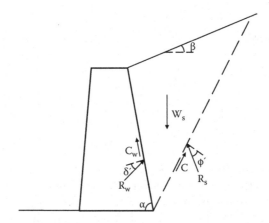

Figure 3.14 Definition of parameters in Table 3.2.

$$\sigma_a = K_a \left[\int_0^z (\gamma \, dz) + q - u \right] - 2c'\sqrt{K_a(1 + a/c)} + u$$

$$\sigma_p = K_p \left[\int_0^z (\gamma \, dz) + q - u \right] + 2c'\sqrt{K_p(1 + a/c)} + u$$

(3.8)

where

K_a and K_p are the active and passive earth pressure coefficients, given in Figures C.1.1 through C.2.4 of EN 1997-1 Annex C.

γ is the bulk unit weight of the soil.

z is depth below the soil surface.

q is a surcharge applied uniformly at the ground surface.

u is the pore water pressure at depth z.

c' is the effective cohesion intercept of the soil.

a is the adhesion (c'_w above) between the soil and the wall (normally assumed to be zero).

Annex C of EN 1997-1 also provides a numerical procedure for determining passive earth pressures coefficients, as a function of

- The effective angle of friction, ϕ', of the soil
- Wall/soil friction, δ'
- The inclination of the soil surface relative to horizontal, β, taken as positive when the soil surface rises away from the wall
- The inclination of the wall/soil interface to the vertical, θ, taken as positive when the soil overhangs the wall

Adhesion (a or c'_w) between the soil and the wall is taken as $c' \tan \delta'/ \tan \phi'$.

The normal earth pressure on the wall resulting from applying normal stress to the surface of the soil is given as

$$K_n = \frac{1 + \sin \phi' \sin(2\,m_w + \phi')}{1 - \sin \phi' \sin(2\,m_t + \phi')} \exp(2v \tan \phi') \tag{3.9}$$

where

$$2\,m_w = a\cos\left(\frac{\sin \delta'}{\sin \phi'}\right) + \phi' + \delta'$$

$$\tag{3.10}$$

$$2\,m_t = a\cos\left(-\frac{\sin \beta_0}{\sin \phi'}\right) + \phi' + \beta_0$$

$\beta_0 = \beta$ for $c' = 0$ with surface load vertical or zero

$$v = m_t + \beta - m_w - \theta \tag{3.11}$$

From this, the earth pressure coefficients for vertical loading on the surface (K_q), cohesion (K_c), and soil self-weight (K_γ) can be found from

$$K_q = K_n \cos^2 \beta$$

$$K_c = (K_n - 1)\cot \phi' \qquad\qquad (3.12)$$

$$K_\gamma = K_n \cos\beta \cos(\beta - \theta)$$

These equations can readily be solved using a spreadsheet. Active earth pressure coefficients can be obtained by entering negative values for ϕ', δ' and c'.

Calculations of earth pressure acting on retaining structures are more sensitive to some inputs than others. Key decisions are

- Whether active, at-rest or passive conditions are relevant
- What the groundwater levels are on either side of the wall
- How much surcharge needs to be allowed for on the retained side of the wall
- Whether the soil surfaces on either side of the wall are horizontal, or slope
- What wall friction and adhesion can be justified

The presence of a sloping ground surface increases (or decreases) the earth pressure acting on an embedded wall when the ground rises (or falls) with increasing distance from the wall face. This is taken into account in most earth pressure theories through the inclusion of an additional parameter (β) to represent the slope angle in the formulation of the active earth pressure coefficient K_a.

For example, Müller-Breslau's (1906) solution for the horizontal component of K_a against a vertical wall is

$$K_{a,h} = \frac{\sin^2(90° + \phi)}{\sin(90° - \delta).\left[1 + \sqrt{\dfrac{\sin(\phi + \delta).\sin(\phi - \beta)}{\sin(90° - \delta).\sin(90° + \beta)}}\right]^2} \qquad (3.13)$$

where ϕ is the soil's angle of shearing resistance and δ the angle of friction between soil and wall. The degree to which rising ground adversely affects earth pressures can be quantified by calculating the ratio

$$R_a = \frac{K_{a,h}}{K_{a,h,\delta=0}} \qquad\qquad (3.14)$$

for different angles of slope and soil resistance, as illustrated in Figure 3.15. When the slope angle β is less than about $1/2\phi'$, the value of R_a increases almost linearly with β (depending of the value of ϕ'). But when the slope angle β exceeds $1/2\phi'$, the value of R_a increases rapidly as β approaches the value of ϕ. As a simple rule of thumb, it seems wise therefore to limit the slope angle to less than 1/2 if at all practicable.

The presence of a sloping excavation decreases the passive earth pressure coefficient and therefore the force supporting the embedded part of the wall, when the ground falls away with increasing distance from it. For example, Müller-Breslau's solution for the horizontal component of K_p against a vertical wall is

$$K_{p,h} = \frac{\sin^2(90° - \phi)}{\sin(90° + \delta).\left[1 - \sqrt{\dfrac{\sin(\phi + \delta).\sin(\phi + \beta)}{\sin(90° + \delta).\sin(90° + \beta)}}\right]^2} \tag{3.15}$$

The degree to which a falling excavation adversely affects earth pressures can be quantified by calculating the ratio

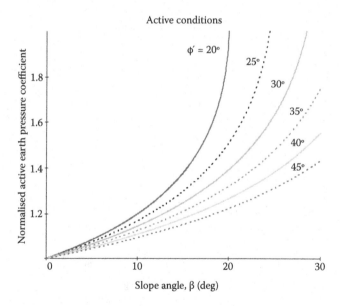

Figure 3.15 Effect of slope angle on normalised active earth pressure coefficient: ratio $R_a = K_{ah}(\delta' > 0)/K_{ah}(\delta' = 0)$ versus slope angle, β, for different angles of soil shearing resistance, ϕ', from 20° to 45°. Angle of wall friction assumed equal to 2/3 times δ'.

$$R_p = \frac{K_{p,h}}{K_{p,h,\delta=0}} \qquad (3.16)$$

for different angles of slope and soil resistance, as illustrated in Figure 3.16. The adverse effect of a sloping excavation increases dramatically as angle β approaches ϕ, and is greater for soils with larger angles of shearing resistance.

Figure 3.17 shows the effect of three common assumptions in the literature regarding wall friction on the values of the earth pressure coefficients K_a and K_p. The values of K_a and K_p have been calculated using the equation given in Annex C of Eurocode 7 Part 1 (2004), taking account of the angle of wall friction δ'—and then normalised by corresponding values of K_a and K_p calculated with $\delta' = 0°$. The normalised values are given by

$$\psi_a = \frac{K_a}{K_{a,\delta'=0}}$$
$$\qquad (3.17)$$
$$\psi_p = \frac{K_p}{K_{p,\delta'=0}}$$

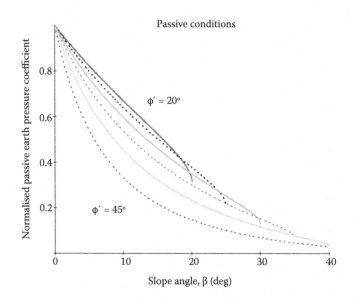

Figure 3.16 Effect of slope angle on normalised passive earth pressure coefficient. Ratio $R_p = K_{ph}(\delta' > 0)/K_{ph}(\delta' = 0)$ versus slope angle, β, for different angles of soil shearing resistance, ϕ, from $20°$ to $45°$. Angle of wall friction assumed equal to 2/3 times ϕ.

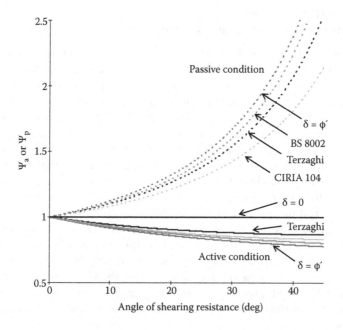

Figure 3.17 Effect of wall friction assumption on earth pressure coefficients.

The graph illustrates two important points:

- Wall friction has a much greater effect on values of ψ_p than on ψ_a.
- The value of ψ_a is relatively insensitive to the soil's effective angle of shearing resistance, ϕ'; whereas the value of ψ_p is highly sensitive.

The relative conservatism of the recommendations with regard to wall embedment (which depends on the ratio ψ_a/ψ_p) has been found to be (most to least conservative): CIRIA > Terzaghi > EAU (German Committee for Waterfront Structures) > BS 8002.

In selecting an appropriate value of wall friction for the design of an embedded retaining structure, the engineer should bear in mind the advice given in Chapter 2, Section 2.3.3, and that

- Wall roughness and vibration needs to be assessed in the context of soil particle size. A wall may be considered smooth when sheet piling is supporting a gravel, for example, but rough when a bored pile wall is supporting a clay.
- The more significant effect of wall friction on K_p than on K_a.
- Precedent practice most often suggests $\delta' = 2/3\phi'$, when ϕ' is determined using routine methods based on particle size and *in situ* testing.

3.3.5 Wall movements required for active and passive conditions

It can be seen in Figure 3.12, that when a wall retains normally consolidated granular soil, larger displacements are required to reach the passive state than to reach the active state. This is because the soil mass is compressible, and in the passive state the volume of soil involved in the failing wedge is much larger than in the active state. Further, the change in applied stress is larger in the passive state than in the active state. Therefore, in the passive case, more wall displacement is required before the shear stress is applied to the farthest part of the shear surface from the wall. In practice, loose sands and soft clays are more deformable than dense sands and stiff clays. For these more deformable materials, the wall movements necessary to mobilise full passive resistance may be unacceptably high.

If movements are restricted then active or passive conditions (or both) may not be reached, and the pressures on the wall will be different from those that might be calculated using theoretical rigid-plastic solutions. For example, pressures on the retained side of an embedded wall may be higher than 'active', whilst pressures below the excavated side of an embedded wall will be less than passive, thus requiring more support from anchors, or struts.

As an example, Rowe and Peaker (1965) considered that a wall movement of 5% of the wall height represented the limit of acceptability, and on this basis determined the available passive resistance and compared it with the peak passive resistance from theoretical considerations. For loose sands, they found in large-scale experiments on sheet-pile walls that the maximum passive pressure occurred for wall movements of between 25% and 40% of the wall height. For dense sand movements at maximum passive resistance were equal to approximately 5% of the wall height. Table 3.3 compares the values of passive earth-pressure coefficient on theoretical grounds with those recommended by Rowe and Peaker, based upon experimental results, to restrict wall movements in loose sand to 5% of the

Table 3.3 Comparison of theoretical K_p values with those recommended by Rowe and Peaker (1965) to restrict wall movements to 5% of wall height

| δ' (°) | Passive earth pressure coefficients, K_p | | |
	Theory	*Recommended*	*Ratio, theory/recommended*
0	3.4	2.5	1.4
10	4.5	3.0	1.5
20	5.6	3.6	1.6
30	6.7	4.3	1.6

Note: Results for loose sand, $\phi' = 33°$. Theoretical values from CP2:1951. Vertical wall, with horizontal surface to backfill.

wall height. It can be seen that in loose sand deposits, a reduction factor of about 1.5 is required on the passive earth pressure calculated for a rigid plastic material, in order to limit wall movements to 5% of the wall height.

In normally consolidated soil, the *in situ* earth pressure coefficient will be quite low, and typically between 0.4 and 0.6 (see Table 3.1). For an anchored sheet-pile wall, normal anchor movements can be expected to be large enough to allow active conditions to develop, and reserve of strength has traditionally been built into the passive pressures by making the wall longer than required simply for equilibrium. However, in heavily over-consolidated soils, the *in situ* effective stress ratio $\left(\sigma'_h/\sigma'_v\right)$ near to the ground surface may be so high that passive failure conditions are approached. Under these conditions, the pressures on the excavated side of an *in situ* embedded wall will probably be similar to those at passive failure, whilst on the retained side pressures will remain higher than active. In this situation, a margin of safety needs to be applied to the active, driving pressures, or perhaps to both active and passive pressures, when carrying out simple limit equilibrium analyses.

The general principles based on Terzaghi's observations are still widely used in calculating earth pressures, but this application is not completely satisfactory because (as described earlier in this chapter) structures do not generally behave in the same way as his models. Limit equilibrium earth pressure calculations generally use coefficients of earth pressure derived from analytical methods (see Appendices A and B), and these methods assume that

   a. Wall movements can be divided into active and passive zones, relative to the soil.
   b. Within each zone, the wall is rigid and rotates sufficiently to mobilise full active pressure, or an assumed proportion of passive pressure.
   c. Pressure distributions are triangular (hydrostatic), increasing linearly with depth within each soil type.

In the next sections, we examine some effects of departures from idealised behaviour.

3.3.6 Strain softening and progressive failure

As Terzaghi noted, real soils do not act in the way that is assumed for the sake of simplicity. In particular, classical limit equilibrium analyses imply a soil which behaves as a rigid plastic material (Figure 3.18a) so that the peak strength of the soil is mobilised at the same instant throughout the soil mass, after an infinitesimally small displacement of the wall. Most soils exhibit strain hardening, followed by plastic failure which may or may not be associated with a drop in shear strength (strain softening) as strains

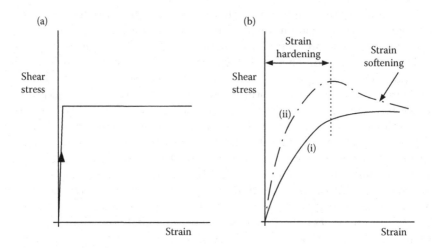

Figure 3.18 Idealised and observed stress-strain behaviour of soil. (a) Idealised, and
(b) observed.

are further increased (Figure 3.18b). Curve (i) in Figure 3.18b is typical of
loose sands and soft, normally consolidated clays, and these materials do
not display significant strain softening. Curve (ii) in Figure 3.18b is typical
of the behaviour of dense sands or stiff, over-consolidated clays. For these
materials, once peak strength is achieved, further strains lead to a reduction in available strength, termed 'strain softening' or 'brittleness'.

Since real soils are compressible and can strain soften, there are situations
where the peak strength cannot be mobilised over the entire shear surface at
any one time. Rowe and Peaker (1965) consider an analogy of a series of rigid
blocks, interconnected by springs to represent the soil compressibility (Figure
3.19a). Force applied to the wall will be transmitted to block A, but block B
will not receive any thrust until the peak shear strength has been mobilised
at the base of block A. Once the shear strength at the base of block A is fully
mobilised, then block B will contribute to the resistance of the soil mass. For
this to happen, there must be more movement of block A, which for a strain-
softening material will imply a reduction in the shear strength at 'a', throwing
more load onto block B. Figure 3.19b shows a possible distribution of shear
stress along the shear surface at some stage when Q_p reaches a maximum.

By the time the shear surface reaches block F and emerges at ground
level, the shear strength at 'a' (block A) will be reduced from its peak value.
The maximum force, Q_p, that the soil can resist will be much less than is
calculated on the basis of the available peak shear strength, if strain softening occurs. If strain softening does not occur, then theory should be able
to provide a reasonable prediction for Q_p. This is borne out by the experimental results in Figure 3.20, which shows that data from tests on loose

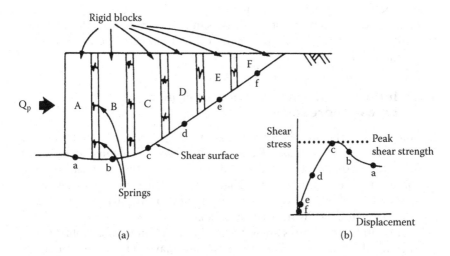

Figure 3.19 Rowe and Peaker's analogy to allow for the effects of progressive failure. (a) Rigid soil blocks connected by springs. (b) Mobilization of shear stress.

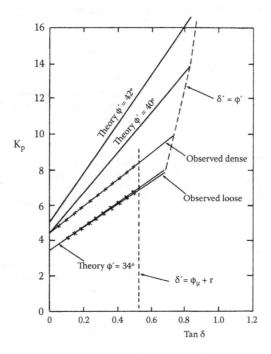

Figure 3.20 Comparison of maximum thrust on model walls with theoretical values. (From Rowe, P.W. and Peaker, K., *Géotechnique* 15, 57–78, 1965.)

sand coincide with predictions made using $\phi' = 34°$ (a reasonable value for a sand in a loose state) whilst for dense sands the predicted values (theory $\phi' = 40°$ and $42°$) are much greater than the observed values, particularly for a rough wall.

3.3.7 Influence of type of wall movement on earth pressure

As noted above, the initial observations made by Terzaghi were for a rigid wall rotating about its base. Figure 3.21 shows the idealisations so far considered, i.e. the 'at-rest', active and passive conditions. As can be seen by comparing this figure with the shape of the deflected walls shown, for example, in Figures 3.10 and 3.11, two significant departures from ideal behaviour occur as walls translate and flex.

Most retaining structures are not rigid, but bend under the applied loads. Often, they do not rotate about the base, but may well translate or rotate about the top or some other part of the structure. These departures from the most commonly assumed behaviour can cause significant deviations from pressure distributions calculated on that basis.

In the late 1920s and 1930s, work continued on refining and extending the available analytical solutions of earth pressure. But objections to the

Idealised cases

(a)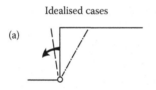

Active: rigid structure rotates away from soil about its base. Eventual soil failure involves a small mass of soil, which is partly supported by the shear stresses on the failure plane. Pressures are low.

(b)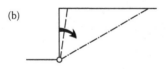

Passive: rigid structure rotates toward the soil about its base. Eventual soil failure involves a large mass of soil, with shear strength acting against the wall. Pressures are high.

(c)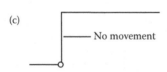

— No movement

Earth pressure at rest: structure is rigid, does not move, and can be placed in the soil without allowing any lateral soil movement. Lateral pressures existing in the soil before wall installation are applied to the wall.

Figure 3.21 Idealised wall movements. (a) Active case, (b) passive case and (c) earth pressure at rest.

classical solutions arose, based partly on a reassessment of their assumptions, and partly on the fact that observations of excavations reported by, for example, Meem (1908) and Moulton (1920), indicated pressure distributions which did not increase linearly with depth, as implied by Coulomb.

Coulomb had determined that the pressure behind a wall in the active condition increased 'hydrostatically' (i.e. linearly with depth), by differentiation of the expression for the active thrust with respect to the height of the wall. He had failed to realize, however, that this was not the only possible solution and that his result implied a particular mode of wall movement, namely rotation about the base of the wall. For the soil in the wedge adjacent to a wall to come from K_0 conditions to the active condition, it is necessary to reduce the thrust on the wall which acts to support the soil. This change in normal stress, accompanied by an increase in shear stress, implies a volumetric increase in the soil which is determined only by the geometry and the properties of the soil. Thus, to achieve the Coulomb active condition, the wall must move in a prescribed manner, namely by rotating about its base.

Terzaghi (1936) concluded that other modes of wall movement could yield very different pressure distributions, partly on the basis of reasoning, partly as a result of analysis and partly as a result of observations of retaining structures. If, for example, rotation occurs about the top of the wall, then after a small movement, the soil close to the base of the wall (Figure 3.22) will have mobilised its full shear resistance, and will attempt to move downward. Because the soil above has not yet reached failure, it

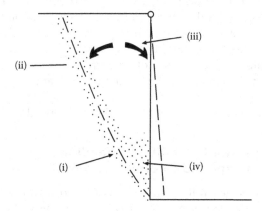

Figure 3.22 Arching during rotation about the top of a rigid retaining wall. (i) Yield at base mobilises full shearing resistance at that level. (ii) Zone rupture occurs as upper retained material follows the lower material downward. (iii) Because of insufficient yield at the top of the wall, the horizontal stress at the top of the wall increases. (iv) Soil 'hangs' at the top of the wall, reducing pressure at the base.

will be partially suspended by the shear forces on the final shear surface and the top of the back of the wall. The soil will 'arch' between the wall and the shear surface, and the centre of pressure will be moved upward relative to the position it would have had if the wall had rotated about its base.

A number of relatively complex analyses have been made of the condition of rigid rotation about the top of the wall. Ohde (1938) used the arc of a circle to represent the failure surface, while Terzaghi (1941) produced his 'General Wedge Theory' based on a logarithmic spiral shear surface. Neither of these methods is currently in use, because of their complexity, but they demonstrate that, for rotation about the top of the wall,

a. The centre of pressure is high—Ohde computed a theoretical value of 0.55h above the base of a wall of height h, while for the Coulomb active condition the centre of pressure is at 0.33h.

b. The total thrust on the wall remains approximately equal to the Coulomb active value, provided the wall yields sufficiently—Terzaghi (1943) obtained a maximum 11% difference between the Coulomb and General Wedge Theory forces, for a soil with an internal effective angle of friction, $\phi' = 38°$, and with an effective angle of wall friction, $\delta' = 0$.

When the wall translates, arching similarly occurs, but only in the initial stages of wall movement. Terzaghi (1936) identified two stages during wall yield:

a. When limited yield occurs $((\Delta/H)_{average} = 0.0005$, for a dense sand), the value of the total thrust rapidly falls to its Coulomb value; however, at this stage the pressure distribution on the back of the wall is far from hydrostatic, and the centre of pressure is high.

b. With further yield the centre of pressure drops to its Coulomb value of 0.33h, when the yield (Δ/H) is about 0.005 at the top of the wall.

These ideas are expressed in Figure 3.23, after Terzaghi (1936).

3.3.8 Effect of wall flexibility on bending moments applied to embedded walls

Classical solutions generally implicitly assume rotation of a rigid wall about its base. In practice, many reinforced concrete walls behave in an approximately rigid manner, but steel sheet-pile walls can be much more flexible. Experience in the first half of the twentieth century showed that, for anchored steel sheet-pile walls, designs based on the classical earth pressures were over-conservative in terms of the thickness of the steel section required to support the bending moments applied by the soil.

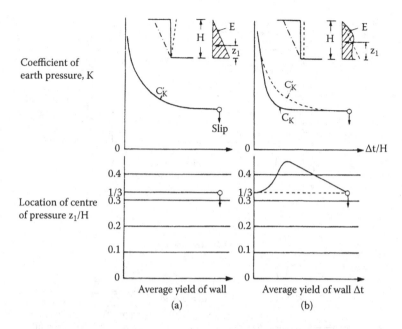

Coefficient of
earth pressure, K

Location of centre
of pressure z_1/H

Figure 3.23 Idealised relationships between the average yield of a wall and the coefficient
of earth pressure. (a) Wall rotates about its base. (b) Wall translates away
from the soil. Shaded areas represent the resulting distribution of earth pressure immediately after the total lateral pressure has become equal to the
Coulomb value. (From Terzaghi, K., *J. Boston Soc. Civil Eng.* 23, 71–88, 1936.)

Early explanations considered vertical arching, in much the same way
as Terzaghi's General Wedge Theory. Rowe (1952), however, pointed out
that for anchored sheet-pile walls' vertical arching is unlikely because the
anchor will yield sufficiently to restore hydrostatic pressures in the soil.
It is now accepted that bending moments in the sheets are affected by the
deflected shape of the wall below dredge level and that this is a function of
the flexibility of the wall relative to the soil.

Figure 3.24a shows the conventional assumed earth pressures on an
anchored sheet-pile wall. The pressure from the supported soil causes wall
movement, and it is assumed that full active and passive pressures can be
mobilised on the wall. Thus, the centre of passive pressure is at one third of
the embedded depth from the base of the wall, and the wall can be simplified into the beam shown in Figure 3.24b, where the maximum bending
moment will be a function of (approximately) the square of the span, L.

Figure 3.24c shows the simplified 'assumed-rigid' passive pressure distribution for a rigid wall rotating about its base. This is the assumption
made to obtain the pressure distribution in Figure 3.24a. Figure 3.24c also
shows the type of pressure distribution observed by Rowe (1952) on model

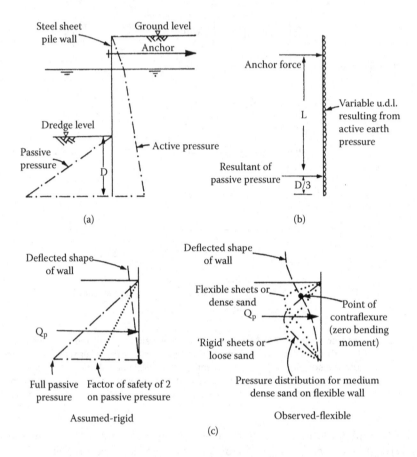

Figure 3.24 Mechanism of moment reduction due to wall flexibility. (a) Assumed (design) earth pressures. (b) Equivalent simply supported beam. (c) Wall deflection and passive pressure distributions below dredge level for rigid and flexible walls.

flexible walls on sand. Generally, there is a point of contraflexure in the wall some distance below dredge level. For very dense sands, the point of contraflexure may be at, or slightly above, dredge level. For loose sands, it will be lower. Because the deflections at the bottom of the wall are small, passive pressures are not obtained. Therefore, the pressure distribution for a medium-dense sand might be parabolic, as shown. For this condition, the resultant force acts near to D/2 from the base of the wall, and therefore the equivalent span, L, and hence the maximum bending moment applied to the steel sheets, are both reduced.

The deflected shape of the wall is a function of the stiffness of the sheets relative to the stiffness of the soil. As the wall becomes more flexible relative

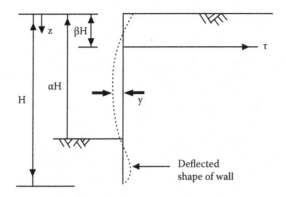

Figure 3.25 Terms used in Rowe's analysis.

to the soil, the position of the resultant passive force, Q_p, moves up, pro-gressively reducing the applied maximum bending moments.

   Rowe (1952) carried out tests on model walls, 500–900 mm high, of differing metal thickness, supporting various soils in loose and dense con-ditions. For similitude between the model and the prototype (Figure 3.25), there must be geometrical similitude. If slopes of the deflected shape of the wall are to be equal at corresponding points,

$$\left(\frac{dy}{dz}\right)_{model} = \left(\frac{dy}{dz}\right)_{prototype} \tag{3.18}$$

but if slope = dy/dz then bending moment

$$M \propto \frac{d^2y}{dz^2} \quad \text{since} \quad \frac{M}{EI} = \frac{d^2y}{dz^2} \tag{3.19}$$

and shear force

$$S \propto \frac{d^3y}{dz^3} \tag{3.20}$$

and load

$$P \propto \frac{d^4y}{dz^4} \tag{3.21}$$

Therefore, from the above

$$\int\left(\frac{Mdz}{EI}\right)_{model} = \int\left(\frac{Mdz}{EI}\right)_{prototype} \tag{3.22}$$

i.e. at any depth, for similitude

$$\left(\frac{Mz}{EI}\right)_{model} = \left(\frac{Mz}{EI}\right)_{prototype} \tag{3.23}$$

If $z = H$, and as, for a triangular pressure distribution, horizontal stress αH, then

- shear force αH^2;
- bending moment αH^3

i.e.

$$M = \zeta H^3$$

and, for similitude,

$$\left(\frac{\zeta H^3 \eta H}{EI}\right)_{model} = \left(\frac{\zeta H^3 \eta H}{EI}\right)_{prototype} \tag{3.24}$$

so that (H^4/EI) must be the same for both model and prototype. Rowe termed this 'ρ' and thus was able to use the results of his model tests to provide design curves for moment reduction in full-scale structures (see Section 10.3.4).

A recent contribution to the effects of wall flexibility has been made by Diakoumi and Powrie (2009), using simplified kinematically admissible strain fields in the Mobilised Strength Design approach (Osman and Bolton 2004) to analyse embedded retaining walls propped at the crest. For the assumptions and simplifications made by Diakoumi and Powrie, the method shows that as the wall flexibility or soil stiffness increase, the bending moments may fall by about 20% of the values calculated using the Free Earth Support method (described in Chapter 10), whilst the prop force may reduce by about 15%. In contrast, Rowe's moment reduction factors suggest reduction of maximum bending moment by as much as 70% when wall flexibility is high.

At this stage, caution should be exercised in applying large moment reductions. As long ago as 1953, Skempton, mindful of the fact that Rowe's moment reduction factors were derived from model tests, suggested that when used in design the amount of reduction should be as follows:

Sands: use 1/2 of the moment reduction suggested by Rowe.
Silts: use 1/4 of the moment reduction suggested by Rowe.
Clays: use no moment reduction.

3.3.9 Stress relief during *in situ* embedded wall construction

The construction of an *in situ* embedded wall, whether from bored piles or diaphragms, involves excavation of the soil and replacement with reinforced concrete. Excavation inevitably reduces the *in situ* total stresses in the soil to either zero (if the hole is unsupported), or some value smaller than originally present, for example, when a bentonite slurry is used for temporary support. In heavily over-consolidated cohesive soils, the placement of fluid concrete in the ground does not bring the horizontal total stresses back to their original values—lateral stress relief occurs in the vicinity of the wall. These total stress reductions are known as 'installation effects'.

The construction of walls in inner-city sites routinely requires that lateral wall displacements must be minimised in order to prevent damage to adjacent buildings. If lateral wall movements are kept small then it is possible that the wall will have to support much higher than active pressures, perhaps approaching earth-pressure-at-rest (K_0) values. Thus, the price to be paid for restricting ground movements adjacent to a construction excavation will be the need to design a stronger and stiffer wall and support system.

In normally consolidated soil, *in situ* horizontal total stresses may be significantly higher than active stresses. Taking the Rankine and Jaky values for K_a and K_0 respectively, i.e.

$$K_a = (1 - \sin \phi')/(1 + \sin \phi')$$
$$K_0 = 1 - \sin \phi' \tag{3.25}$$

yields, for a cohesionless soil with $\phi' = 30°$, $K_a = 1/3$ and $K_0 = 1/2$, suggesting a 50% increase in earth pressure. But as noted in Section 3.3.2 above, in heavily over-consolidated clays, such as the London Clay, K_0 may rise to well in excess of 2 at the surface (Bishop et al. 1965; Skempton 1961; Simpson et al. 1979). If it is assumed that, as a result of the rigidity of a wall and its support system, such stresses remain at the end of construction, then prop forces and bending moments will be calculated that are many times those derived on the basis of active pressure (Potts and Fourie 1985). Yet, despite the fact that few designers have, until recently, taken *in situ* earth pressure into account, walls have not failed.

This difference between predicted and observed behaviour is probably due to the influence of wall installation processes on *in situ* stresses. Gunn and Clayton (1992) have noted that retaining walls may be divided into two types: filled walls, where the wall is constructed above ground and backfill is subsequently placed against it, and embedded walls, where the wall is constructed within the soil mass and the ground is subsequently removed from the front of it. Wall construction (installation) effects are only relevant in the case of embedded walls, although the construction method (particularly the method of backfill placement) is important for back filled walls, as will be described below.

We have seen that embedded walls may be of either a displacement or a replacement type. Displacement type walls are typically placed by driving either steel or pre-cast concrete sections into the soil. For the section to be driven, it must be relatively slender—it is not likely that the driving of commonly used steel sheet-pile sections will lead to a significant increase in *in situ* horizontal stress conditions. However, the excavation of diaphragm wall panels or bored piles is certain to result in significant total stress reduction, because during formation of the wall sections, a hole (which may be unsupported or may be supported by bentonite slurry) must be excavated in the soil. The total horizontal stress on the boundary of this hole will reduce from the initial *in situ* horizontal total stress in the undisturbed soil to either zero (if the hole is unsupported) or to a value which approximates to the pressure exerted by a fluid with the same bulk density as bentonite. The total stress acting on the soil is then increased to a value approximating to the pressure applied by wet concrete, at least in the upper 10 m of the wall (Figure 3.26). If the soil subsequently swells against the concrete, before bulk excavation takes place, horizontal pressures may rise somewhat.

A field experiment on a small-diameter bored pile in London Clay, reported by Mililitsky (1983), also showed little increase in the measured total stress over a period of 150 days following concreting (Figure 3.27). But this may have resulted from the rather long time taken in excavating the bore, placing the instrumentation, and subsequently concreting. Field observations of the Bell Common Tunnel, both during construction and afterward, showed not only that 30% of the measured ground surface settlements adjacent to the wall occurred during the process of the installation of the secant-bored pile, but also that this was associated with a significant reduction in horizontal total stresses, particularly in the London Clay, where K_0 was initially higher (Figure 3.28). Measured bending moments over a period of five years after construction were much lower than was predicted during design, perhaps partly as a result of this effect.

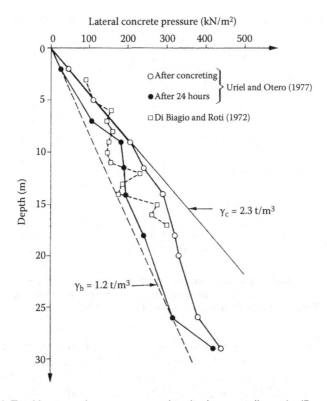

Figure 3.26 Total horizontal stress measured in diaphragm wall panels. (From Clayton, C.R.I. and Mililitsky, J., *Ground Engineering* 16(2), 17–22, 1983.)

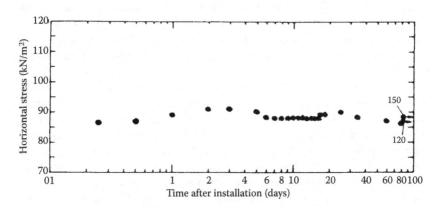

Figure 3.27 Measured total horizontal stress as a function of time since concreting, for a small-diameter bored pile installed in the London clay. (From Mililitsky, J.M., *Installation of Bored Piles in Stiff Clays: An Experimental Study of Local Changes in Soil Conditions.* Ph.D. thesis, University of Surrey, 1983.)

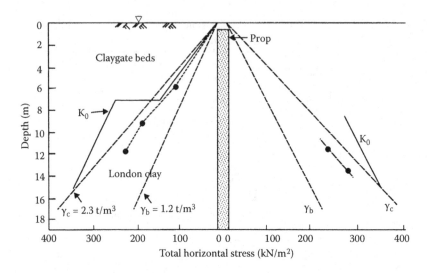

Figure 3.28 Total horizontal stress measured adjacent to the Bell Common retaining wall, after installation of secant piles and before main excavation. (From Tedd, P. et al., *Géotechnique* 34 (4), 513–532, 1984.)

Numerical analyses of retaining walls, using finite element methods, have shown the important implications of both construction detail and installation effects. Potts and Fourie (1985) and Fourie and Potts (1989) have demonstrated that for a propped cantilever wall in clay, finite element analyses can provide predictions of equilibrium depths of embedment which agree with those calculated using simple limit equilibrium solutions, regardless of the initial value of horizontal total stress assumed for the soil. But, in contrast, the values of prop force and maximum bending moment will be much higher than are predicted by limit equilibrium methods, when the assumed initial horizontal *in situ* stresses in the finite element analyses are high.

Figure 3.29 shows how this effect can be visualized on the basis of simple limit equilibrium analyses. For normally and lightly over-consolidated soils, *in situ* horizontal stresses are closer to the active than the passive state. Conventional analyses of a uniform soil deposit, using a factor of safety on passive pressure, predicts that the prop force will be 302 kN/m run of wall, while the bending moment will be 1916 kNm/m run. For a heavily over-consolidated soil, however, the *in situ* horizontal stress close to the ground surface will approach the passive value. Full passive pressures will therefore remain after excavation at the front of the wall, and the pressures behind the wall will not fall to active (see, for example, the measured earth pressures given by Carder and Symons 1989). For this scenario, the prop force and maximum bending moments will be doubled if moment

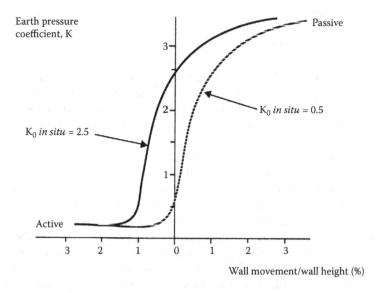

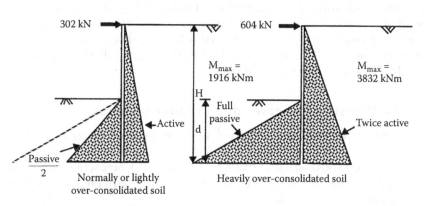

Figure 3.29 Example limit equilibrium calculation to show effects of initial K_0 value on calculated prop force and maximum bending moment. (From Gunn, M.J. and Clayton, C.R.I., *Géotechnique* 42, 137–141, 1992.)

equilibrium is maintained as before, because the horizontal stresses to be supported must be twice the active value. However, in reality, the stresses on the wall will be a function of installation procedures and, in addition, of wall flexibility.

Finite element analyses reported by Kutmen (1986), Higgins et al. (1989) and Gunn et al. (1992) have all shown the sensitivity of the calculated wall bending moments to the details of the construction procedure. In a coupled-consolidation analysis, modelling in a simple way the complete construction process, Gunn et al. (1992) have demonstrated the following:

- The impact of installation effects will depend upon the position of the groundwater table. When the groundwater table is high, installation has little effect, since most of the total active pressure is as a result of the water pressures.
- The greater the restraint imposed at the top of the wall, for example by propping or anchoring, the larger will be the effects of installation. This results from the fact that if the wall is relatively unrestrained then the effective horizontal stresses will fall to their active values.

Higgins et al. (1989) modelled the installation effects of the Bell Common Tunnel using high-quality soil data obtained after the wall was constructed. Wall construction was modelled as an undrained event, which was followed by the application of seepage forces to determine long-term conditions. A number of slightly different (but plausible) assumptions were made for strength and stiffness parameters, according to whether the soil was expected to experience compression or extension stress paths. The results showed that analyses based upon non-linear elastic soil properties, measured in high-quality laboratory tests, gave predictions as good as those made using parameters derived from back-analysis of structures in similar ground conditions. Modelling all construction phases, including wall installation, brought the computed behaviour closer to that observed, although agreement between observed and calculated displacements and bending moments still remained poor.

3.3.10 Pressure increases due to external loads

Most retaining walls are subjected to some kind of external loading, whether this is from surrounding structures acting on an inner-city basement, goods stacked on a quay wall, or from traffic. Compaction of backfill also applies additional loading to soil and increases earth pressures. This is covered in the next section.

The soil supported by many types of retaining structures may be subjected to external loads, i.e. loads not derived from the self-weight of the soil itself. For example, a quay wall in a dock will obviously have traffic driven over it, and freight placed upon it. A bridge abutment will be subjected to both the vertical loading of passing vehicles, and also to horizontal breaking forces. Some temporary and permanent retaining structures are built specifically to provide support for preexisting permanent structures, for example, adjacent buildings or power pylons, while temporary excavations for new foundations are made.

External loads normally act to increase the horizontal stresses on a retaining wall. A number of methods exist to predict their effect, but there is little field data against which to try these methods or to check that they are of sufficient reliability for design purposes.

The simplest case of an external load is where a uniformly distributed load is placed over the entire ground surface behind or in front of a retaining wall. It can be used to demonstrate the problem which faces the designer, in trying to use hand calculation methods to predict the increased horizontal total stress to be supported by a wall. Three approaches are possible:

- Simple 'elastic' solutions
  - With implied horizontal wall displacements
  - Rigid wall
- Simple 'plastic' solutions
  - Active
  - At rest
  - Passive
  - Numerical modelling of wall, soil and construction process

Elastic stress distributions, readily available in texts such as Poulos and Davis (1974), can be used to obtain both the vertical and horizontal stress increases resulting from a wide variety of load geometries. Simple 'elastic' solutions assume that the soil is a linear elastic material, and will normally (in order to make use of existing solutions) also assume that the soil is a semi-infinite half-space (implying that the wall and excavation do not exist) and is homogeneous and isotropic. The horizontal stress is calculated directly from the elastic equations, and the result will be a function of Poisson's ratio, which is one of the least-well-known parameters in soil mechanics.

In the simplest case—of a uniformly distributed load of great lateral extent behind a rigid retaining wall—there will be no horizontal strains in the soil. From Hooke's law

$$\Delta\sigma_h = \Delta\sigma_v \frac{\nu}{(1-\nu)} \tag{3.26}$$

where ν is Poisson's ratio (which might be of the order of 0.25 under drained conditions, and 0.5 for undrained (short-term) loading of clay).

Equations to allow the lateral stress increase created by loads of more complex geometry, and where the wall is allowed to deflect, will only be available for certain cases (see Appendix A). If the wall is assumed rigid, and the loading geometry is simple, then the horizontal stress distribution calculated from elastic solutions may need to be doubled, according to Mindlin's 'Method of Images' (Figure 3.30). If the horizontal wall displacements implied in elastic analysis are thought to be realistic, then the value given by elastic theory may be used directly. But for other cases, no simple hand solution is available.

Simple 'plastic' approaches in fact use a combination of elastic and plastic methods, implying that the soil is simultaneously both far from failure and at

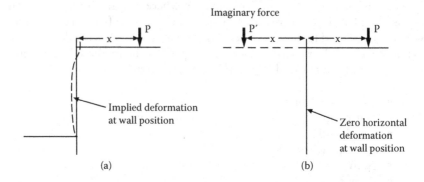

Figure 3.30 Principle of Mindlin's 'Method of Images'. (a) Horizontal deformation of flexible wall caused by force P. (b) Horizontal deformation reduced to zero by addition of imaginary force P.

failure. The vertical stress at different elevations down the back of the wall is calculated using elastic stress distributions (see Appendix A), and the horizontal total stress increase down the back of the wall is then obtained by multiplying these values by a relevant earth pressure coefficient (i.e. K_a, K_p or K_0), depending upon whether the wall is moving, and in what direction relative to the soil. Active and passive coefficients are often used, depending on whether the surcharge is on the retained or excavated side of the wall, and in principal K_0 would seem appropriate in the case of a rigid wall. In the simplest case, for a uniform surface surcharge, elastic solutions predict that the vertical stress increase at any depth will be equal to the applied surcharge pressure. In other words, the pressure on the retaining structure will be modified in the same way as if there were an extra layer of soil placed on the ground surface. Therefore, the horizontal total stress at any depth will be given by

σ_h = (surcharge pressure times K)
  + (effective vertical stress due to soil self-weight times K)
  + (pore pressure)

The horizontal force due to an external load may also be taken into account for failure conditions by using graphical techniques described in Appendix A. This will often be done when ground or groundwater conditions are too complex to allow earth pressure coefficients to be used. As a result of this approach, a number of methods of calculation have been proposed which are based loosely upon graphical techniques, but aim only to estimate the influence of the external force, rather than both the soil and any external forces. These techniques are described in Appendix A. They can lead to rather irrational results. For example, if a Coulomb wedge analysis is carried out for the general case of a line load behind a wall, it will be found that the force on the wall is increased by a uniform amount for all positions of the line load away from the wall, up to a certain point. Beyond that point the line load has no effect.

It will be evident, from the discussion above, that for a given loading geometry there may be several possible ways for calculating the horizontal forces and stresses on a wall. Each method will yield a different distribution of horizontal stress, and a different increase in horizontal force, on the wall. If hand calculation methods must be used then it is recommended that the calculations be carried out using all the available techniques, so that the full range of stress and force increase can be appreciated.

However, if the effect of external loading is critical to the design, it is recommended that numerical methods are used. The various options available are discussed later in this book, where the methods of modelling the soil, the wall and its construction are also considered. Ideally, a constitutive model invoking both elasticity and plasticity should be used to estimate the effects of external loading, and wall installation and time effects (i.e. due to dissipation of excess pore pressures) should be included. Unfortunately, it will not normally be economically feasible to carry out such an analysis, especially when three-dimensional modelling is required in order to simulate the loading and excavation geometry. For almost all purposes, it will be possible only to model plane strain conditions, so that considerable simplification of the design problem will probably be required before numerical analysis can begin.

3.3.11 Compaction pressures—granular backfill

In practice, many earth-retaining structures are built before the soil to be retained is placed. Compaction can significantly increase the earth pressure on a retaining wall, particularly if the wall is rigid and cannot slide, and is of small or modest height.

A typical situation, of a highway bridge abutment, is shown in Figure 3.31. The fill, placed after the construction of the abutment, must be well compacted

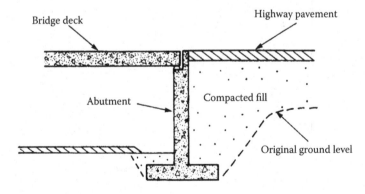

Figure 3.31 Simplified section through highway bridge abutment.

in order to prevent future self-settlement of the fill and consequential surface settlements at the surface of the pavement. Pavement settlements in this situation are very noticeable to road users because bridges are normally designed for minimal settlements, and any settlement in the adjacent fill therefore gives a very bad ride as vehicles cross from the fill to the bridge deck. The price paid for compaction by heavy rollers is, however, an increase in the lateral pressures on the abutment.

The application of a roller, which may weigh 5–10 tonnes, to the fill, causes a temporary increase in the vertical stresses within the fill. If the roller were infinitely long and wide, it would be reasonable to assume that, adjacent to an unyielding wall, the horizontal pressures set up by the vertical stress increase would be related to the vertical stress increase caused by the roller, multiplied by the coefficient of earth pressure at rest, K_0. It could similarly be argued that, during unloading, no lateral strain conditions would apply.

Such a situation can be modelled in a laboratory K_0 'triaxial test', where as the vertical load on the specimen is increased or decreased, the diameter of the soil specimen is monitored and the cell pressure is adjusted to maintain the diameter of the specimen at its original dimension at all times. Figure 3.32 shows results of such a test on sand. (Note that the axes are reversed compared with the normal conventions for plotting a stress path.) Upon first loading, as the vertical stress on the specimen is increased, the cell pressure must also be increased to prevent the specimen from barrelling. It is found that the ratio of horizontal to vertical effective stress (K_0) remains constant (OA). If the vertical stress is reduced from A, little reduction in the horizontal stress (the cell pressure) is initially required to

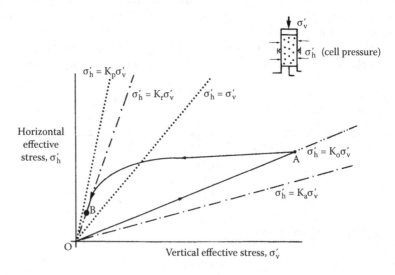

Figure 3.32 K_0 triaxial test result.

maintain no lateral strain conditions. Eventually, however, once the vertical stress approaches the horizontal stress, increasing reduction of horizontal stress becomes necessary. Finally, as the vertical stress is further reduced, the passive failure state is approached, and the unloading curve moves down the line $\sigma'_h = K_r.\sigma'_v$ to point B. The value of K_r will depend upon the angle of friction of a granular soil. Broms (1971) and Lambe and Whitman (1969) suggest that $K_r = 1/K_0$, where K_0 (in this case) is the ratio of effective stresses upon first loading (as distinct from the ratio of effective stresses in the undisturbed ground, described in Section 3.3.2 above).

Relatively little is known about compaction pressures. Broms (1971) proposed a theory which allows an understanding of the mechanism, for a granular soil, as follows. Consider an element of soil (Figure 3.33b) at a depth z below the temporary ground surface, during filling and compacting behind a rigid unyielding retaining wall.

For a shallow element (i.e. z less than about 1 m) the stress path followed in Figure 3.32 can be simplified to that shown in Figure 3.33a. Initially, before the roller is passed over the surface of the fill, it is assumed that $\sigma'_h = K_0\sigma'_v$ (see point A' in Figures 3.33a and 3.33b). When the roller is positioned immediately above the soil element, the vertical stresses are increased, and the horizontal stress is estimated on the basis of the assumption of no lateral yield, i.e.

$$\sigma'_{hm} = K_0.\sigma'_{vm} \qquad (3.27)$$

(see point B' in Figures 3.33a and 3.33c). In reality, this would only be true for an infinitely wide and long load at ground surface.

As the roller moves off the fill, the vertical stress at a depth z below the ground surface decreases. Initially, very little horizontal pressure reduction will take place, and so the assumption is made that $\sigma'_h = \sigma'_{hm}$ until the vertical stress is reduced below a critical value at point C'. Once this occurs, horizontal pressures are assumed to reduce linearly with σ'_v until the original vertical stress $\left(\sigma'_{vi}\right)$ is once more reached. From C' to D' (Figure 3.33a)

$$\sigma'_h = K_r.\sigma'_v \qquad (3.28)$$

where K_r is the coefficient of earth pressure at rest for unloading. It can be seen (Figure 3.33a) that the residual horizontal effective stress $\left(\sigma'_{hf}\right)$ is much higher than the initial horizontal effective stress $\left(\sigma'_{hi}\right)$.

For a deeper soil element, the initial and final vertical stresses are higher than before (Figure 3.33e). The vertical effective stress on the soil element increases under the roller load from A″ to B″, but upon unloading the full maximum horizontal load $\left(\sigma'_{hm}\right)$ is retained. Because the roller is not of infinite extent, the increased vertical stress at greater depth will be smaller than at shallow depth.

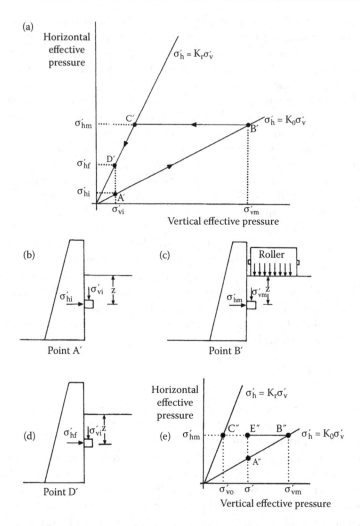

Figure 3.33 Broms' simplified compaction pressure theory. (a) Horizontal and vertical stresses during a single cycle of compaction. (b) Initial stresses on soil
element before application of roller. (c) Stresses when roller is applied. (d)
Stresses after removal of roller. (e) Stresses after for a deeper soil element,
after removal of roller.

A critical depth (z_c) will exist, where the stress state after compaction will
return exactly to point C'' (Figure 3.33e), and the depth can be calculated,
since the residual vertical stress

$$\sigma'_{vc} = \frac{\sigma'_{hm}}{K_r} = \frac{K_0 . \sigma'_{vm}}{K_r} \tag{3.29}$$

and also

$$\sigma'_{vc} = \gamma.z_c \tag{3.30}$$

so that

$$z_c = \frac{K_0}{\gamma}.\frac{\sigma'_{vm}}{K_r} \tag{3.31}$$

Figure 3.34a shows the assumed effect of placing and removing a roller at the surface of a fill. Before the roller is applied to the fill, the lateral earth pressure is equal to $K_0.\sigma'_v$ (curve 1). The application of the roller leads to an increase in vertical stress which decreases with depth, and the method assumes that the maximum horizontal stress is now $K_0.\sigma'_{vm}$, where σ'_{vm} equals $\sigma'_v + \Delta\sigma_v$, and $\Delta\sigma_v$ is the increase in vertical stress at any depth due to the roller (curve 2). Once the roller is removed, material below the critical depth retains its increased horizontal stress, and material above the critical depth reduces its horizontal stress to $K_r.\sigma'_v$ (curve 3). The final pressure distribution is then shown by the shaded area.

In reality, compaction is carried out methodically on thin layers of fill placed against the back of the retaining wall. The residual lateral pressure distribution is then given by the locus of point A as the surface of the fill moves upward relative to the back of the wall. A simplified distribution based upon this is given in Figure 3.34b, where it can be seen that for greater depths of fill, the earth pressure at rest will not be increased by compaction pressures. This distribution is obtained from the assumptions that

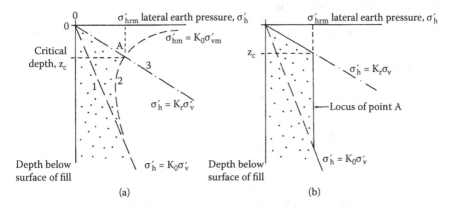

Figure 3.34 Earth pressure distribution after compaction. (a) Lateral earth pressure caused by a single pass of a roller. (b) Lateral earth pressure resulting from compaction of multiple layers of fill. (After Broms, B., *Lateral Pressure Due to Compaction of Cohesionless Soils*. In Proc. 4th Int. Conf. Soil Mech. Found. Engg, Budapest, pp. 373–384, 1971.)

a. $K_0 = 1 - \sin \phi'$

b. $K_r = 1/K_0$

c. $\sigma'_{vm} = \sigma'_v + \Delta\sigma_v$ (where $\gamma.z$ is the vertical stress due to the weight of soil above, and $\Delta\sigma_v$ is the temporary increase in vertical stress at depth z due to the roller and can be calculated from simple elastic stress distribution theory)

d. $z_c = \dfrac{K_0}{K_r} \cdot \dfrac{\sigma'_{vm}}{\gamma}$

A simplified analysis presented by Ingold in 1979 is essentially the same as that of Broms, but with the following modifications. Ingold substitutes K_a for K_0 and K_p for K_r, in considering the simplified stress path followed during compaction. There is little doubt that in most circumstances the horizontal stress due to a roller will not be as high as $K_0.\sigma'_{vm}$ because the soil is at least partially able to yield in the horizontal direction when the vertical roller load is applied. If lateral yield of the wall takes place before the roller is applied, then initially the horizontal stress at some depth z is given by

$$\sigma'_h = K_a.\sigma'_v = K_a.\gamma.z \tag{3.32}$$

If the vertical stress is now increased by rolling, by $\Delta\sigma_v$, and provided lateral yield occurs locally in the soil

$$\sigma'_{hm} = K_a(\gamma.z + \Delta\sigma_v) \tag{3.33}$$

When the roller is removed, the vertical stress returns to its initial value, $\sigma'_v = \gamma.z$. At the critical depth, z_c

$$\sigma'_{hm} = K_p.\sigma'_v = K_p.\gamma.z \tag{3.34}$$

and therefore

$$\gamma.z_c = K_a.\sigma'_{hm} \tag{3.35}$$

It follows from the above equations that

$$\gamma.z_c = K_a^2.(\gamma.z + \Delta\sigma_v) \tag{3.36}$$

and at shallow depth, where $\Delta\sigma_v \gg \gamma z$, this expression may be simplified by putting $\gamma.z = 0$, to yield

$$z_c = \dfrac{K_a^2.\Delta\sigma_v}{\gamma} \tag{3.37}$$

Ingold notes that a useful approximation for the vertical stress increase set up by a roller is obtained by using the expression derived by Holl (1941) for an infinitely long line load on an elastic half space.

$$\Delta\sigma_v = \frac{2.p}{\pi.z}$$

(3.38)

where p is load per unit length, z is the depth below the surface and $\Delta\sigma_v$ is the vertical stress increase immediately below the line load.

By comparison with field observations, Ingold has shown that this equation is a good approximation to measured vertical stresses below static rollers. For vibrating rollers, the line load may be taken as the sum of the static load per unit length and the centrifugal vibrator force per unit length. (If this is unknown, the total force per unit length may be taken as twice the static load per unit length.)

Thus at shallow depth, assuming $\gamma.z$ to be negligible

$$\sigma'_{hm} = \frac{2p.K_a}{\pi.z}$$

(3.39)

and the critical depth becomes

$$z_c = K_a \sqrt{\frac{2.p}{\pi.\gamma}}$$

(3.40)

and thus, the maximum residual horizontal earth pressure, after removal of the roller, is

$$\sigma'_{hrm} = \sqrt{\frac{2p.\gamma}{\pi}}$$

(3.41)

A further expression can be derived for the point below which compaction pressures are insignificant (point B in Figure 3.35), where the maximum compaction pressure equals the active pressure, i.e.

$$\sqrt{\frac{2p.\gamma}{\pi}} = K_a.\gamma.h_c$$

(3.42)

or

$$h_c = \frac{1}{K_a}\sqrt{\frac{2p}{\pi.\gamma}}$$

(3.43)

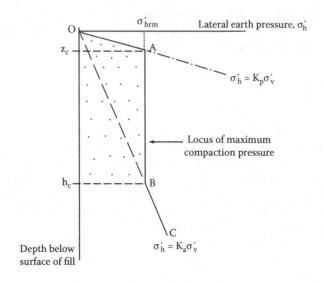

Figure 3.35 Simplified lateral pressure distribution due to compaction. (After Ingold, T.S., *Géotechnique* 29 (3), 265–283, 1979.)

Ingold's method has the great merit of simplicity, but it is difficult to decide on the amount of lateral yield that the soil can undergo. From this point of view, it would appear sensible, particularly in the case of more rigid walls, to use K_0 and K_r values rather than K_a and K_p values. If the roller is assumed to come right up to the back of the wall (as in the analysis above, where $\Delta\sigma'_v$ is taken as $2p/\pi.z$), this assumption has no effect on the value of σ'_{hrm}, but will increase the critical depth, z_c.

Where the roller is prevented from reaching the back of the wall, the increase in vertical stress will be less than that given by Holl's equation, and can be calculated from the integrated form of the Boussinesq equation for a point load. The increase in vertical stress at a wall face a vertical distance z below a line load of intensity, p, length, l, situated a distance, a, back from the face of the wall is given by

$$\Delta\sigma_v = \frac{p}{2.\pi.z}\left[\frac{3x}{R} - \frac{x^3}{R^3}\right]_{x=a}^{x=a+l} \tag{3.44}$$

where

$$R = (x^2 + z^2)^{1/2} \tag{3.45}$$

For an unyielding wall, this stress must be doubled according to the 'method of images' (Mindlin 1936) (Figure 3.30). Ingold (1980) adopted the assumption of an unyielding wall, and also considered an infinitely long roller, to calculate vertical stresses at the wall face when compaction plant is prevented from coming up to the face of the wall, so that

$$\Delta\sigma_v = \frac{p}{\pi.z}\left[\frac{3x}{R} - \frac{x^3}{R^3}\right]_{x=a}^{x=a+l} \qquad (3.46)$$

where

$$R = (x^2 + z^2)^{1/2} \qquad (3.47)$$

and putting $(a + l) = \infty$

$$\Delta\sigma_v = \frac{p}{\pi.z}\left[2 - \frac{3a}{R} - \frac{a^3}{R^3}\right]_{x=a}^{x=a+l} \qquad (3.48)$$

where

$$R = (a^2 + z^2)^{1/2} \qquad (3.49)$$

Where the roller is prevented from coming close to the wall, two effects occur:

a. The horizontal pressures are reduced.
b. A critical depth may not occur, since the horizontal pressure induced by the roller may not exceed the yield pressure upon unloading.

Unlike the previous situation, the assumption of either at-rest or active coefficient of earth pressure has a significant effect on the calculated pressures, and it would therefore seem sensible to calculate the horizontal pressures on the basis of

$$\sigma_{hm} = K_0(\gamma.z + \Delta\sigma_v) \qquad (3.50)$$

and to determine the critical depth (if any) from the intersection of this curve with that from the expression $\sigma_h = K_r.\gamma.z$.

A further method of predicting compaction pressures has been proposed by Murray (1980). The method is similar to that proposed by Broms, except that the increase in horizontal stress due to the roller $\left(\Delta\sigma'_{hm}\right)$ is calculated directly from elastic theory, by integrating the Boussinesq equation for a point load at the ground surface, i.e.

$$\Delta\sigma_h = \int_a^{a+l} \frac{p}{2\pi R^2} \left[\frac{3zx^2}{R^3} - \frac{R(1-2v)}{(R+z)} \right] dx \tag{3.51}$$

The calculated horizontal pressure must be doubled in order to make lateral strains at the wall zero, as before. Therefore,

$$\Delta\sigma_{hm} = \frac{p}{\pi z} \left[\frac{x^3}{R^3} - \frac{(1-2v)}{(R+z)} \cdot x \right]_a^{a+l} \tag{3.52}$$

where

$$R = (x^2 + z^2)^{1/2} \tag{3.53}$$

The critical depth can be found from the intersection of the curve defined in Equation 3.52 above with the line of $s_h = K_r.\gamma.z$, or by using the equation given by Murray,

$$z_c = \left[\frac{p}{\pi\gamma} \left[\frac{\left[\left(\frac{x}{R}\right)^2 + 2\frac{x}{R} \right]_a^{a+l}}{\left(\frac{K_r}{K_0} - 1\right)} \right] \right]^{1/2} \tag{3.54}$$

Poisson's ratio may be taken as its elastic value, $K_0/(1 + K_0)$ and $K_r = 1/K_0$. Like Ingold before him, Murray makes the simplifying assumption that wall pressures generated by the self-weight of the fill are negligible, and therefore that

$$K_0.\gamma.z + \Delta\sigma_h \approx \sigma_h \tag{3.55}$$

Murray has pointed out the importance of the minimum distance of the roller from the back of the wall in determining wall pressures due to compaction, but for design, it would seem prudent to assume the worst, i.e. that the roller will come hard up at the back of the wall. Experimental evidence given by Murray (1980) based on Carder et al. (1977) indicates workable agreement between Ingold and Murray's method and observed stresses on a small rigid retaining wall.

3.3.12 Compaction pressures—cohesive backfill

The compaction of clay can produce even greater horizontal pressure than is the case in granular soil. For some years, it has been appreciated that this might be so and, indeed, the South African Code of Practice in 1989 noted that 'clayey materials used for backfilling will produce higher pressures than sandy materials'.

Apart from the magnitude of the observed earth pressures, there are a number of fundamental differences between the processes involved during and after the compaction of cohesive soil, as compared with those occurring during the compaction of granular soils. Free-draining granular soil is compacted under drained (i.e. $\Delta u = 0$) conditions, and, in addition, volumetric strains of any magnitude will not occur after compaction. It is therefore reasonable to use the effective stress equations developed above, and to assume that (providing the wall does not move) subsequent pressure changes will be negligible. Neither of these conditions is true for cohesive backfill, and therefore the Broms/Ingold equations cannot be applied.

Observations of the pressures exerted by clays on relatively rigid retaining walls (Carder et al. 1980; Symons et al. 1989; Mawditt 1989) suggest that there are at least three overlapping stages which need to be considered:

- Compaction
- Relaxation
- Pore pressure equilibration

Figure 3.36 shows, diagrammatically, the total horizontal stress variations to be expected for each stage, for two materials, one undergoing swelling during pore pressure equilibration and the other undergoing consolidation. The last of these stages is considered in the next section.

As far as compaction pressures are concerned, the first point to note is that clay fill will only begin to develop significant pressures against a wall once its air-void content is reduced, by compaction, to less than about 15%. Typically, a well-compacted engineered fill will have about 5% air void content, so that most clays placed as part of civil engineering projects will be in this category.

Results from field, pilot-scale and laboratory studies have all demonstrated that compaction can lead to high total lateral stresses (point 1, Figure 3.36).

In a field trial, Sowers et al. (1957) observed lateral pressures of the order of 100–150 kPa, which reduced by about 30% in the first 24 hours, when a sandy silty clay was compacted some 15% dry of its plastic limit. At present, there are few reported measurements of compaction pressures on full-scale structures (Symons and Murray 1988). Laboratory experiments by the first author have shown that the total lateral stress is

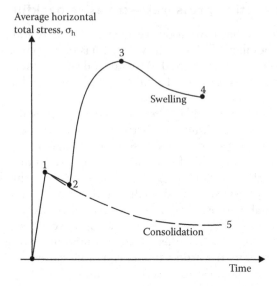

Figure 3.36 Variation in total horizontal stress in cohesive backfill.

proportional to the compacted undrained shear strength of the clay, and is a function of its plasticity. For a high-plasticity clay (LL = 73%, PL = 25%) lateral stresses of the order of 0.8 c_u were observed, while an intermediate plasticity clay (LL = 38%, PL = 16%) gave 0.25 c_u. Compaction of the same clays behind a 3-m-high rigid concrete wall forming part of the pilot-scale retaining-wall facility at the UK Transport and Road Research Laboratory produced lateral pressures of the order of 0.4 and 0.2 c_u, respectively, on completion of filling (Table 3.4). In embankments, clays are typically placed at moisture contents which give compacted undrained

Table 3.4 Examples of pressures exerted by compacted clays

| | Average total stress (kPa) | |
	High plasticity clay	Intermediate plasticity clay
End of compaction	47	18
After relaxation/consolidation	41	8
Max. during swelling	98	–
Long term—2 years	69	–
Water contents (%)		
Moisture content as placed	28	19
Liquid limit	78	43
Plastic limit	29	17

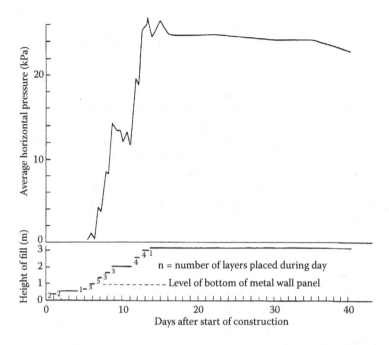

Figure 3.37 Development of pressures during compaction of London Clay against a 4-m² retaining wall panel. (From Clayton, C.R.I. et al., *Can. Geotech. J.* 28 [April], 282–297, 1991.)

shear strengths of about 50–150 kPa, so that total lateral stresses of the order of 10–60 kPa may be expected over the upper part of a wall on completion of construction.

Clayton et al. (1991) have reported the results of pilot-scale retaining wall experiments, where the total stresses applied by a compacted stiff high-plasticity clay (London Clay) against two relatively rigid walls were measured. Figure 3.37 shows the development of total stress with time during compaction of the fill, and for a period of about 30 days thereafter.

For clays placed relatively dry, a relaxation in lateral stress after completion of filling has been observed (Figure 3.36, points 1 and 2) (Sowers et al. 1957; Symons et al. 1989). This may explain why large-scale trials involving many layers of fill and a significant construction period have shown lower total lateral stresses at the end of construction than might be expected from small-scale laboratory experiments of short duration.

3.3.13 Swelling of backfill

In the long term, after compaction, there will be equilibration of pore pressures in a compacted clay fill with those at its boundary. If positive excess pore pressures exist after compaction, then the clay will consolidate and the lateral stresses will reduce with time (points 1–5, Figure 3.36). The pressures on completion of backfilling will then be the greatest that a retaining wall must support. If negative excess pore pressures remain in the clay fill after compaction, then swelling will take place (points 2, 3 and 4, Figure 3.36) as and when rainfall or groundwater penetrates the fill. This would normally be the case when stiff or very stiff clay of high plasticity is used as backfill in a temperate climate, and considerations such as this led to the limitations on plasticity of cohesive backfill given in the UK Department of Transport Specification (1986). The total thrust exerted on a retaining wall in the long term is likely to exceed that on completion of compaction. Due to the stress paths followed during swelling, the total thrust at an intermediate stage is likely to be even greater (as at point 3, Figure 3.36, and see also Figure 3.38).

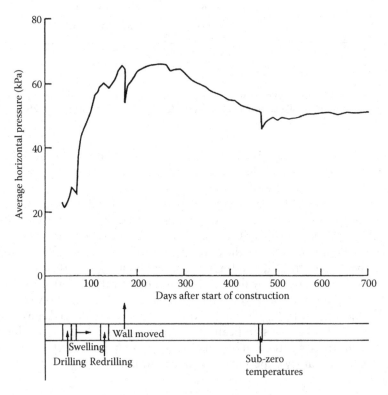

Figure 3.38 Development of pressures during swelling of London Clay against a 4-m² retaining wall panel. (From Clayton, C.R.I. et al., *Can. Geotech. J.* 28 [April], 282–297, 1991.)

Table 3.4 summarises the average lateral stresses measured on a rigid concrete retaining wall during two pilot-scale experiments conducted at the Transport and Road Research Laboratory (UK). *In situ* determinations of lateral stresses within a 6-m-high embankment of London Clay (Mawditt 1989) have shown horizontal total stresses of up to 180 kPa near its centre and up to about 70 kPa close to retaining walls.

During this period of swelling, there is likely to be significant surface heave, as Figure 3.39 illustrates.

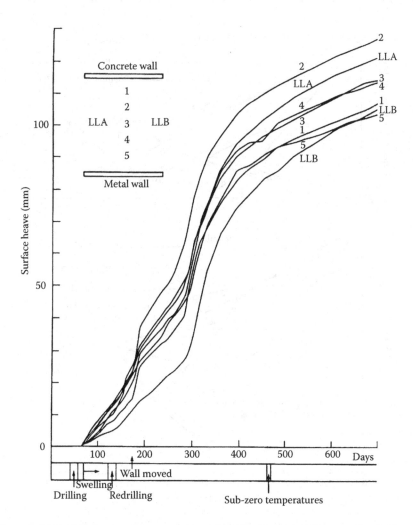

Figure 3.39 Surface heave of compacted London Clay fill retained by a rigid wall. (From Clayton, C.R.I. et al., *Can. Geotech. J.* 28 [April], 282–297, 1991.)

In the long term, after pore pressure equilibration, the horizontal effective stresses remain much greater than the vertical. The limited large-scale experimental evidence currently available suggests that

$$\sigma'_h = K_p.\sigma'_v \tag{3.56}$$

where K_p is the passive earth pressure coefficient. Figure 3.40 shows the pressure distributions at the end of the London Clay experiment, about two years after compaction. The effective strength parameters measured

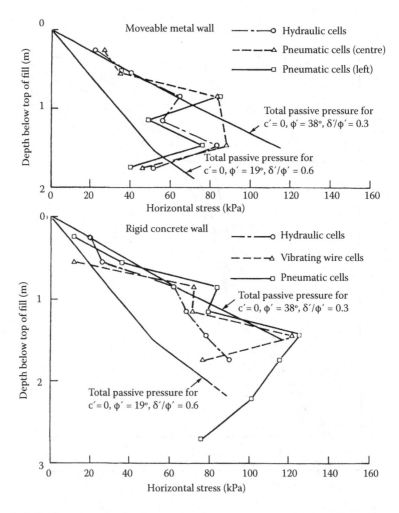

Figure 3.40 Horizontal total stress versus depth in compacted London Clay two years after placement. (From Clayton, C.R.I. et al., Can. Geotech. J. 28 [April], 282–297, 1991.)

for the remoulded clay, in conventional triaxial compression tests, are typically quoted as between $\phi' = 20°$ and $\phi' = 25°$, with $c' = 0$. Yet values of the order of 38°, with $\delta'/\phi' = 0.3$ are necessary to predict the very large residual total horizontal stresses. At the time that the experiment was halted, swelling of the fill was still continuing, suggesting that still higher pressures might be expected at greater depths (Figure 3.40). It would appear from the back-figured effective strength parameters that residual pressures should be calculated from the expected pore pressure distribution, coupled with parameters obtained from passive stress relief tests (Burland and Fourie 1985). Even assuming $\delta' = \phi'$, it is necessary to use $\phi' = 32°$ to fit the data.

Clearly, the differences in behaviour between clay fills which swell and those which do not, following compaction, are very significant. Some attempt to estimate the moisture content limits for swelling behaviour is possible. Figure 3.41 shows postulated stress paths for two idealised (elastic perfectly plastic) soils with the same plasticity but different initial moisture contents, during pore pressure equilibration. Soil 'A' has been placed 'wet' and has positive pore pressures at the end of compaction. Soil 'B' is 'dry', with negative pore pressures at the end of compaction (points a). The stress paths are shown for soil elements at the same depth below the surface of the fill, i.e. the vertical total stress is the same and remains constant for both the wet and the dry soils. The final points (b) on the effective stress paths have the same

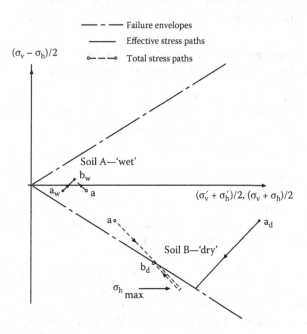

Figure 3.41 Total and effective stress paths for an elastic perfectly plastic soil swelling or consolidating against a 'smooth' rigid retaining wall.

value of vertical stress, because it is assumed that the long-term equilibrium pore pressure is zero in both cases. The effective stress paths before yield have been constructed using elastic equations, with Poisson's ratio equal to 0.25.

Figure 3.41 captures some of the most important elements of behaviour which have been observed in pilot and full-scale studies. Soils placed wet will consolidate, and total horizontal stresses will decrease. Those placed dry will swell and produce a maximum total horizontal stress as they start to yield in the intermediate term, and subsequently show a decrease in lateral stress. It can be concluded that, for a given depth of burial, there will exist a placement moisture content for which the pore pressure after compaction will be equal to its long-term value. If large lateral swelling pressures are to be avoided, then clay should be placed wet of this value.

Calculations to establish the minimum moisture contents to avoid swelling have been reported by Clayton et al. (1991). The results are shown in Figure 3.42; Figure 3.42a gives the limiting moisture contents, while Figure 3.42b indicates the undrained shear strength of the clay at the time of compaction. The data indicate that

a. Low plasticity clays (PI < 30%) are unlikely to swell.
b. Soil at greater depths will swell less than that at shallow depth.
c. Clays with a PI greater than 50% cannot be placed using conventional plant without swelling occurring, because trafficability requires a minimum undrained strength of more than 50–100 kPa.

3.3.14 Shrinkage and thermal effects in propped walls and integral abutments

The forces in props used to support deep excavations, for inner city developments and cut and cover tunnels, for example, are not constant. If placed during excavation, they will naturally be expected to increase as the excavation deepens. But they are also affected by

- Shrinkage (if formed of cast *in situ* concrete)
- Creep under load
- Temperature (due to hydration of cement in the period immediately after concrete props are poured, and due to subsequent daily and seasonal changes in ambient temperature)

The changes in prop loads caused by these mechanisms can be considerable.

All concrete undergoes shrinkage as it ages, as a result of a number of mechanisms. The controls on this are complex, and therefore (as noted by the American Concrete Institute [ACI 1992]) there is considerable scatter in the observations of shrinkage as a function of time. ACI 209R suggests a simple equation:

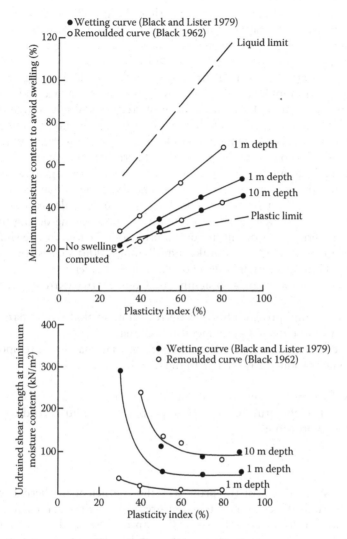

Figure 3.42 Minimum moisture contents and maximum undrained shear strengths to avoid swelling of compacted clay.

$$\varepsilon_{shr} = \frac{\varepsilon_{shr\infty} \cdot t}{(B + t)}$$

(3.57)

where t is the age of the concrete in days, B is 20 days, and the ultimate shrinkage strain is of the order of 0.1%. As an example, for encastré concrete props spanning 30 m at the top of an embedded wall, or an integral bridge deck, this is equivalent to 15 mm inward movement at the top of each wall. The ACI equation suggests that about 50% of this movement will occur within 20 days of concrete placement, and 75% within 60 days. Such movements can be expected to lead to reductions of earth pressure, and to small ground movements adjacent to the wall.

Concrete that exhibits high shrinkage generally undergoes high levels of creep under constant loading. Evidence suggests that shrinkage and creep are closely related, although the mechanism of creep is still not entirely understood. For a given concrete, the lower the relative humidity, the higher the creep. Strength of concrete has a considerable influence on creep and within a wide range, creep is inversely proportional to the strength of concrete at the time of application of load. The modulus of elasticity of aggregate also has an effect on the amount of creep that occurs. Experiments have shown that creep continues for a very long time. However, the rate decreases continuously, with approximately 75% of 20-year creep occurring during the first year. Long-term creep strains at normal utilization factors can be similar to those due to shrinkage. Fanourakis and Ballim (2006) suggest that the BS 8010: (1985) model can provide a good estimate of creep.

Temperature effects are also significant. Consider two extreme situations:

- The ground provides little or no restraint, so that free expansion or contraction occurs as temperatures change.
- The ground provides sufficient restraint that changes in temperature do not produce any change in the length of the props.

In the first case, there will be no increase in prop load as a result of daily or seasonal temperature change. The prop will expand according to the following expression:

$$\Delta L = \alpha L \Delta t \tag{3.58}$$

where L is the length of the prop, α is the coefficient of thermal expansion, and Δt is the change in temperature. Taking the coefficient of thermal expansion for steel as $11 \times 10^{-6}/°C$, an unrestrained 25-m-long prop subjected to a 50°C temperature change will undergo a length increase of 13.7 mm, or 0.055% of its length. For an example, see Powrie and Batten (2000a).

It is the difference between the free thermal strain described above, and the actual strain that will lead to changes in forces in props and bridge abutments. These will increase with increasing temperature, and increasing restraint of the ground. A fully restrained prop or bridge deck will undergo an increase in load that can be calculated from

$$F_{fully\ restrained} = \varepsilon EA = \alpha \Delta t EA \tag{3.59}$$

For the props described above, the increase in load under fully restrained conditions amounts to about 5200 kN. In practice, the restraint offered by the retained ground will be between the two extremes. On the basis of case records, Twine and Roscoe (1997) suggest an effective restraint of 40%–60% for temporary props supporting rigid walls in stiff ground. At Canary Wharf and Canada Water stations, London, Powrie and Batten (2000) and Batten and Powrie (2000) measured prop load changes of the order of 1700–2900 kN, and calculated average effective restraints of 52%–63%, depending upon prop location. Figure 3.43 shows a typical relationship between calculated prop force and temperature, at a time when there was no excavation or construction activity, and when variations in pore water pressures were minimal. It can be seen that temperature variations result in large fluctuations in prop loads. These will produce variations in the stresses between the wall and the soil, affecting the bending moments applied to the wall.

Similar effects occur in integral bridge abutments. These structures have no bearings, and as the result of the rigid connection between the top of the abutment wall and the bridge deck (Figure 3.44a), daily and seasonal temperature changes lead to change in the length of the deck, and induce cyclic strains in the retained soil. Because they are shallow structures (in the UK and USA minimum bridge clearance is of the order of 5 m), the soil provides less restraint than in the case of the deep propped excavations discussed above. Numerical modelling (Springman, Norrish and Ng 1996, Lehane 1999 and Clayton, Xu and Bloodworth 2006), suggests that, for the range of stiffness to be expected, the soil behind the abutments provides little restraint to the deck, and that the horizontal strain levels predicted by

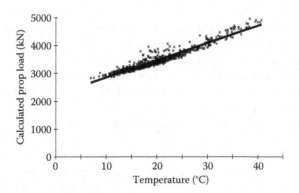

Figure 3.43 Relationship between temperature and calculated prop force. Canada Water station, London. (From Powrie, W. and Batten, M., *Géotechnique* 50, 2, 127–140, 2000.)

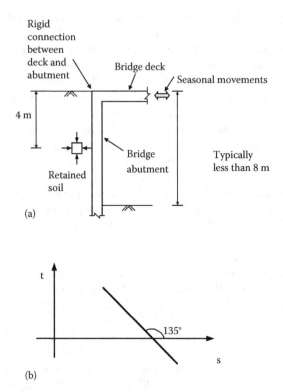

Figure 3.44 Schematic diagram of an integral bridge abutment. (a) Location of representative soil element. (b) Total stress path for constant vertical total stress. (From Clayton, C.R.I. et al., *Géotechnique* 56, 8, 561–571, 2006.)

the simple methods described below give reasonable estimates of the strains imposed in the field. The total stress path applied to the soil retained by a smooth integral bridge abutment is shown in Figure 3.44b. The vertical stress can be expected to remain approximately constant, controlled by the self-weight and depth below the surface, whilst the horizontal stress cycles as the abutment rotates.

Abutments are normally constructed in one of two ways:

a. As diaphragm or bored pile walls, extending to a significant depth below the base of the retained soil, and typically supporting *in situ* over-consolidated clay (termed as an 'embedded' abutment)
b. As backfilled walls, with foundations a relatively small depth below the base of the retained soil, typically supporting compacted selected granular fill (termed as a 'frame' abutment)

For a rigid frame abutment rotating about its base, with retained fill of height H, the horizontal strain in the backfill is of the order of δ/H, where δ is the displacement induced at the top of the abutment, by the deck expansion. For example, a concrete-decked integral bridge constructed in the London area can be expected to experience an annual change in effective bridge temperature (EBT) of 43°C, resulting in a change in deck length of 0.52 mm/m length (Emerson 1976). Assuming that the temperature changes result in equal displacements at each abutment, a 60-m-long deck retaining 8 m of soil and subjected to an EBT change of 43°C will impose a change of horizontal strain of 0.2% on the soil behind its integral abutments. Shorter decks, taller abutments and smaller temperature changes will produce proportionately smaller strain changes.

Two types of behaviour have been observed in the backfill to integral abutments:

1. In some materials (e.g. *in situ* clays supported by an embedded abutment), cycling is controlled solely by the stiffness of the soil and the imposed strain, with no build-up of horizontal total stress with time.
2. In other materials (e.g. backfilled sands and gravels retained by frame abutments), the horizontal stresses progressively increase, until the active state is reached each time that the wall moves away from the soil, and the passive state is reached each time the wall moves toward the soil.

In the first case, a rough and conservative estimate of the lateral earth pressure induced by thermal cycling can be made by calculating the thermal movement at the top of the wall (as described above), calculating the horizontal strain in the soil (the thermal movement divided by the distance from the top of the wall to the estimated point of rigid wall rotation), and multiplying this by the estimated horizontal Young's modulus of the soil. Young's modulus will vary with position down the wall, typically increasing with effective stress (and therefore depth).

A more sophisticated prediction of earth pressures developed by soil supported by embedded abutments requires numerical modelling of soil–structure interaction (Bloodworth, Xu, Banks and Clayton 2011). The stiffness of the soil used in such a model should take into account anisotropy, strain-level dependence and non-linearity. Two approaches are possible:

a. Use a non-linear constitutive model that faithfully reflects the degradation of soil stiffness with strain.
b. Incorporate soil stiffness values that are appropriate for the strain levels expected, given the geometry of the structure and its foundations, and the predicted temperature-induced movement range.

In the second case, for frame integral abutments, the change in earth pressure can be estimated from the vertical effective stress down the back of the wall, combined with active and passive earth pressures. It is clearly possible that a bridge deck and abutment backfilling might be completed in either the summer or the winter months. However, Springman, Norrish and Ng (1996) have demonstrated that the initial direction of loading has no influence on the behaviour of granular soil during subsequent cyclic loading. The use of limiting active and passive pressures on frame integral abutments may seem conservative, but there is growing evidence from field monitoring that these are indeed achieved (Barker and Carder 2006).

3.4 SUMMARY

In conclusion, Coulomb's calculations and initial observations made by Terzaghi gave results for a rigid wall rotating about its base. The general conclusions based on the observations are still widely applied in calculating earth pressures, particularly in limit equilibrium methods of analysis. Such earth pressure calculations generally use coefficients of earth pressure derived from analytical methods, and assume that

a. Active and passive zones can be identified, on the basis of wall movements relative to the soil.
b. Within each zone, the wall is rigid and rotates sufficiently to mobilise full active pressure, or an assumed proportion of passive pressure.
c. Pressure distributions are triangular (hydrostatic), increasing linearly with depth within each soil type.

In reality, most retaining structures may well translate or rotate about the top or some other point on the structure. These departures from the most commonly assumed behaviour can cause significant deviations from pressure distributions calculated on that basis. In addition, wall installation, soil placement and creep and thermal movements in props and bridge decks may have a significant role in controlling earth pressures. The design of earth-retaining structures therefore requires a good knowledge of earth pressure theory, a feel for soil behaviour, an appreciation of construction methods and real wall behaviour and an understanding of the limitations of available theoretical and design methods, coupled with experience.

Chapter 4

Water and retaining structures

The importance of groundwater in the design and performance of earth retaining structures is hard to overemphasize.

- Groundwater level (in combination with soil conditions) has a great influence on the selection of an appropriate type of wall, and the way it must be constructed.
- High groundwater reduces effective vertical stresses and therefore the effective stress component of earth pressure.
- Differences in water levels behind and in front of the wall result in out-of-balance loads that can be much greater than those resulting from the effective earth pressures on the wall.
- In addition, water can produce erosion and induce instability in the ground, leading to complete collapse of earth-retaining structures.
- Groundwater may be contaminated, giving rise to problems (e.g. disposal) during construction, or attacking the wall materials in the longer term.

This chapter therefore describes the way in which groundwater occurs, how seepage and water pressures can be estimated, the loading that water can exert on retaining structures, and how water can induce internal instability in soils. Finally, some basic information is given on groundwater control.

4.1 TYPICAL GROUND WATER CONDITIONS

Water occurs in soil under almost all climatic conditions, but for the purposes of retaining wall design, it is the positive pore water pressures exerted on a wall that are of most significance. The profile of pore pressure with depth can be quite complex, depending on how the groundwater is fed to the subsoil, and the variation of soil permeability with depth.

4.1.1 Definition of water table

For geotechnical purposes, the water table at any location is identified as the level at which the water pore pressure is zero (Figure 4.1). Positive pore water pressures exist below the water table, with negative pore water pressures, and matric suctions, above the water table. The level of the water table varies from place to place, and seasonally. In a dry climate, or in a permeable rugged terrain, the water table may be tens of metres below the ground surface. In a temperate climate, in relatively impermeable soil such as clay, the water table will be close to the surface, and typically within one or two metres of it.

Because the groundwater table marks the level of zero pore pressure, u, and seepage rates, v, are slow in soil, the total head (h_t, from Bernoulli's equation, 4.1 below)

$$h_t = h_E + h_p + h_V = z + \frac{v^2}{2g} + \frac{u}{\gamma_w} = z \qquad (4.1)$$

at the water table is equal to the elevation head, z, i.e. the elevation of the point under consideration above an arbitrary reference datum. Total head controls flow. So for groundwater to flow toward river valleys, there must be a head gradient and lowering of the water table as the river is approached. This is illustrated in Figure 4.2.

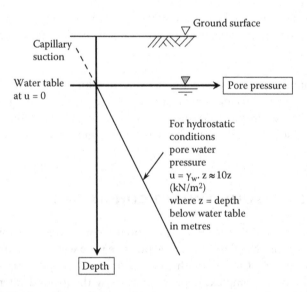

Figure 4.1 Definition of groundwater table.

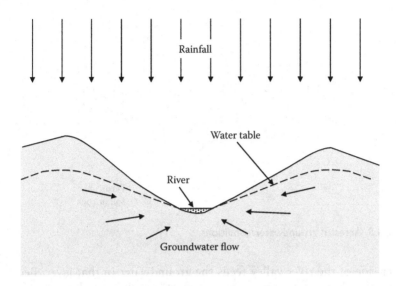

Figure 4.2 Effect of groundwater flow on water table position.

4.1.2 Hydrostatic groundwater

As Figure 4.1 shows, the water pressure generally increases with depth below the water table. When there is no flow in any direction, 'hydrostatic' conditions are said to exist. Under these conditions, the water pressure is easily calculated, since

$$u = \gamma_w . z \tag{4.2}$$

where

   u = pore water pressure
   γ_w = bulk unit weight of water
   z = the depth below the water table

Since the bulk unit weight of water is equal to its density times the gravitational constant, g, under hydrostatic conditions, the pore pressure increases at a rate of about 10 kN/m^2 per metre of depth below the water table. If groundwater flows downward, then the pore water pressure gradient is less than $10z$. When it flows upward, it is greater than $10z$.

4.1.3 Artesian groundwater

Artesian groundwater conditions can occur when groundwater becomes trapped below an impermeable layer, termed an aquiclude. An example is shown in Figure 4.3. Rainfall on the permeable soil or rock outcrop on

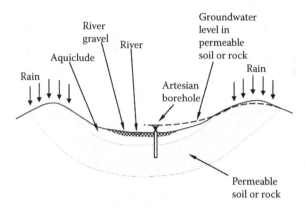

Figure 4.3 Artesian groundwater conditions.

either side of the river valley feeds the groundwater in that layer. Because of the aquiclude, water pressure in permeable rock beneath the valley can increase above that in the river. When a borehole is drilled through the aquiclude, water will rise above ground level.

Artesian conditions exist in many parts of the world. In the nineteenth century, for example, London relied on artesian groundwater for much of its water supply, and indeed the famous fountains in Trafalgar Square are supposed to have been fed from artesian water in the permeable Chalk, which underlies the London clay aquiclude. Artesian conditions produce higher than expected pore pressures, and can contribute to base heave (described later in this chapter).

4.1.4 Perched groundwater

Rainwater falling on permeable soil percolates downward without developing excess pore pressures. If the pore pressure is zero everywhere, then by definition the total head is equal to the elevation, and the hydraulic gradient is unity. When an impermeable layer or lens of soil is encountered the water must travel laterally; in order to do so, it requires a head gradient, and it therefore mounds above the impermeable material. This condition is known as 'perched' groundwater.

Figure 4.4 gives an example of perched groundwater. During ground investigation, this condition might appear to be associated with groundwater strikes at a number of levels. This situation, although common, is difficult to investigate thoroughly.

A similar condition can arise when services such as water pipes and sewers fracture close to a retaining structure, in ground that is above the groundwater table. In designing urban retaining walls, it is as well to

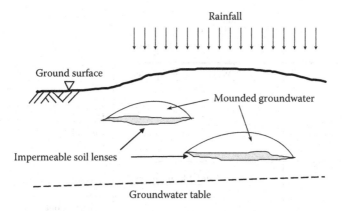

Figure 4.4 Perched groundwater.

remember that wall movements may damage adjacent existing infrastructure. In the case of pipes and sewers, this may in turn lead to additional loads on the wall.

4.1.5 Underdrainage

The widespread domestic and industrial use of groundwater, which is often of good quality and can be used for drinking water supplies, has lead to significant reductions in groundwater levels in many cities around the world. Under the ground conditions that produce artesian groundwater, rainfall infiltration is unable to recharge the aquifer, as it remains perched above the aquiclude. As before, two groundwater tables exist, but with under drainage, the groundwater level in the aquifer is reduced, and may fall below the level of the top of the aquifer.

Figure 4.5 shows schematically the situation in some parts of London. The groundwater in the relatively permeable Chalk was originally artesian, but after many years of abstraction, the stiff London clay is now under drained. Approximately, hydrostatic conditions exist in the upper parts of the London clay, and the Thames Gravels that overly it in places. In the lower part of the London clay, there is downward flow, as the water attempts to recharge the Chalk aquifer.

4.1.6 Rising groundwater

The underdrainage of the Chalk below London, and of a number of other cities in the United Kingdom and elsewhere, were associated with rapid growth of population and industry during nineteenth century urbanization. In 1900, London was the world's largest city, with a population at that time

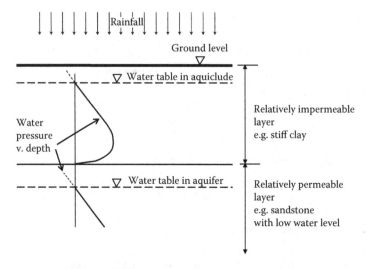

Figure 4.5 Under drained groundwater.

of 6.5 million (which would not qualify as a megacity today). The water demands of the population and of industry were fed both from surface and groundwater. In the 1960s and 1970s, as Britain moved toward a post-industrial economy, industrial water demands reduced and as a result, groundwater levels in the Chalk aquifer started to rise. By this time, ground-water levels had been reduced by as much as 65 m. Since then, rates of groundwater rise have reached 3 m/year in central London. Rising ground-water is a serious concern in a number of UK cities (Figure 4.6).

Projects involving the construction of deep excavations (for example, the new British Library basements) need to take into account final, higher, groundwater levels, which may not only increase the loads on deep retain-ing structures, and apply uplift forces to basement slabs, but may also lead to decreases in strength and stiffness as a result of decreases in effective stress (Simpson et al. 1989).

4.1.7 Conditions above the groundwater table

The level of the groundwater table is partly a function of the permeability of different soil layers, as we have seen above, but it is also significantly affected by the climatic conditions of the area. Rainfall introduces new water, but surface evaporation and plant transpiration remove it. When rainfall exceeds evapo-transpiration (for example in winter, or during mon-soons), the groundwater table rises, and when evapo-transpiration exceeds rainfall the groundwater falls.

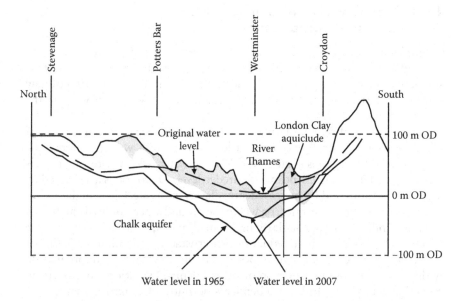

Figure 4.6 Rising groundwater in London. (Redrawn from UK Groundwater Forum/ Thames Water. Available at http://www.groundwateruk.org/html/issues9.htm.)

Above the groundwater table, the pore water pressures are by definition negative. The ability of a soil to sustain negative pore water pressure, termed as 'suction', depends upon its pore size, which is a function of its grain size. Clays can sustain high negative pore water pressures (of the order of hundreds of kPa) because they are fine grained. The clay remains saturated, because air cannot penetrate the fine pore spaces between the soil particles.

On the other hand, sand and fine gravel will only support a suction of a few kPa, after which air will penetrate the soil. In this situation, there are two pore pressures that affect the strength and stiffness of the ground—the pore air pressure (u_a, which is generally close to zero) and the pore water pressure (u_w, which is negative). It is generally accepted (for example, see Fredlund and Rahardjo, 1993) that under these conditions, it is the net normal stresses (($\sigma_v - u_a$) and ($\sigma_h - u_a$)), and the matric suction ($u_a - u_w$) that control soil behaviour. An assumption of zero air pressure and of zero matric suction generally leads to a conservative estimate of strength and stiffness.

The application of high suction to a clayey soil leads to significant volumetric reduction, and this causes near surface desiccation cracking to occur. These cracks can fill with water after rapid rainfall. In common with tension cracks, which can occur as a result of lateral stress reduction, it is often wise to assume that desiccation cracks immediately behind the back of a wall exert a hydrostatic fluid pressure.

4.1.8 Groundwater and soil attack on structural components

Groundwater and the ground itself should not be assumed to be benign. Ground conditions can be acidic, attacking the lime in concrete, or contain sulphates, causing destruction of concrete as a result of the creation of new substances (ettringite and thaumasite) that can break up and weaken the cement paste.

The corrosion of steel Sheet-piles is a function of their environment. Sheet piling corrosion generally occurs at a low rate in fresh water, or above the splash zone. Higher rates occur just below the water level, where water levels fluctuate, and immediately below the dredge level. Corrosion is accelerated by high temperatures and by increased stress levels in the steel. The highest rates of corrosion occur in seawater, or where the groundwater salinity is high.

Timber piling can, particularly in warm climates, be attacked by marine borers such as ship worm, teredo, or other crustaceans. Traditionally, timber has been protected from marine borers or fungus attack by impregnation with a creosote-coal tar solution or other chemical solutions, but environmental concerns now limit the chemicals permitted for marine usage.

4.1.9 Groundwater contamination

Contaminated groundwater or leachate behind the wall can not only of themselves produce corrosion, but can also promote the growth of bacteria that are aggressive to steel, or produce a highly acidic groundwater. During construction, contaminated groundwater can be a health hazard for construction workers, and may require special disposal.

4.2 SEEPAGE AND WATER PRESSURE CALCULATIONS

For a full treatment of seepage, the reader is referred to Harr (1962) and Cedergren (1989). The flow of water through soil is normally evaluated on the basis of the Darcy equation, which is valid for hydraulic gradients (i) of less than unity:

$$q = k.i.A \tag{4.3}$$

where

    q is the rate of flow per unit area of soil
    k is the coefficient of permeability (in units of length/time)
    i is the hydraulic gradient (dimensionless)
    A is the cross-section area of flow

The hydraulic gradient is the change in total head per unit length of flow. Total head is determined on the basis of Bernoulli's equation:

$$h = z + \frac{v^2}{2g} + \frac{u}{\gamma_w} \qquad (4.4)$$

where

h   is the total head
z   is the elevation head above datum
v   is the velocity of flow
g   is the gravitational constant
u   is the pore water pressure
γ_w is the unit weight of water

For flow of water in soil, the velocity term $v^2/2g$ is considered insignificant and therefore

$$h = z + \frac{u}{\gamma_w} \qquad (4.5)$$

or

$$u = \gamma_w(h - z) \qquad (4.6)$$

The two-dimensional steady-state flow of water through soil is governed by the Laplace equation, in the x–y plane

$$k_x \cdot \frac{\partial^2 h}{\partial x^2} + k_y \frac{\partial^2 h}{\partial y^2} = 0 \qquad (4.7)$$

For many purposes, it is impractical to determine the difference between k_x and k_y or the coefficients of permeability in two orthogonal directions, and it is assumed that $k_x = k_y$ and thus that

$$\frac{\partial^2 h}{\partial x^2} + \frac{\partial^2 h}{\partial y^2} = 0 \qquad (4.8)$$

In reality, the horizontal permeability of a soil will normally be many times greater than the vertical permeability.

Equation 4.4 or 4.5 must be solved taking into account the relevant boundary conditions for the particular problem. A complete mathematical solution is normally impractical for real problems, but a number of approximate methods are available:

  i. Flow net sketching
  ii. Electrical analogy
  iii. Finite difference analysis
  iv. Finite element analysis

Many software packages now exist that allow a flownet to be derived, seepage calculations to be carried out, and heads and pore pressures to be determined. In the absence of a suitable software package for a numerical solution, the most practical method is that of flownet sketching. Two families of curves are defined, termed 'flow lines' and 'equipotentials'. The first corresponds to the route of an element of water as it travels through the soil. Two flow lines form the boundaries for what is termed a 'flow channel'. In contrast to flow lines, equipotentials join points with the same hydraulic head.

The following rules are derived from the mathematics of the Laplace equation, and are used during flownet sketching, for an isotropic (i.e. $k_x = k_y$) flow region.

  a. Flow lines and equipotentials cross each other at 90°, and encompass square or equilateral areas.
  b. Impermeable surfaces are flow lines. An example for a Sheet-pile cofferdam is shown in Figure 4.7. Boundary 1 has been treated as impermeable, because it is a plane of symmetry. Boundary 2 has been

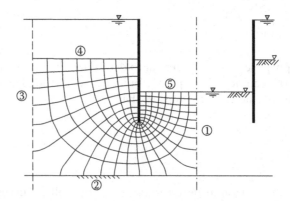

Figure 4.7 Initial attempt at flownet sketching for a sheet-pile cofferdam. $N_f = 8$, $N_d = 22$.

treated as impermeable, because the underlying soil is impermeable relative to the upper material. Boundary 3 does not exist in nature. To allow a solution, an arbitrary position has been chosen. For the solution to be correct, this must be sufficiently far from the wall not to significantly affect the results of calculation.

c. Discharge and entrance surfaces are equipotentials, if submerged. Examples of these are boundaries 4 and 5 in Figure 4.7.

d. If not submerged, discharge and entrance surfaces are not equipotentials or flow lines, but have a total head that is equal to the elevation head at every point along them.

e. If the water table is horizontal, it is an equipotential.

f. If the free surface of the groundwater (defined by the positions at which the pore pressure is zero) is not horizontal, it is a flow line. In addition, it has a total head at every point which is equal to the elevation head.

The solution for a particular seepage problem is achieved by trial and error sketching of a flownet. A simple method of trial and error sketching is to plot the boundaries of the flow domain using ink on tracing paper. After making the first attempt to sketch the flownet in pencil (see Figure 4.7 for an example), the tracing paper is turned over, and the second attempt is obtained by making improvements to the first attempt. Once the second attempt is complete, the first may be rubbed out, and a further attempt is made. Trials continue until all the conditions described above (i.e. (a) to (f)) are achieved.

When the necessary conditions have been achieved by a flownet, the correct position of the flow lines and equipotentials has been found. In such a case, it can be shown that the same difference in total head exists between any two adjacent equipotentials, and that the same amount of discharge occurs between any pair of equipotential lines. The following information is then available from the completed flow net:

i. The direction of flow at any point in the deposit
   — Given by the streamline direction
ii. The total head at all points within the flow domain
   — By interpolation between equipotential lines
iii. The pore water pressure at any point
   — From Equation 4.3
iv. The rate of flow (q) at all points in the flow domain
   — The head gradient (i) is the head drop between two adjacent equipotential lines, divided by the distance (unscaled) between them. From Darcy's law $q = k.i$
v. The total flow (Q, per unit length) through the flow region

— The total flow can be calculated from the coefficient of permeability, k, and the number of flow channels and equipotential drops (see Figure 4.7 for example), and using Equation 4.6

$$Q = kH \frac{N_f}{N_d} \tag{4.9}$$

Sufficiently accurate estimates of total flow (especially considering the difficulty of determining the coefficient of permeability with sufficient certainty) can readily be obtained from flownet sketching. Much greater care is required when estimates of pore water pressure distribution are required, when it is recommended that numerical (computer) modelling is to be used.

4.2.1 Wall loading from water

As noted at the start of this chapter, water can impose direct loads on a retaining structure, and can also lead to decreased effective stresses between the back of the wall and the retained soil. The simplest situation that the designer must deal with is the presence of a horizontal water table with no flow.

Retaining structures frequently intercept the groundwater table. In many cases, the structure will have to sustain the full water pressures if (as is usual) groundwater lowering is not to be carried out in the long-term, throughout the life of the structure. The influences of a static groundwater table are as follows:

a. Bulk density may be increased below the water table.
b. Effective vertical stresses are reduced by groundwater pressure.
c. In both the active and passive cases, effective horizontal pressures are reduced by groundwater pressure.
d. Total active horizontal pressures are increased by groundwater pressure, but total passive horizontal pressures are reduced.

Therefore, a rise in groundwater above that assumed in design leads not only to an increase in the load to be supported by a retaining wall, but also to a reduction in any resistance provided at the front of the wall. An example for active pressure is given below.

Groundwater example calculation 4.1

Obtain the earth pressure acting on the retaining structure shown in Figure 4.8, and calculate the active force on the wall for the following cases:

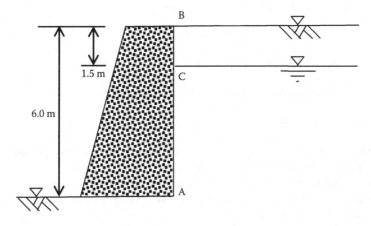

Figure 4.8 Groundwater example 4.1.

a. No water table
b. With a static water level at 1.5 m below the ground surface, and hydrostatic groundwater conditions, for the parameters given below.

$c' = 0$, $\phi' = 30°$, $\delta' = 0$ (i.e. smooth wall)

$\gamma = 18$ kN/m³ above ground water level

$\gamma = 20$ kN/m³ below ground water level

For $\delta' = 0$, $\phi' = 30°$, $K_a = (1 - \sin \phi')/(1 + \sin \phi') = 0.33$

a. Without water

$\sigma_{vB} = 0$ $u = 0$ ∴ $\sigma_{hB} = 0$

$\sigma_{hA} = \gamma.h.K_a = 18 \times 6.0 \times 0.33$ $(u = 0) = 35.6$ kN/m²

∴Resultant force $= (35.6 \times 6.0)/2 =$ **106.9 kN/m run**

b. Water level at 1.5 m depth

$\sigma_{vB} = 0$ $u = 0$ ∴ $\sigma_{hB} = 0$

$\sigma_{vC} = 1.5 \times 18 = 27.0$ kN/m²

∴$\sigma_{hC} = 0.33 \times 27.0 =$ **8.9 kN/m²**

$\sigma_{vA} = 27.0 + (6.0 - 1.5) \times 20 = 117.0$ kN/m²

$$\sigma'_{vA} = 117.0 - 4.5 \times 10 = 72.0 \text{ kN/m}^2$$

$$\therefore \sigma'_{hA} = 72.0 \times 0.33 = \mathbf{23.8 \text{ kN/m}^2}$$

$$\therefore \sigma_{hA} = 23.8 + 4.5 \times 10 = \mathbf{68.8 \text{ kN/m}^2}$$

$$\therefore \text{Resultant force} = 1.5/2 \times 8.9 + 4.5 (8.9 + 68.8)/2$$

$$= \mathbf{181.5 \text{ kN/m}^2}$$

See Figure 4.9, for total and effective stress distributions on the wall. The resultant force is increased by almost 70%.

The following sections show examples of flownets for flow adjacent to retaining structures, and use these to derive earth pressures on the basis of classical wedge analyses.

Groundwater example calculation 4.2

Evaluate soil action against a retaining structure with a lowered water level, using a vertical drain, for the potential failure surface AC (inclined at 60° to the horizontal), in Figure 4.10.

Estimated soil parameters

$$\phi' = 30°$$
$$\delta' = 0°$$
$$\gamma = 17.6 \text{ kN/m}^3$$
$$\gamma_{sat} = 20.2 \text{ kN/m}^3$$

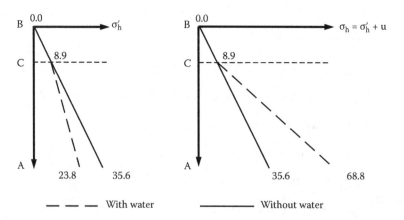

Figure 4.9 Comparison of pressures on wall for groundwater example 4.1 (pressures given in kPa).

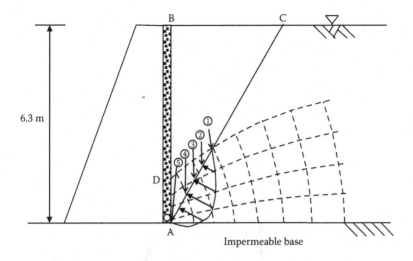

Figure 4.10 Seepage example 4.2. Flow net for wall with drainage.

Draw a flownet (a preliminary attempt is shown).
Calculate the weight of wedge ABC.

$$W = \gamma \times Area_{BCD1} + \gamma_{sat} \times Area_{AD1} \approx 180 \text{ kN/m run}$$

Evaluate the pore water pressure on the potential failure surface.

$u_1 = 0 \text{ kN/m}^2$
$u_2 = 3.3 \text{ kN/m}^2$
$u_3 = 6.0 \text{ kN/m}^2$
$u_4 = 5.25 \text{ kN/m}^2$
$u_5 = 0 \text{ kN/m}^2$

Therefore the resultant force from the pore pressure acting normal to AD

$$U = \Sigma[(u_n + u_{n+1})/2 \times \text{length}] \text{ of each segment} \approx 10 \text{ kN/m run}$$

Because of the presence of the vertical drain, there is no pore pressure on the back of the wall (AB).

Drawing a force polygon for the soil mass ABC (Figure 4.11), one can obtain the (horizontal) force on the structure, $E_a \approx 130 \text{ kN/m run}$.

The inclination of AC in Figure 4.10 is varied, and the pore water pressures reevaluated, to plot new force polygons and find the maximum value of the force E_a.

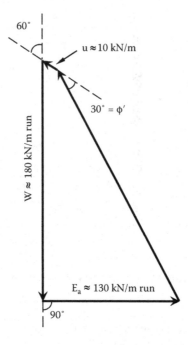

Figure 4.11 Force polygon for seepage example 4.2.

Groundwater example calculation 4.3

Calculate the total force acting on the retaining structure shown in Figure 4.12, for the failure surface shown (inclined at 60° to the horizontal), due to heavy rainfall.

Estimated soil parameters

$\phi' = 30°$
$\delta' = 0°$
$\gamma = 20.2 \text{ kN/m}^3$

Calculate the weight of wedge ABC

$$W = \frac{1}{2}. \ 6.3 \times 3.6 \times 20.2 \approx 230 \text{ kN/m run}$$

Evaluate the pore water pressure at various positions along the failure surface, BC

For a datum taken at the base of the wall

- head at AC = 6.3 m, since pore pressure u = 0
- head at point 1 = $\frac{7}{8} \times 6.3 = 5.51$ m

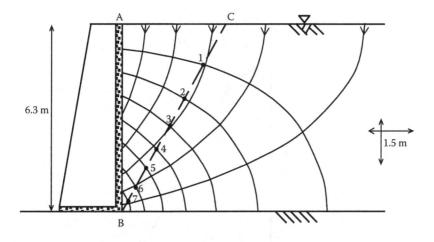

Figure 4.12 Seepage example 4.3. Flow net for steady rainfall.

- elevation at point 1 relative to datum = 4.99 m
- $u_1 = \gamma_w. (5.51 - 4.99) \approx 5.2$ kN/m^2
- similarly

$u_2 = 9.0$ kN/m^2

$u_3 = 10.1$ kN/m^2

$u_4 = 9.9$ kN/m^2

$u_5 = 8.9$ kN/m^2

$u_6 = 7.0$ kN/m^2

$u_7 = 4.1$ kN/m^2

Calculate the resultant force from the pore pressures above

$$U = \tfrac{1}{2}(0 + 5.2)1.5 + \tfrac{1}{2}(5.2 + 9.0)1.31 + \tfrac{1}{2}(9.0 + 10.1)1.06$$

$$+ \tfrac{1}{2}(10.1 + 9.9)0.92 + \tfrac{1}{2}(9.9 + 8.9)0.74$$

$$+ \tfrac{1}{2}(8.9 + 0.7)0.70 + \tfrac{1}{2}(7.0 + 4.1)0.58$$

$$+ \tfrac{1}{2} \times 4.4 \times 0.43 = 49 \text{ kN/m run}$$

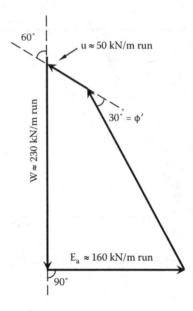

Figure 4.13 Force polygon for seepage example 4.3.

Draw the force polygon for soil mass ABC in Figure 4.12 and obtain the horizontal force on the structure. (From Figure 4.13.)

$E_a \approx 160$ **kN/m run**

In cases such as Example 4.1, where the groundwater surface is lowered close to the retaining wall, the upper flow line is not known *a priori*. It may be found during flownet sketching by trial and error, in order to satisfy the boundary conditions. It can be seen that all these solutions involve the use of the Coulomb wedge analysis, which is described in Appendix A. The use of such a method is likely to be relatively time-consuming.

4.2.2 Tidal lag effects

Situations such as that shown in Figures 4.7 and 4.14 are often associated with tidal water. For example, the water level outside the cofferdam in Figure 4.7 may vary, whilst the level inside is maintained constant by pumping from sumps. Figure 4.14 shows a typical situation in which sheet piling is used for river protection and for dock quays. Here, if the dock water is tidal, the groundwater level behind the wall will also fluctuate, and because the ground is not completely permeable, the groundwater levels will be continuously chasing those in the dock. This has two effects:

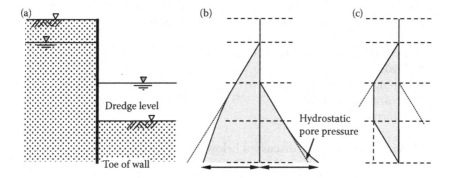

Figure 4.14 Out-of-balance water pressure on a quay wall. (a) Wall layout. (b) Water pressures on wall. (c) Net pressure.

- There will be a time lag between the groundwater levels and the dock water levels.
- The fluctuations in the groundwater levels will be smaller than the fluctuations in the dock water level.

The effect of this is that the wall will be continuously subjected to out-of-balance water pressures. Figure 4.14 shows a worst case for wall stability, where the water in the dock is at its lowest, and the groundwater level is following it down. A seepage analysis would give the pattern of pore water pressures on each side of the wall that are shown in Figure 4.14b. Subtracting the pressures on the dock side from those on the retained soil side of the wall yields net pressures that can, for a wall driven into permeable soil such as sand or gravel, be approximated to those shown by the full line in Figure 4.14c. The net water pressure at the toe of the wall is zero, because of seepage.

In the case of a wall driven into impermeable soil, where there is a negligible flow around the wall, the full head is retained on each side. The net water pressure can be approximated by the dashed vertical line, below the dredge level, in Figure 4.14c.

4.3 WATER-INDUCED INSTABILITY

The creation of a retaining structure in an area of high groundwater is normally associated with a change in groundwater conditions. For example, a basement excavation needs to be kept reasonably dry during construction, and this is often done by pumping from sumps within the excavation. This has the effect of lowering the groundwater table inside the excavation, thus promoting flow. Flow into the base of an excavation can induce instability of the soil, especially near to the wall, where the hydraulic gradients are highest. But instability can occur for a number of reasons:

- Reduction in passive resistance
- Hydraulic uplift
- Piping and fluidisation
- Hydraulic fracture
- Internal erosion

These mechanisms are discussed below.

4.3.1 Reduction in passive resistance

Figure 4.14 shows that as a result of seepage around a retaining wall there will be an increase in pore water pressures near the toe of the wall, on the dredged side, where passive pressures are generated. This increase is in excess of the hydrostatic pore pressures (shown by the straight dotted line in Figure 4.14b) which are allowed for when calculating passive pressure. As a result, the effective stresses are lower than assumed, as are the passive resistances acting to support the wall.

This effect has been calculated, using the classical log spiral method, by Soubra and Kastner (1992). Full charts are given in Chapter 10, Part II. Figure 4.15 gives an extract to illustrate the magnitude of the effect, which can be considerable.

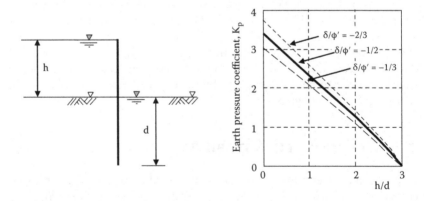

Figure 4.15 Effect of seepage pressure on passive resistance, for $\phi' = 25°$. (From Soubra, A.H. and Kastner, R., Influence of seepage flow on the passive earth pressures. Proc. ICE Conf. on Retaining Structures, Cambridge, pp. 67–76, 1992.)

4.3.2 Hydraulic uplift (base heave)

Base heave can occur, for example, at the bottom of a braced excavation for the installation of a service pipe, due to two factors:

- Reduction in vertical stress can produce sufficient shear stress, $(\sigma_h - \sigma_v)/2$, that the shear strength of the soil beneath the base of the excavation is exceeded. This condition can be a problem when excavating in soft clays.
- The uplift due to ground water, normally when trapped below an impermeable layer, is sufficient to overcome the weight of the soil.

An example of this second condition is shown in Figure 4.16. Assuming that

- The gravel/clay interface to be 3 m below the base of the excavation, and
- The groundwater table to be 4 m above the base of the excavation,
- The bulk unit weight of the sand and the clay to be 18 kN/m³,

then
The uplift pore pressure

$$\approx (4 + 3)\, 10 = 70 \text{ kPa}$$

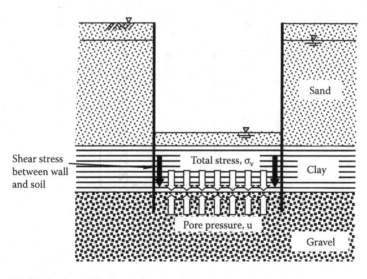

Figure 4.16 Hydraulic uplift.

The downward vertical total stress under the clay, inside the excavation

= 3.18 = 54 kPa

Under these conditions uplift can only be prevented by the shear stresses (shown by the black downward arrows in Figure 4.16) acting between the walls and the soil below the base of the excavation. These are only likely to help in a narrow excavation.

4.3.3 Piping

Figure 4.17 shows an element of soil subjected to upward water flow. If side friction is ignored, there will be stability provided that the weight of the soil element is greater than the force resulting from the pore pressures on the upper and lower faces of the element. The weight of the element

$$W = \gamma_{bulk} . \overline{AB} . \overline{CA} . 1 \text{ per unit length} \tag{4.10}$$

The resultant force due to the difference in water pressure

$$U = (u_{AB} - u_{CD}) . \overline{AB} \tag{4.11}$$

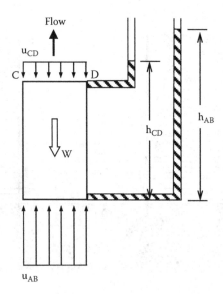

Figure 4.17 Instability due to piping, during upward seepage.

where

$$u_{AB} = h_{AB} \cdot \gamma_w \tag{4.12}$$

$$u_{CD} = \left(h_{CD} - \overline{AC}\right) \cdot \gamma_w \tag{4.13}$$

$$\therefore \quad U = (h_{AB} - h_{CD}) \cdot \gamma_w \cdot \overline{AB} + \gamma_w \cdot \overline{AC} \cdot \overline{AB} \tag{4.14}$$

For conditions of limiting stability, $W = U$, and therefore

$$\overline{AC}(\gamma_{bulk} - \gamma_w) = (h_{AB} - h_{CD})\gamma_w \tag{4.15}$$

or

$$\frac{(h_{AB} - h_{CD})}{\overline{AC}} = i_{crit} = \frac{(\gamma_{bulk} - \gamma_w)}{\gamma_w} \tag{4.16}$$

Since $\gamma_w = 9.81$ kN/m³ and typically $\gamma_{bulk} = 16\text{–}22$ kN/m³, for piping to occur

$$i_{crit} \approx 1 \tag{4.17}$$

The condition of upward seepage with a hydraulic gradient greater than 1 can occur below the bottom of an excavation. An example of a possible situation is the seepage around sheet piling, as shown in Figure 4.18, where the soil inside the excavated area is at risk. The deeper the excavation in such a case, the greater the risk, because as the excavation is increased the flow length becomes smaller and the hydraulic head difference between the upstream and downstream boundaries becomes greater.

Experience shows that reductions in passive resistance and unstable conditions for equipment and labour may occur when the hydraulic gradient at the exit (i.e. at the base of the excavation) is of the order of 0.5–0.75. Figure 4.19 shows the penetration of sheet piling required to prevent piping in an isotropic sand deposit, according to NAVFAC-DM7.

The computation of a factor of safety against piping at the base of an anchored sheet-pile wall can be carried out according to the method described by Terzaghi (1943). In this method, the equilibrium of a rectangular soil element, with a depth equal to the depth of embedment of the sheet piling and a width equal to one-half of that depth, is considered. The factor of safety is defined as the ratio between the head difference between the upstream and downstream boundaries required to cause instability

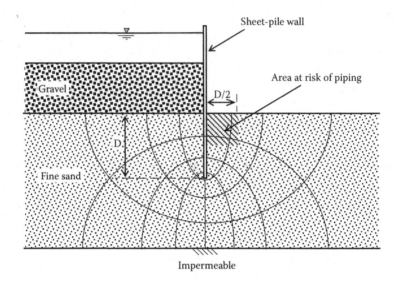

Figure 4.18 Example of soil vulnerable to piping.

divided by the actual head difference. A factor of safety of 3–4 is suggested. Example 4.4 carries out a calculation by this method.

Example calculation 4.4—Factor of safety against piping

Calculate the factor of safety against piping failure by the Terzaghi method, for the geometry in Figure 4.20, where

> head difference, $H = ai = 4$ m
> Sheet-pile penetration, $he = 4.8$ m
> $\gamma_{bulk} = 20.2$ kN/m³
> $oe = 2.4$ m

Calculate the buoyant weight of soil in element *heop*

$$W_{buoy} = \gamma \,.\overline{he}. \,\overline{eo} - \gamma_w .\overline{he}. \,\overline{eo}$$

$$= (20.2 - 10.0)\, 4.8 \,. 2.4$$

$$\approx 118 \text{ kN/m run}$$

Calculate the average head perpendicular to *eo*

> At e, $h_e = 2.00$ m
> At b, $h_b = 1.50$ m
> At c, $h_c = 1.38$ m

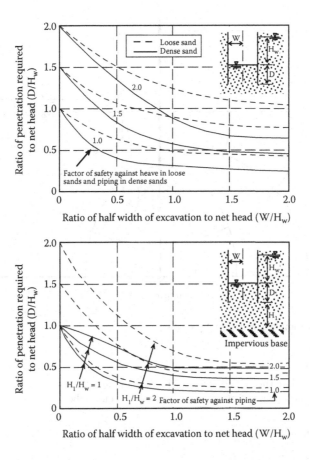

Figure 4.19 Penetration of sheet piling required to prevent piping in isotropic sand (NAVFAC-DM7 1982). Top—penetration required for sheeting in sands of infinite depth; Bottom—penetration required for sheeting in sands of limited depth.

At d, $h_d = 1.20$ m
At o, $h_o = 1.10$ m

$$\overline{h_{eo}} = 1.44 \text{ m}$$

Proportion of total head loss

$$m = \frac{\overline{h_{eo}}}{h} = \frac{1.44}{4} \approx 0.36$$

$$m = 0.36$$

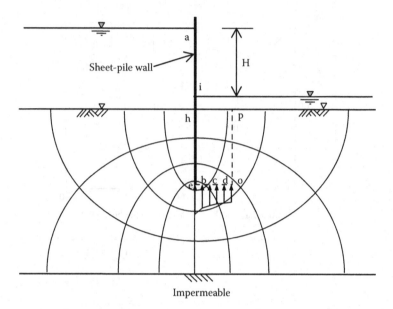

Figure 4.20 Example calculation 4.4 - piping.

Calculate the critical head at the base of element heop
As shown above, the critical condition occurs when

$$\frac{h_{hp} - h_{eo}}{4.8} = \frac{\gamma_{bulk} - \gamma_w}{\gamma_w}$$

or

$$\gamma_w(h_{hp} - h_{eo}) = 4.8(\gamma_{bulk} - \gamma_w)$$

$$\therefore \overline{eo} \cdot \gamma_w(h_{hp} - h_{eo}) = W_{buoy}$$

$$(h_{hp} - h_{eo}) = \frac{117.5}{10.0 \cdot 2.4} \approx 4.9 \text{ m}$$

$$h_{crit} = \frac{4.9}{m} = \frac{4.9}{0.36}$$

$$= 13.6 \text{ m}$$

Calculate the factor of safety

$$F = \frac{h_{crit}}{h} = \frac{13.6}{4.0}$$

$$= 3.4 \therefore O.K.$$

In cases where it is not possible to achieve a satisfactory depth of Sheet-pile penetration, or where the required depth is clearly uneconomical, a number of other measures may be considered, such as

a. The use of well-points from original ground level, to lower the groundwater level in the area of the excavation (see Section 4.4). This is suitable for relatively homogeneous soil conditions, or where permeability decreases with depth.
b. The use of pressure relief wells in the base of the excavation (see Section 4.4). This technique may be suitable where a thin relatively impermeable soil layer overlies permeable soil relatively close to the base of the excavation.
c. The use of a filter layer in the base of the excavation. This method provides weight and prevents the upward movement of soil particles with the inflowing water (see Section 4.3.5).

If the groundwater level is higher than ground level on the upstream side of the wall, a clay carpet or bentonite slurry may be used to create a relatively impermeable barrier and reduce water inflow. The effectiveness of each option must be evaluated for the individual geometry of a particular case on the basis of flownet sketching, or other seepage analysis.

4.3.4 Hydraulic fracture

Hydraulic fracture can occur when the pore pressure in the soil exceeds the total stress in any direction. For a soft, normally-consolidated clay (i.e. with $K_0 < 1$), the horizontal effective stress (and therefore the horizontal total stress) is less than the vertical effective stress. Once the pore pressure rises to the level of the horizontal total stress then the soil will crack in the vertical plane. Water flow then accelerates.

For a wide excavation (see Section 4.3.2), base uplift occurs when the pore pressure exceeds the vertical total stress, plus some component of resistance obtained from shear between the soil and the retaining walls (Figure 4.16). In a narrow excavation in soft clay underlain by sand or gravel, hydraulic fracture can be expected to start somewhat before base heave conditions are arrived at.

4.3.5 Internal erosion

In order for flow to occur, hydraulic gradients must exist across and within a soil mass. As was seen in the case of piping, water flow leads to body forces on the soil mass and on individual particles within the soil. Thus, at the flow exit region there is a tendency for particles to be washed out. This is particularly likely in non-cohesive or laminated soils subjected to high hydraulic gradients. Seepage erosion presents problems because

    i. The loss of volume of the soil may lead to surface settlements adjacent to a retaining structure, and
    ii. Migration of soil particles may lead to blocking of drains, and subsequently to an increase in pore pressures behind the wall.

To avoid such phenomena, filters are used between the soil and the drainage systems. Filters are traditionally composed of clean sands or gravels, with a selected grading. The grading is selected so that

    a. The voids between the smallest particles of the filter must be greater than the voids between the smallest soil particles, so that the filter has a higher permeability than the soil; and
    b. The voids between the smallest filter particles must be smaller than the biggest soil particles, so that the soil does not progressively flow into the filter under high hydraulic gradients.

These requirements are normally based on the particle size distributions of the soil. Tests performed by Terzaghi, later extended by the US Corps of Engineers, form the basis of the commonly used filter criteria, that

$$D_{15(\text{filter})} < 5.D_{85(\text{soil})} \tag{4.18}$$

and

$$4 < \frac{D_{15\,(\text{filter})}}{D_{15\,(\text{soil})}} < 20 \tag{4.19}$$

where D_{15} and D_{85} are the particle sizes which have 15% and 85% of the soil by weight finer, normally obtained from a particle size distribution curve. It is also good to have a clean well-graded soil as a filter, with less than 5% by weight of particles finer than 75 μm, and no particles larger than 75 mm in diameter.

The US Corps of Engineers criteria are satisfactory for the filtration of granular soils. In the case of clays, such criteria cannot be applied because the D_{15} size of the soil may well be finer than 0.001 mm, which implies that a number of filter layers will be required. According to Cedergren (1989), in practice clays are satisfactorily filtered by materials with

$$D_{15(\text{filter})} < 0.4 \text{ mm} \tag{4.20}$$

which are well-graded, but with a coefficient of uniformity (D_{60}/D_{10}) of less than 20.

As an alternative to graded granular filters, coarser granular soils are sometimes used with a covering of filtration geotextile. The contact between a granular filter and pipes or conduits leading water away from it must also be protected from erosion. Filter fabrics can be used in this location, or, for perforated tubes with circular holes,

$$D_{15(\text{filter})} > D_{\text{hole}} \tag{4.21}$$

and for tubes with unsealed joints

$$D_{15(\text{filter})} > 1.2 \text{ (length of joint)} \tag{4.22}$$

Opinions differ concerning the filter criteria that are necessary. According to Terzaghi and Peck (1967), 'the quantity of water that percolates through a well-constructed backfill is so small that there is no danger of the drains becoming obstructed by washed-out particles. Therefore, it is not necessary that the grain size of the materials in the draining layers should satisfy the requirements for filter layers'.

Cedergren (1989) and Bowles (1968) have different opinions on the matter. Possibly these different opinions reflect different regional experience, not just due to building techniques and practice but also due to different soil types. Therefore, any decision about the need to strictly obey filter criteria should be based on an evaluation of the performance of existing retaining structures, particularly in rainy periods. A typical case is the tropic humid regions, where porous soils eventually become very erodible. A drainage system certainly must consider these effects, taking into account the frequency of failures of retaining structures.

4.4 GROUNDWATER CONTROL

As has been noted, pressures on a retaining structure can be reduced by lowering the groundwater table, using a drainage system. Although the water pressure acting directly on the back of the wall can be eliminated, other effects may remain. The lowering of the watertable induces water flow adjacent to the structure, and depending on the configuration of the drainage system, water pressures may still act along the critical active failure surface of the retained soil mass. This implies that earth pressures will be higher than in the case of a completely drained soil.

A number of serious problems are related to groundwater flow and retaining structures:

a. The lowering of the groundwater table by flow around a retaining structure may lead to settlements and damage the adjacent structures, especially if those are founded on soft cohesive or compressible soils.
b. Flow beneath a retaining structure may be such that there is an excessive hydraulic gradient near the downstream region of flow. This may lead to boiling, especially in fine sands or silts, and the result may be a loss of passive resistance.
c. Drainage systems may become blocked by fine soil particles, if not properly designed. Once the groundwater returns to its original position, higher pressures will be imposed on the back of the wall.
d. Ice may penetrate the back of the wall, causing very high internal pressures. Alternatively, it may block exits from drainage systems, causing temporary increases in lateral pressure.
e. Groundwater may be polluted, requiring special means of disposal.

4.4.1 Wall drainage

An example of flownet for the condition of steady rainfall, with water ponding at the ground surface, was presented in Example 4.3 in Figure 4.12. In an earth pressure calculation which takes this condition into account, it is necessary to evaluate the pore pressures for a number of trial surfaces, and then determine the surface which gives the maximum (active) pressure. Comparison between the conditions on the wall for soil fed by groundwater and drained by a filter at the back of the wall (Example 4.2, Figures 4.10 and 4.11) and a wall subject to flow following intense rainfall (Example 4.3, Figures 4.12 and 4.13) shows that the total force applied to the wall is increased by heavy rainfall. This increase, due to the groundwater level being maintained at ground level during a storm, may be greater than 30%–40% of the active force in dry conditions, according to Huntingdon (1957). In the case where the groundwater is lowered by drainage, the increase in force due to water is seldom greater than 10% of the dry active force. Consequently, the influence of heavy rainfall on the stability of a retaining structure is a critical problem.

Terzaghi (1936) suggests drainage measures to minimize these effects. These drainage measures need to guard against internal erosion, and incorporate the filtration rules set out in Section 4.3.5 above.

The simplest drainage system for a retaining structure consists of 'weepholes', which are holes precast or drilled horizontally through the wall, typically at 1.5 m centres both horizontally and vertically. These holes should be of the order of 100 mm in diameter, and may be separated from the retained soil by a layer of free-draining granular soil.

A more efficient system consists of separating the structure from the soil using a permeable filter blanket, on the back of the wall, as shown in Figure 4.12. This layer intercepts the water percolating through the soil and carries it to a perforated horizontal pipe at the bottom of the wall. The top of the drain is sometimes sealed with a layer of impermeable material, such as clay, in order to minimise the penetration of surface run-off.

To eliminate pore water pressure in the critical zone, Terzaghi suggested the use of inclined drains, typically placed below the backfill to a wall. The flownets due to seepage give horizontal equipotentials and vertical flow lines, and since the pore water pressures are equal to zero at the ground surface and may be assumed to equal zero at the drain, it is easily proved that they will be zero at any point within the soil mass, above the drain. The construction of an inclined drain may sometimes be difficult or impractical. In the case of a cantilever reinforced concrete wall, almost the same effect can be achieved by using a horizontal drain over the top of the heel. Although the pore pressures on potential failure surfaces are not eliminated, they will be lower for a horizontal drain than for a vertical one.

When drain material is to be placed on the back of the wall it should be free-draining, i.e. without fines (GP, GW, SP or SW according to the Unified Soil Classification system), to avoid high earth pressures on the wall due to frost. A sand layer may be needed between a coarse drain and the supported fill, in order to avoid contamination of the drain.

Although flow nets of the type given in Example 4.3 as shown in Figure 4.12 are obtained regardless of soil permeability, in practice they are observed only in materials with intermediate permeability ($k \approx 10^{-6}$ m/s) (Terzaghi 1936). With more permeable soil, rainfall is seldom of sufficient intensity to establish steady-state flow with free water at ground surface. With less permeable soils, water disperses by run-off before it can penetrate a significant depth of soil. Even when the rainfall is of very great duration, it is unlikely that steady-state seepage will result, because the soil will tend to swell. As a consequence of this, the soils that deserve the most care are those of intermediate permeability, such as silty fine sands and porous residual soils. When clays are to be retained, the lateral stresses created by swelling against the back of the wall as a result of water ingress may well be much greater than the effects of groundwater.

4.4.2 Soil dewatering

The use of drainage blankets against the back of the wall, or in the soil, is generally only possible for backfilled walls. For permanent walls, the use of other techniques, such as well points, deep wells or ejector systems is often impractical and too costly, except where they are intended to reduce wall loads during construction (e.g. before final floor or permanent prop construction, in a bottom-up basement, or multi-propped structure).

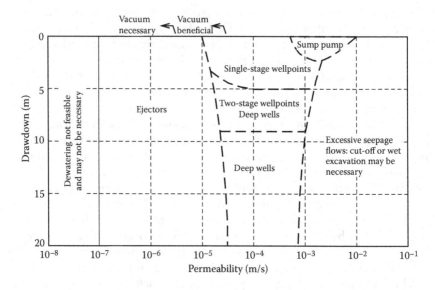

Figure 4.21 Approximate ranges of application of pumped-well groundwater control techniques CIRIA Report 515. (From Preene, M. et al., Groundwater control—design and practice. Construction Industry Research and Information Association report C515, 204 pp. CIRIA, London, 2000.)

However, for temporary retaining structures, such as a sheet-pile wall, such systems are often used, and are worth considering. *CIRIA Report C515* (Preene, Roberts, Powrie and Dyer 2000) gives guidance on dewatering, and a summary of the application range for different dewatering techniques is given in Figure 4.21. *CIRIA Report C532* (Masters-Williams et al. 2001) addresses the issue of water pollution arising from construction sites.

Chapter 5

Global and local instability

Although a retaining structure may be able to support the calculated adjacent soil and groundwater loads without failure, wider or more local instability may still occur. Retaining structures are often used to create locally steeper slopes, which would otherwise not be stable. Critical, deeper, failure surfaces may develop that do not intercept the wall support system.

The 'Coulomb wedge' (Coulomb 1776) and log spiral methods (Terzaghi 1941) of analysis, which are traditional methods of analysis carried out for a uniform wall backfill to determine the soil loading that must be carried by a retaining structure, are particular cases of limit equilibrium analysis. As Figure 5.1 shows, as early as 1776, Coulomb recognised that the designer should also be searching for other critical mechanisms of failure. These may involve non-planar failure surfaces, and surfaces that do not intersect with the base of the retaining wall, and may be affected by surcharge loads.

The limit equilibrium methods used to estimate the force on a retaining structure from the retained fill represent particular cases of local instability analysis. Global and local stability checks, such as those described in this chapter, are normally carried out in the search for other critical mechanisms of failure. This is particularly important when earth-retaining structures are being used to improve the stability of existing but potentially unstable slopes.

5.1 TYPES OF INSTABILITY AFFECTING RETAINING STRUCTURES

An example of instability is shown in Figure 5.2, where the wall has been able to support the ground immediately behind it without undergoing bearing capacity failure or unacceptable forward sliding, yet is unsatisfactory because of a deeper-seated instability in the ground beneath it. This situation is sometimes termed 'external instability', or 'global instability'. An early and serious case of such a failure occurred in Gothenburg in 1916 and is illustrated in Figure 5.3.

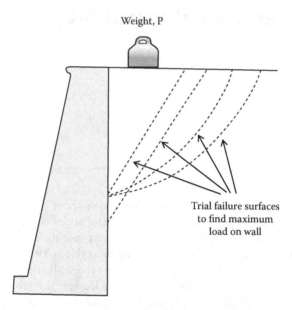

Weight, P

Trial failure surfaces
to find maximum
load on wall

Figure 5.1 Coulomb's examples of potential failure surfaces. (Redrawn from Coulomb, C.A., Essai sur une application des regles de maximis et minimis a quelques problemes de statique, relatifs a l'architecture. Memoires de Mathematique et de Physique présentés a l'Academic Royale des Sciences, Paris, 1773, 1, 343–382, 1776.)

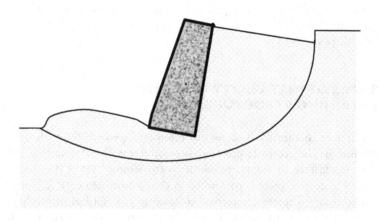

Figure 5.2 Global instability of a gravity retaining wall.

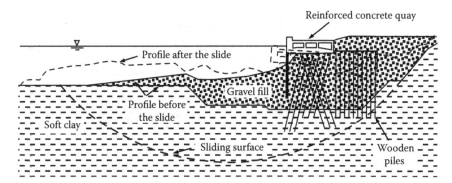

Figure 5.3 Gothenburg Harbour quay wall failure, 1916. (From Petterson, K.E., *Géotechnique*, 5, 4, 275–296, 1955.)

Analyses of local stability are also required in order to search for mechanisms of failure that may be overlooked in routine earth pressure analysis, for example,

- In tiered mechanically stabilised earth walls (Figure 5.4)
- Where loads are placed on backfill behind walls
- When there are weaker layers within the retained soil

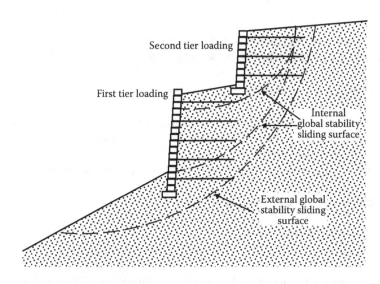

Figure 5.4 Global and local instability of a tiered mechanically stabilised earth wall. (Redrawn and modified from www.keystonewalls.com/media/technote.pdfs/globlstb.pdf.)

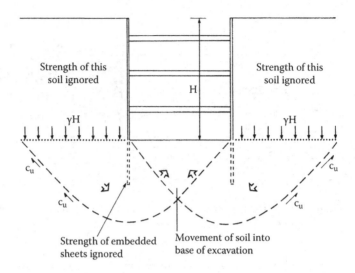

Strength of this soil ignored

Strength of this soil ignored

H

γH

γH

c_u

c_u

c_u

c_u

c_u

Strength of embedded sheets ignored

Movement of soil into base of excavation

Figure 5.5 Instability in the base of a braced excavation.

This type of local failure is sometimes termed 'internal global instability'. Local or internal global instability can also occur in weak clay soils by 'base heave', a form of 'negative' bearing capacity caused by unloading the ground inside, (for example) a braced excavation (Figure 5.5).

5.2 CLASSIFICATION OF INSTABILITY AND SELECTION OF PARAMETERS

Instability can occur at three different times:

- During construction, before completion of the wall
- After completion of wall construction, when loads (such as on a dock quay) are first applied
- In the long term, well after completion of wall construction

Global instability can also occur as a

- First-time slide
- Reactivation of instability on a preexisting shear surface

These issues are important because the designer must choose appropriate soil strength parameters during analysis (see Chapters 1 and 2). The parameters to be used are not the same in each and every case, and indeed may often be different for earth pressure calculations and for calculations of global and local instability. For example, Figure 5.3 shows a quay wall retaining gravel fill, where failure has occurred in the soft clay below. Calculations to determine earth pressures on the wall would use effective strength parameters for the gravel, but global stability might in this case need to be estimated on the basis of the undrained strength of the clay.

5.2.1 Short-term or long-term parameters?

Long-term, peak effective angles of friction (ϕ', with $c' = 0$) should be used for all analyses in sands and gravels, because of their rapid drainage rate. This applies whether for earth pressure calculation or for global stability estimates. The most difficult issue is whether drained (long-term, c' ϕ') or undrained (short-term, c_u) parameters are appropriate for calculating the strength of shear surfaces passing through clays. (See discussion in Chapter 1, Section 1.6). If in doubt, one option is to carry out separate calculations using undrained and drained parameters, and adopt the parameters that give the least favourable results.

If a wall fails during, or shortly after construction, then a number of scenarios are possible:

1. Forward sliding on the base of a gravity wall, or rotation on a slightly deeper shear plane, may occur during construction as a result of backfilling and compaction (see Chapter 3). Calculation of compaction pressures should, of course, use parameters relevant to the type of backfill. Global stability should be checked using parameters relevant to the shear surface under consideration. If the wall fails by forward sliding, and has been founded on clay, the resistance of the wall/soil interface to sliding should be calculated using the preexisting undrained shear strength, as determined from undrained triaxial tests carried out on samples from the founding depth that have been obtained during ground investigation, with a reduction factor to allow for the smoothness of the wall soil interface and for any swelling of the clay that may have occurred during construction. Deeper shear surfaces, passing, for example, through a layer of clay at some depth below the wall, can use the undrained shear strength without taking into account the reduction of strength along the wall/soil interface.

2. If an embedded wall fails during excavation of the soil at the front of the toe (such as in Figure 5.5) then, depending on the time taken for construction to occur, some drainage of excess pore water pressure may have occurred. But it is by no means sure whether the mean total stress in front of the wall will have increased or decreased, as the vertical unloading during excavation at the base may be compensated for by the horizontal total stress increase as the soil supports the toe of

the wall. Assuming no time for drainage, an analysis using undrained strength parameters is normally used in a clay soil. This is the common assumption when designing braced excavations, for example.

3. The addition of loads shortly after wall construction, for example, by stacking cargo behind a quay wall, can produce three types of failure. First, there may be a bearing capacity failure immediately below the loaded area; second, loading may lead to deformation of ducting carrying tie-back anchors, and the strain induced by this may cause the anchor stresses to exceed their ultimate values; and third, an overall global instability may be triggered, involving the wall, any anchors and the surcharge loading behind the wall (Figure 5.3). Bearing capacity calculations should be carried out using undrained analysis (with undrained shear strength, c_u) for clay, and drained analysis (with effective angle of friction, ϕ') for sands and gravels. A judgement will need to be made whether to use undrained shear strength for global stability analysis of clays. It is probably wise to make an assessment using both undrained and drained analyses for clays.

4. Long after wall construction ground, water levels may have readjusted to their new boundary conditions as a result of changes in the ground profile, and the impermeability of the wall. Any excess pore pressures set up either by loading or by unloading, during regrading of the ground, will have dissipated leading either to swelling or to consolidation. Under these conditions the shear strength of cohesive soil is likely to have decreased (as a result of unloading and swelling), but may have increased locally (as a result of loading and consolidation). The available strength must be calculated using effective strength parameters c' and ϕ'. As noted in Chapter 2, great caution needs to be exercised when adopting values of effective cohesion intercept (c') for analyses. In soft or firm normally consolidated clays, it is prudent to assume $c' = 0$. In stiffer heavily overconsolidated clays, it is sensible to restrict c' to 1–2 kPa (Chandler and Skempton 1974), unless there is considerable evidence (e.g. from back analyses of field case records) to support the adoption of a higher value.

5.2.2 Peak, residual or critical state parameters?

Estimates of global stability for first-time slides (i.e. slides on a surface that has not experienced failure in the past) can be carried out using peak effective strength parameters for any kinematically admissible shear surfaces. If effective shear strength parameters are poorly known, then critical state parameters may be used to obtain a conservative estimate of stability.

The development of a 'first-time' slide produces complex patterns of undulating failure surfaces and the peak effective stress angle of friction

(ϕ') obtained from effective stress laboratory testing, used with a conservative value of effective cohesion intercept ($c' = 0-2$ kPa) can safely be used to analyse potential failure surfaces. However, as displacements on failure surfaces become greater, the failure surface is smoothed, and clay particles align, giving a polished and striated (slickensided) appearance to exhumed surfaces. The effective angle of friction in medium and high plasticity clays is gradually reduced, eventually reaching a 'residual' value, ϕ'_r. The residual effective angle of friction, which can be less than $10°$ in 'fatty' clays (i.e. compressible clays of high plasticity), must be used in stability analysis to assess the effects of reactivating preexisting shear surfaces.

Such a low value of angle of friction is only relevant on the actual plane or planes upon which large movements have previously occurred, so these need to be identified and their positions mapped. Very often, these pre-existing shear surfaces are shallow (less than 10 m deep) and non-circular. Non-circular failures require a particular form of analysis (see below).

As Lupini, Skinner and Vaughan (1981) show, rolling of particles during shearing prevents alignment of clay particles in low plasticity soils, so that the residual angle of friction for low-plasticity clays, is much the same as the peak value. Analysis of a reactivated slide in granular material should be carried out using the critical state angle of friction, equivalent to assuming a low relative density on the shear surface itself.

5.3 BASE HEAVE AND LOCAL FAILURE CALCULATIONS

Base instability can be a serious problem in soft or firm clays, if the strength of the soil does not increase significantly below the base of the excavation. This type of failure is analogous to a foundation bearing capacity failure, but with a negative load (Figure 5.5), caused by the imbalance between the weight of overburden outside the excavation and inside it.

Local failure also can be said to occur when excavations reach such a depth that lateral pressures from the base of the wall coupled with stress relief overstress the soil, leading to an inward movement of the sheets.

For walls which are continuous below excavation level, the estimation of passive stress relief stresses can be carried out using conventional calculation methods, in combination with predicted long-term (equilibrium) pore water pressures in the passive zone. The use of conventional triaxial compression effective strength parameters will yield conservative estimates of the factor of safety against failure. Because there is evidence that such parameters may lead to underestimates of the soil strength, if better estimates of the soil behaviour are required, then special 'passive stress relief' tests must be considered (Burland and Fourie 1985).

5.4 LIMIT EQUILIBRIUM ANALYSIS
OF OVERALL INSTABILITY

In situations where earth retaining structures are underlain by weak soils (e.g. Figure 5.3), or the ground adjacent to the excavation slopes upward, the overall stability of the excavation should always be analysed. One approach to the analysis of such a failure is to use classical circular arc slope stability methods, or commercial computer programs based upon them. Alternatively, wedge stability analyses can be made, involving an active wedge outside the excavation and a passive wedge inside.

In areas where there are preexisting failures, it is necessary to carry out overall instability analyses using residual parameters on predefined shear surfaces, as has been discussed above. However, the use of residual parameters in conventional circular stability analyses will yield overconservative estimates of stability.

Overall stability analyses should also be carried out to check that the fixed lengths of ground anchors will be placed sufficiently far back from the excavation. Circular and non-circular failure surfaces should be used, and the assumption is often made that they can only propagate beyond the remote ends of the fixed lengths of the anchors. This assumption may not always be sound, as the contribution of an anchor to the calculated stability on a particular sliding surface may not be great, in terms either of its shear resistance or as a result of the force it applies normally and tangentially to that surface. For further discussion, see Chapter 11, where the design calculations for anchoring and nailing systems are described.

5.4.1 The method of slices

Slope-stability problems in engineering works are generally analysed using limit equilibrium methods. Many computer programs using such methods are available in practice and the most common ones call on the method of slices. In this method, the failure mass is broken up into a series of vertical slices (Figure 5.6) and the equilibrium of each of these slices is considered. This procedure allows both complex geometry and the variable soil and pore pressure conditions of a given problem to be considered.

Traditionally, the factor of safety is defined in terms of the ratio between the average shear strength available on the shear surface and the average shear strength mobilised for stability, i.e.

$$F = \frac{\overline{\tau}_{available}}{\overline{\tau}_{mobilised}} \qquad (5.1)$$

where F = 1 at failure.

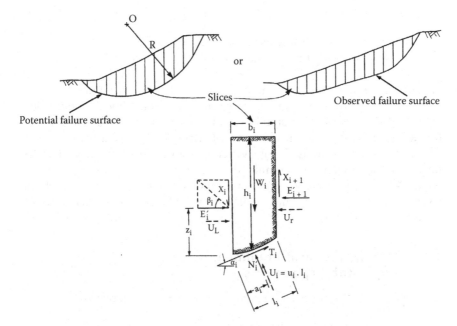

Figure 5.6 Complete system of forces on a single slice of a slipping mass of soil.

To obtain a solution to the problem, slope-stability analyses also examine the overall equilibrium of the mass of soil which is being considered. A number of different solutions can be derived, depending on the simplifications adopted in the analysis. Methods used to derive the basic equations are

a. Force equilibrium of a single slice
b. Moment equilibrium of a single slice
c. Force equilibrium of the total mass of soil above the slip surface
d. Moment equilibrium of the total mass of soil above the slip surface

For example, Bishop's method (Bishop 1955) combines (a) and (d) to achieve a solution, Janbu's method (Janbu 1973) uses (a) and (c), and Morgenstern and Price's method (Morgenstern and Price 1965) uses a combination of (a), (b) and (c). A solution cannot be obtained for the complete stability of the slipped mass without making simplified assumptions, and therefore, it is common to make assumptions concerning the interslice forces E and X (Figure 5.6), for example, in deriving a factor of safety for a circular failure surface. From the definitions of factor of safety

$$\text{available shear strength} = \frac{\text{peak shear strength}}{\text{factor of safety}}$$

For moment equilibrium about the centre of the circle

Moment of available shear strength on shear surface
= moment of weight of soil mass

i.e. disturbing moments must equal restoring moments, at failure.

A summary of the most significant features of each method recommended is presented below. Global stability analyses will normally be carried out for earth retaining structures in terms of effective stresses, using estimates of the long-term, stabilised pore water pressure, since this will give the lowest factor of safety for an unloading case (e.g. excavation in front of a wall). Where load increase takes place on clays, (e.g. for fill placed behind a sheet-pile wall) short- or intermediate-term stability analysis may give the lowest factor of safety.

5.4.2 Methods of limit equilibrium slope stability analysis

The methods that have most commonly been used in practice are

- Fellenius' (1936) method of slices
- The modified or simplified Bishop method (1955)
- Janbu's generalised method of slices (1973)
- Morgenstern and Price's method (1965)
- Spencer's method (1967)
- Sarma's method (1973)

The first two methods do not satisfy all the moment and force equilibrium equations and can only accommodate circular slip surfaces. The last four methods may be used to calculate the factor of safety along any shape of slip surfaces. The first five of these methods were compared by Fredlund and Krahn (1977) and by Duncan and Wright (1980) in order to evaluate their accuracy and reliability. The results of such studies have shown that those methods which satisfy all the conditions of equilibrium, and Bishop's modified method, give accurate results which do not differ by more than 5% from the 'correct' answer, obtained by the log spiral method. The main conclusions of these comparative studies can be found in La Rochelle and Marsal (1981). These authors prefer the simplified Bishop and Janbu methods due to the simplicity of computer programming and the low cost of running such programs. When using commercial computer code and computers with fast processors, these considerations are not important, but they are relevant when using hand calculations to check computer predictions.

In practice, the design engineer must choose between circular and non-circular methods of analysis. Because of their relative simplicity, it is common to carry out routine slope-stability analyses using a circular failure surface. In relatively homogeneous soil conditions, this assumption will be justified, since experience shows that the analysis can make good estimates of the factor of safety when failure is imminent.

Non-circular analyses are necessary when

a. A preexisting shear surface has been found in the ground, and is known to be non-circular.
b. Circular failure is prevented, perhaps by the presence of a stronger layer of soil at shallow depth.

Under either of these conditions, the use of a circular shear surface may overestimate the factor of safety against failure.

5.4.3 Computer analysis of slope stability

There are now many examples of computer software that can be used for slope stability analyses. Traditional limit slope analysis (described above) is being gradually complemented by more complex, but potentially more flexible, continuum (finite element and finite difference) analysis. It is not the purpose of this chapter to evaluate them, but it may help the reader to give examples. The following were amongst the many programs available at the time of writing:

Program	Comments
FLAC/Slope	Continuum analysis using the explicit 2D finite difference method. The software can deal with heterogeneous soil conditions, surface loading and structural reinforcement. It is produced by Itasca.
PCSTABL	Long-established and widely trusted software, originating from Purdue University. It implements Bishop, Spencer and Janbu limit equilibrium methods. Pre- and post-processing is enabled via STED. GSTABL also uses STABL.
SLOPE/W	Widely used, this software has a modern graphical user interface, and can cope with external loads. It implements a wide selection of methods of limit equilibrium analysis, including Bishop, Spencer, Janbu, Sarma and Morgenstern and Price. Input parameters can be deterministic or stochastic. It is produced by GEO-SLOPE International Ltd.

It is important when using commercial software to examine validation documents, and where none is available, to check the factor of safety on critical failure surfaces by using a number of different codes or analyses, by comparison with design chart solutions, and by hand calculation if necessary.

5.4.4 Hand calculations and design charts

Methods of hand calculation are widely covered in soil mechanics texts (for example, the second edition of this book, and Bromhead 1998). They will not be repeated here. Instead, we consider only infinite slope analysis, and then reproduce key stability charts for undrained and drained analyses. Computer-generated solutions, whether based on limit equilibrium or continuum analysis, need checking. Design charts can be helpful in this respect.

Infinite slope analysis (Haefeli 1948; Skempton and Delory 1957) gives a simple basis on which to make initial assessments of the stability of slopes. It is particularly useful for slopes consisting of granular material, and for residual shear surfaces, both of which tend to be shallow and relatively planar. For flat or long slides where the slip surface is parallel to the slope, such as that shown in Figure 5.7, the influence of the top and toe are relatively small. By resolving parallel and at right angles to the slope, the factor of safety of a slice (length l parallel to the slope, width b) can be written as

$$F = \frac{c'l + (\gamma zb \cos\beta - ul)\tan\phi'}{\gamma zb \sin\beta} \tag{5.2}$$

where

- z is the depth to the slip surface
- γ is the bulk unit weight of the soil
- u is the pore pressure on the slip surface
- β is the inclination of the slope, and of the shear surface

For flow parallel to the ground surface, and with the groundwater surface at a height h_w above the slip surface,

$$u = \gamma_w h_w \cos^2\beta \tag{5.3}$$

since the equipotentials are normal to the slope surface. Therefore,

$$F = \frac{c' + (\gamma z - \gamma_w h_w)\cos^2\beta \tan\phi'}{\gamma z \sin\beta \cos\beta} \tag{5.4}$$

Cross-section through preexisting slope failure in clay

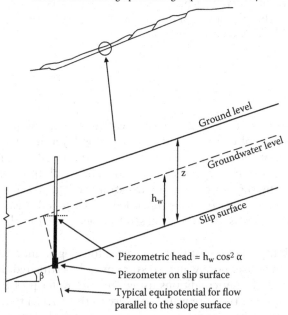

Piezometric head = $h_w \cos^2 \alpha$
Piezometer on slip surface
Typical equipotential for flow
parallel to the slope surface

Figure 5.7 Infinite slope analysis. (From Haefeli, R., The stability of slopes acted upon by parallel seepage. Proc. 2nd Int. Conf. on Soil Mech. Found. Engng, Rotterdam, 1, 57–62, 1948; Bromhead, E.N., *The Stability of Slopes*. Spon, London, 1998.)

Where c' is small or negligible, and the groundwater is below the failure plane,

$$F = \frac{\tan\phi'}{\tan\beta} \tag{5.5}$$

At the other extreme, where the groundwater surface is at ground surface

$$F = \frac{\gamma_w}{\gamma}\frac{\tan\phi'}{\tan\beta} \approx \frac{\tan\phi'}{2\tan\beta} \tag{5.6}$$

So, for example, if a clay has a residual angle of friction of 14° and a pre-existing shear failure occurs on a hillside where, in winter, the groundwater table is at the ground surface, failure (F = 1) can be expected to occur on slopes as flat as 7° to the horizontal. In the summer, were the groundwater to drop below the shear surface, the factor of safety would approximately double.

An assessment of the factor of safety in terms of short-term, undrained shear strength of a clay can be made using Taylor's (1937) chart (Figure 5.8), provided that the strength of the clay is approximately uniform. The chart shows Taylor's stability number

$$N = \frac{c_u}{\gamma HF} \tag{5.7}$$

as a function of slope angle, β. H is the height of the slope. The chart can take account of constraints due to a strong layer at some depth DH below the top of the slope. The full line gives the stability number if the toe breakout position is unconstrained, and the depth of the slip surface is not limited by the presence of a hard stratum. The dashed lines allow the depth of a hard stratum to be included. Since N is decreased, the factor of safety is increased as the hard stratum gets closer to the bottom of the slope.

Other stability charts exist that are useful for long-term, effective stress analyses. In this case, pore water pressure must be estimated. It is normal to express groundwater conditions using the pore water pressure ratio, r_u, which is the ratio of the pore water pressure to the vertical total stress at a point. The average pore water pressure ratio along the failure surface needs

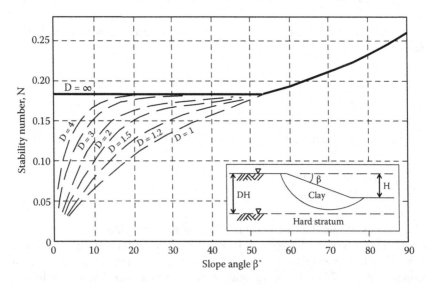

Figure 5.8 Taylor's stability chart for undrained, $\phi_u = 0$, analyses. (From Taylor, D.W., J. Boston Soc. Civil Engrs. 24, 137–246, 1937.)

to be estimated given that the bulk unit weight of soil is about twice the unit weight of water

$$0 < r_u(= u/\gamma h) < 0.5 \qquad (5.8)$$

and is typically in the range 0.2–0.3 in temperate climates.

Bishop and Morgenstern (1960) observed that, approximately, the factor of safety of a slope in soil with modest effective cohesion intercept, c', is a linear function of r_u,

$$F = m - nr_u \qquad (5.9)$$

and they therefore provided charts (Figure 5.9) of m and n for different values of slope (cot β), effective angle of friction, normalised cohesion ($c'/\gamma H$), and associated depth factor (D) (Figure 5.10).

These charts were intended for use in assessing embankment dam stability, so the values of effective cohesion intercept that are provided for are limited. Curves are also given for r_{ue} as a function of slope and angle of friction. When r_{ue} is less than the actual value of r_u in the case under consideration, values of m and n should be obtained from the chart with a greater depth factor.

Bishop and Morgenstern (1960) give the following example of the use of the charts:

Determine the factor of safety for a 42.7-m (140′) embankment with a side slope of 1 on 4, founded on 18.3 m (60′) (thickness) of alluvium with the same properties as the embankment fill, under which lies bedrock. The soil properties are

$\phi' = 30°$
$c' = 28.2$ kPa (590 p.s.f.)
$\gamma = 18.9$ kN/m^3 (120 p.c.f.)
$r_u = 0.5$

From the given data,

$c'/\gamma H = 0.035$
$D = (42.7 + 18.3)/42.7 = 1.43$

From the bottom right hand chart in Figure 5.9a, for $c'/\gamma H = 0.025$ and $D = 1$, and with $\phi' = 30''$ and cot β = 4:1, it can be seen that r_{ue} (≈ 0.43) < r_u (= 0.5). Therefore D = 1.25 is the most critical.

So, from the upper two charts of Figure 5.9b, for D = 1.25 and $c'/\gamma H = 0.025$ values of

m = 2.95, and
n = 2.81

are obtained.

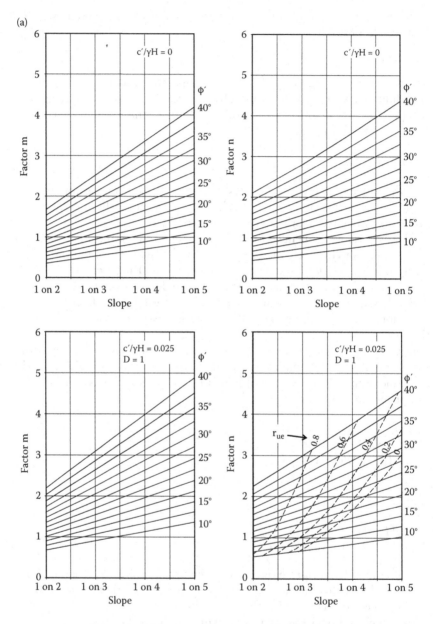

Figure 5.9 (a) Bishop and Morgenstern's (1960) chart for slope stability. Above: m and n for c'/γH = 0. Below: m and n for c'/γH = 0.025 D = 1.00.

(b)

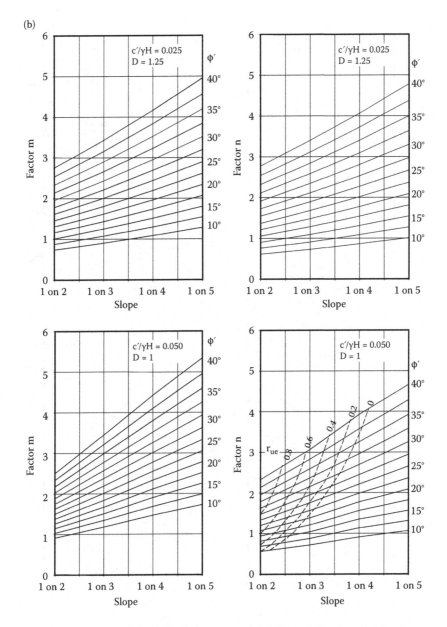

Figure 5.9 (Continued) (b) Bishop and Morgenstern's (1960) chart for slope stability. Above: m and n for c′/γH = 0.025 D = 1.25. Below: m and n for c′/γH = 0.050 D = 1.00.

(c)

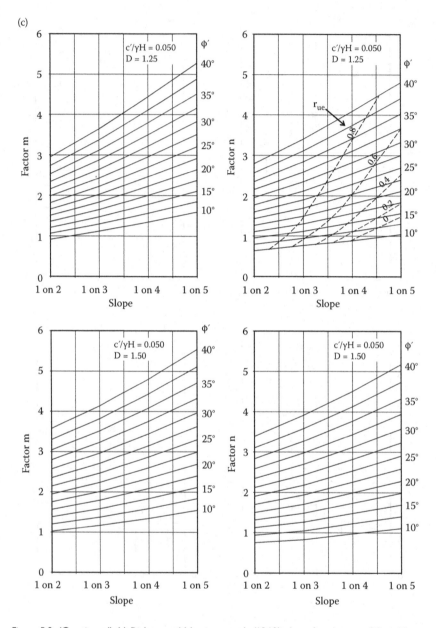

Figure 5.9 (Continued) (c) Bishop and Morgenstern's (1960) chart for slope stability. Above: m and n for c′/γH = 0.050 D = 1.25. Below: m and n for c′/γH = 0.050 D = 1.50.

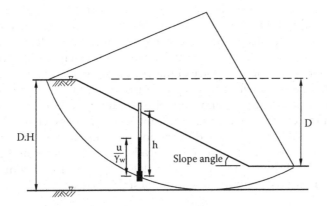

Figure 5.10 Definition of r_u and D for Bishop and Morgenstern's (1960) charts.

From Equation 5.9, with $r_u = 0.50$, it follows that

$$F = 2.95 - 1.405 = 1.545$$

For $c'/\gamma H = 0.05$ and $D = 1$, similarly, the bottom right chart of Figure 5.9b shows that $r_{ue} < r_u$ and therefore that the factor of safety with $D = 1.25$ is more critical than with $D = 1.00$. For $c'/\gamma H = 0.05$ and $D = 1.25$, the chart on the top right hand side of Figure 5.9c shows that

$$r_{ue} \approx 0.72 > r_u \ (= 0.5)$$

Therefore, $D = 1.25$ is the most critical level, and from the upper two charts on Figure 5.9c,

$$m = 3.23$$
$$n = 2.83$$

Applying Equation 5.9 with $r_u = 0.50$,

$$F = 3.23 - 1.415 = 1.815$$

Interpolating linearly for the given value of $c'/\gamma H = 0.035$, we obtain

$$F = 1.545 + 0.4 \times 0.270 = 1.65$$

It can be seen from the above that the groundwater level and hence pore pressures and r_u values have a dominant effect on stability, and on factors of safety calculated using effective stress strength parameters. If realistic assessments of factor of safety are to be made, it is essential that good estimates of pore water pressures are obtained. These estimates must be for the long-term conditions, when excess pore pressures due to loading or unloading have dissipated, and new flow patterns, resulting both from slope reprofiling and the construction of impermeable walls and associated drainage, have been established. When this cannot be done, it will be sensible to adopt conservatively high pore pressure values. In the temperate climate of the UK, for the brown London Clay, a value of $r_u = 0.3$ has sometimes been adopted on the basis of pore pressures in established slopes.

5.5 DETECTING AND STABILISING PREEXISTING INSTABILITY

Because residual effective strength parameters are much lower than peak parameters, particularly for plastic clays, it is essential that the likelihood of preexisting instability is assessed where retaining structures are to be built on, or at the base of, natural slopes.

A number of factors may suggest preexisting instability:

- The combination of existing slope angle and soil type. Slopes in clay of more than about 7° to the horizontal should be regarded with suspicion if winter rain leads to high groundwater.
- Evidence of poorly drained ground and a high groundwater level on slopes, for example, pools and ponds (perhaps only in winter) and water-loving plants such as reeds. These conditions are often created by the movement of the ground, which disrupts natural drainage patterns.
- Hummocky ground, which is often created by multiple rotational slips, or at the toe of slope instability.
- Other evidence of ground movement, including growth of bushes and trees in areas of breaks of slope, and bends in tree trunks.

In addition, valuable evidence can often be obtained from air photography, where hummocky ground and a 'turbulent' texture, caused by differences in drainage, may be seen. A vertical and an oblique air photograph of preexisting slope instability at Stag Hill, Guildford, U.K. (now the site of the University of Surrey) are given in Figure 5.11. A view of preexisting slope instability as seen from ground level is shown in Figure 5.12.

A very useful review of European experience of slope stability problems and parameter selection is given by Chandler (1984).

(a)   (b)

Figure 5.11 Vertical and oblique aerial photographs showing preexisting instability in the London clay, Stag Hill, Guildford. (From Clayton, C.R.I. et al., *Site Investigation*, Second edition, Blackwell Science, 584 pp., 1995.)

Figure 5.12 View from ground level of preexisting instability at Sevenoaks, Kent. (From Clayton, C.R.I. et al., *Site Investigation*, Second edition, Blackwell Science, 584 pp., 1995.)

5.6 STABILISATION OF SLOPES USING RETAINING STRUCTURES

The existing stability of a slope can be of concern not just because it shows signs of failure or is failing. It is common to assess the stability of a slope which shows no signs of a problem, and to find that according to calculation, it has a factor of safety less than required by national legislation, standards, or current practice. Retaining walls may then be able to offer a solution.

However, at the outset, it is important to recognise that a number of methods exist by which the stability of a slope can be improved. Techniques that can be used are

- Regrading (flattening the slope, or cut and fill solutions)
- Shallow or deep drainage
- Prevention of water entering the slope
- Anchoring
- Soil nailing
- Use of geo-grids in fill
- Grouting

These techniques are considered in detail in many standard textbooks. There is a broad agreement that regrading is the most certain method of improving stability, and that drainage can be very effective, although the maintenance of drainage systems needs to be assured. Prevention of water

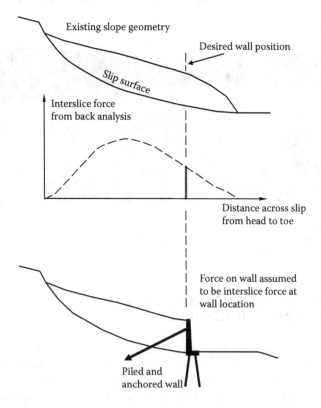

Figure 5.13 Estimating the force on a retaining wall due to slope instability. (From Bromhead, E.N., *The Stability of Slopes*. Spon, London, 1998.)

entering a slope, for example, by infiltration at the top, or from rainfall, is certainly an essential component of many designs. Anchoring tends to be most effective in rock slopes, and reinforcement and strengthening by grouting finds some application.

Retaining walls are not generally the first choice for the stabilisation of preexisting slope failures, simply because the mass of soil to be supported is generally very large. A simple way of obtaining an estimate of the likely passive force on a wall placed in a given position is to extract the interslice forces from a limit equilibrium analysis, as shown in Figure 5.13. If soil is removed down slope, this interslice force must be provided by the wall in order to give an equivalent stability. This type of analysis can also be performed using many commercial computer programs. Note that the magnitude of the interslice force varies up the slope, being zero at both the toe and the rear scarp. Therefore, a key issue is the location of the wall. Another issue is whether the wall will disrupt slope drainage. As can be imagined, a wall of this type needs sufficient rigidity in bending and shear. Therefore, the use of sheet piling is generally only acceptable in stabilising very small slips.

Other viable methods of improving stability include soil nailing and anchoring, which are discussed in Part II of this book, and discrete pile stabilisation (Carder 2005; Smethurst and Powrie 2007).

Part II

Design

Chapter 6

Wall selection

There are many different types of earth retaining structure. The first stage in the design process is therefore to assess the appropriateness of the available types of structure for the given application. Following this, the possible ways in which a given structure might fail to perform satisfactorily should be considered (see Chapter 7). Preliminary design may then be carried out on a number of different types of retaining structure, to assess their viability, and finally detailed design calculations will be undertaken. During the selection of appropriate types of retaining structure, it will be necessary to have an approximate idea of the types of subsoil and their distribution around the proposed wall location.

6.1 REASONS FOR SELECTING A PARTICULAR FORM OF RETAINING WALL

It may be helpful to rank the various considerations that influence the final choice of wall type, as shown in Figure 6.1. This pyramid puts the aspect considered most important at the bottom, and least important at the top. For example, it is usually the case that the function of the wall is paramount; if it will not support the soil, then it does not matter how attractive it is to look at. But beyond function, modern regulations—such as the Construction Design and Management (CDM) Regulations—require that a structure can be constructed safely. And what were once aspirations of sustainability have now become specific requirements, so the ability to dismantle and reuse may need to be considered. Higher layers in the pyramid rely upon those beneath being present, and everything in the pyramid has an influence on overall cost. As the various different types of earth retaining structure are described in the following sections, some comments on these different aspects affecting final choice are included.

Not all types of ground require support. Some materials, for example, unfractured rocks, can stand vertically, or even overhang (see Figure 6.2).

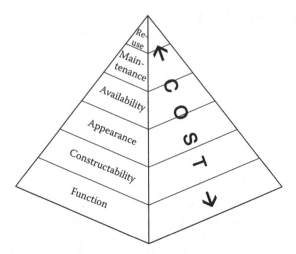

Figure 6.1 Hierarchy of design considerations.

Figure 6.2 Overhanging, unsupported rock face in school playground (Hong Kong 1981).

Other rock types can stand either vertically or at a steep angle, but require some type of facing in order to protect the ground from the effects of weathering (Figure 6.3). However, jointed rock masses generally require some support, depending on the orientation and nature of the discontinuities. These are beyond the scope of this book but see, for example, Simons et al. (2001).

Figure 6.3 'Chunam' rock face protection (Hong Kong).

6.2 GRAVITY WALLS

According to *Eurocode 7*, the defining characteristic of a gravity wall is that 'The weight of the wall itself ... plays a significant role in the support of the retained material'.

Gravity walls can be made of stone, blockwork, or plain or reinforced concrete, and may include a base footing (with or without a heel), ledge, or buttress. The weight of the wall may be enhanced by that of soil, rock, or backfill placed on its footing. A key benefit of gravity walls is their rugged construction, but they are not economical for large retained heights (Teng 1962).

6.2.1 Mass concrete gravity walls

Cross sections through some typical mass concrete gravity walls are shown in Figure 6.4. The dimensions of the wall should be such that the resultant earth pressures on it produce no tensile stress in any part of the wall, since it cannot be assumed that joints between lifts of concrete or masonry blocks have any tensile strength.

Mass concrete walls are probably only viable for small retained heights, say up to 3 m. They can be designed for greater heights, but as the height increases, other types of wall become more economical. The cross-sectional shape of the wall is dictated by stability, the use of space in front of the wall, the required wall appearance, and the method of construction. Economy of material will normally result if either the front or back of the wall is stepped or inclined.

6.2.2 Gabions

A variation on the traditional mass concrete gravity wall is the gabion wall, illustrated in Figure 6.5a. A gabion consists of a box made of

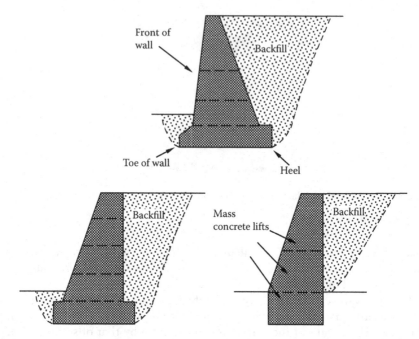

Figure 6.4 Cross sections through typical mass concrete walls: (top) inclined back of wall; (bottom left) vertical back of wall; (bottom right) with footing cast in trench.

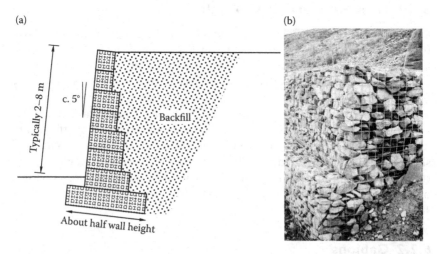

Figure 6.5 Gabion retaining wall. (a) Cross-section through a gabion wall. (b) Detail of gabions and fill.

metal or plastic mesh that is filled *in situ* with coarse granular material such as crushed rock or cobbles, and used as a basic building unit (see Figure 6.5b).

The major advantage of the system is its flexibility, but it has additional advantages when constructing in remote areas. Only the mesh needs to be transported to the site, and local labour and materials can be used to complete the structure. Gabion walls are particularly good at absorbing impact energy and are often used as rock fall barriers.

Gabion walls can be particularly attractive and blend in extremely well with a mountainous natural environment. They are simple to maintain and repair if damaged, and particularly easy to reuse or recycle.

6.2.3 Crib walling

Figure 6.6 shows a cross section through a typical concrete crib wall, with various elements identified. Figure 6.7 shows a large timber crib wall for a sports complex. Crib walling is suitable for walls of small to moderate height (up to 6–9 m) subjected to moderate earth pressure. Timber components are normally used for landscaping and temporary works, with precast concrete being used for most civil engineering construction. The crib components are backfilled with (compacted) granular soil.

The major advantage of the crib wall system is that large movements can be tolerated without damage since it is a flexible structure. Other advantages are that site work is very simple with no need of any major plant or facilities, and the use of a permeable fill improves the drainage of the soil retained behind the wall.

Crib walls are generally considered aesthetically pleasing, especially when surface vegetation has been encouraged to grow in between the stretchers. Not only are they easy to assemble (and ultimately to dismantle and reuse if necessary), but maintenance is also straightforward.

6.2.4 Interlocking block walls

There are a number of proprietary systems that use different shapes of precast concrete blocks that interlock with each other—usually without the need for any cement mortar—to produce a retaining wall, as shown in Figure 6.8.

The interlocking between units can be achieved via a rear lip or protrusions on the upper or lower surfaces (see examples at the top of Figure 6.8). Such systems are, by virtue of their modularity and absence of mortar, not only easy to construct but also easy to dismantle and reassemble—allowing reuse. Their appearance is generally aesthetically pleasing, resembling

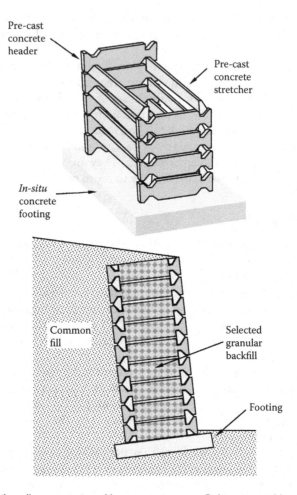

Figure 6.6 Crib wall construction. Above: components. Below: general layout.

dry stone walls. High standards of quality control in manufacture lead to reliable individual units that are unlikely to fail if constructed to the recommended geometry. As an example, the manufacturers state that the 'Porcupine' wall, which uses concrete blocks weighing around 20 kg, can be used to build walls with face angles between 68–73° and heights up to 3 m routinely (and higher, with additional measures).

6.2.5 Masonry walls

For walls of modest height, up to about 4 m, load-bearing brickwork can provide an economical solution that can also be attractive. Design guidance

Figure 6.7 Large timber crib wall at University of Surrey.

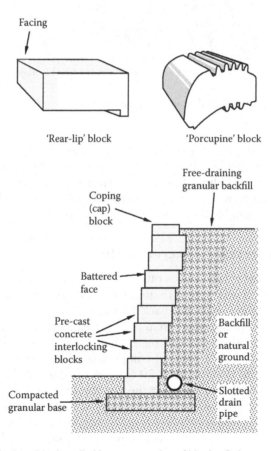

Figure 6.8 Interlocking block wall. Above: examples of blocks. Below: general layout.

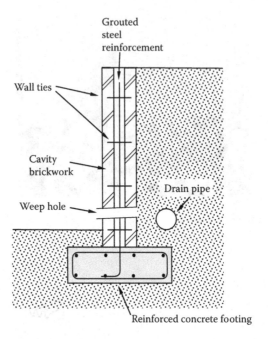

Figure 6.9 Grouted cavity reinforced brick retaining wall.

is provided by the Brick Development Association (Haseltine and Tutt 1991). Mass brickwork walls are generally only suitable for small walls, up to a retained height of about 1 m, but a 330 mm thick 'quetta bond' wall can be used to retain up to 3 m of soil. Double-skinned reinforced and grouted cavity walls are suitable for greater retained heights (see Figure 6.9).

6.2.6 'Semi-gravity' concrete walls

Semi-gravity walls rely more on internal resistance to bending and shear, and less on self weight than gravity walls. By introducing a small amount of reinforcing in the back of the wall (Figure 6.10) as a connection between the vertical stem and the base, and between concrete 'lifts', a more slender stem can be used, resulting in a reduction of the mass of concrete. It is a form of compromise between the simplicity of mass concrete and the low material content of reinforced concrete. This leads to a cost trade-off between the volume of concrete saved and the amount of steel required. From a durability standpoint, mass concrete is easier to maintain and so whole-life costs may be lower, but if reusability is important, the thinner section of reinforced concrete will be easier to break up for recycling.

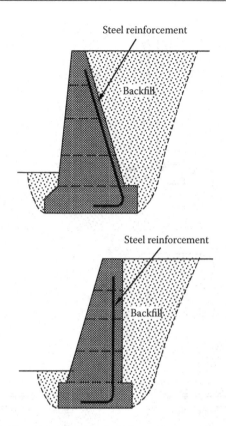

Figure 6.10 Cross sections through typical semi-gravity concrete walls.

6.2.7 Reinforced concrete cantilever walls

Figure 6.11 shows the most common forms of reinforced-concrete cantilever wall. They are made in the form of an inverted T (Figure 6.12, top) or L (Figure 6.12, bottom), with the latter being either forward or backward facing. The cantilever wall is a reinforced concrete wall that uses the cantilever action of the stem to retain the soil mass behind the wall. Stability is achieved from the weight of the soil on the heel portion of the base slab. A shear key may be used to augment sliding resistance. The very simple form of L or inverted T are suitable for low walls (less than 6 m), but for higher walls, it is necessary to introduce counterforts or buttresses.

Moderate heights of cantilever walls are available as precast units, allowing quick assembly on site. Their finish is generally plain, but can be textured to make them more aesthetically pleasing. In general, the quality of a precast concrete would be expected to be higher than that of an *in situ* concrete, but this advantage will be offset by the greater cost of transport

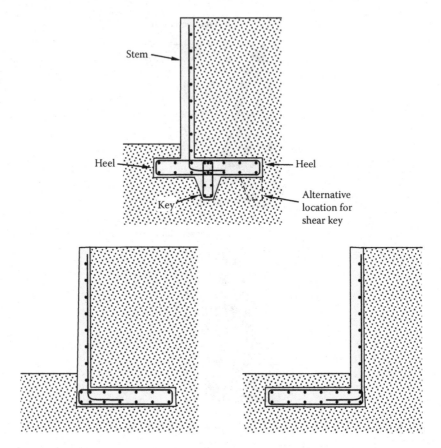

Figure 6.11 Cross sections through typical (inverted) T-shaped and L-shaped reinforced concrete cantilever walls.

and handling. There should be little risk of construction failure, provided that the manufacturers' recommendations with regard to installation and backfill are complied with.

6.2.8 Counterfort walls

Counterfort walls (Figure 6.12) are similar to cantilever walls, but they have counterforts (buried in the retained soil) to connect the wall and base, thus reducing bending moments and shear stresses in the stem. They have in the past been used for high walls or where there is very high pressure applied behind the wall. They are now seldom used, except for very tall walls (10–12 m high).

This sort of wall is more complicated to build because of the counterforts, but the advantages and disadvantages are as much as for other types of semi-gravity wall.

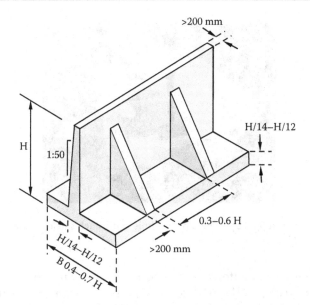

Figure 6.12 Counterfort reinforced concrete wall dimensions (see also Teng 1962).

When excavations of 10–12 m depth are to be made from ground level, then diaphragm walling or secant bored piling are often preferred. These avoid additional excavations, can be constructed from the top down and do not involve the complex geometry of counterfort or buttressed walls, which lead to construction difficulties.

6.2.9 Buttressed walls

Buttressed walls, alternatively known as reverse-counterfort walls, are similar to counterfort walls but the bracing is at the front, subject to compression instead of tension and thereby allowing construction in masonry. As with a counterfort wall, the buttresses make construction more difficult than for other types of semi-gravity wall, but the main advantages and disadvantages are much the same. Buttressed walls were quite common in the UK in previous centuries, but are little used today.

6.3 EMBEDDED WALLS

According to *Eurocode 7* Part 1, Paragraph 9.1.2.2, the defining characteristic of an embedded wall is that 'the bending capacity of [the] wall plays a significant role in the support of the retained material while the role of the weight of the wall is insignificant'. Embedded walls are relatively thin

Figure 6.13 Trench support provided by manhole and trench shields (image courtesy of VP plc).

walls made of steel, reinforced concrete, or timber, which are supported by anchorages, struts, and/or passive earth pressure. Examples include cantilever steel sheet-pile walls, anchored or strutted steel or concrete sheet-pile walls, and diaphragm walls.

6.3.1 Trenching systems

Trenching is very common in urban areas, being used to permit the installation, repair, and/or replacement of buried utilities, such as water, gas, and sewerage pipes. Trench support systems are always temporary, and are typically used for excavations that are quite narrow but can be up to several metres deep. They can comprise prefabricated units (Figure 6.13) that are lifted in and out of the trench as necessary, or individual steel sheets/timber planks with struts and wallings (Figure 6.14). Trench support systems must, by their very nature, be highly reusable. Because they are temporary structures, their appearance is almost irrelevant. Buildability is essential, and made easier with prefabricated units, which are generally safer to use, because they can avoid the need for site workers to enter the excavation during their construction. However, when sites are congested, hand construction may be required.

6.3.2 Sheet-pile walls

Sheet-pile walling is widely used to construct flexible support systems, often for both large and small waterfront structures, or in temporary works. It is often used in unfavourable soil conditions (for example, soft clays) because no foundations are needed. Although sheet-piles are easily driven from ground level (Figure 6.15), construction is straightforward even where

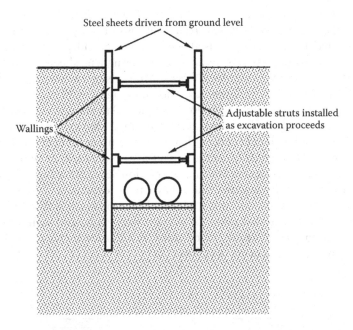

Steel sheets driven from ground level

Adjustable struts installed as excavation proceeds

Wallings

Figure 6.14 Braced excavation for services installation.

Figure 6.15 Driving Larssen steel sheet-piles.

water is present, when other types of structure are difficult to build, since the sheets may readily be driven from pontoons or barges. Over the years, different materials have been used—steel, timber, and precast reinforced concrete. Figure 6.16 shows some typical permanent sheet-pile wall layouts, intended for river and dock works. Methods of anchoring, necessary when retained heights are larger, are shown in Figure 6.17.

Sheet-pile walls are able to follow complex plan shapes with ease, and cause minimal soil displacement during driving. They can be constructed in conditions of low headroom, with a variety of modern installation methods that cause low environmental impact (e.g. by reducing driving noise), and their speed of installation and extraction leads to a high degree of sustainability. On the negative side, they can be expensive if used to provide a permanent solution, and traditional methods of installation can be very noisy. Wall depth is limited by section size, loads, and standard stock lengths, and in certain ground conditions, installation may need to be preceded by water jetting or preaugering.

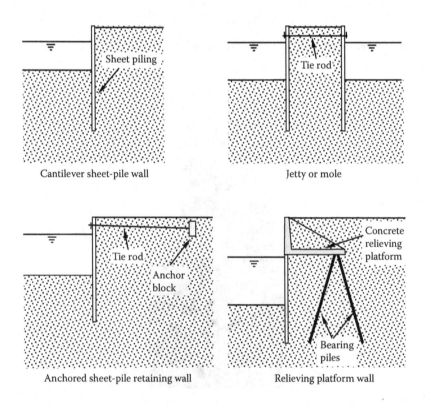

Figure 6.16 Cross sections through some typical permanent sheet-pile structures.

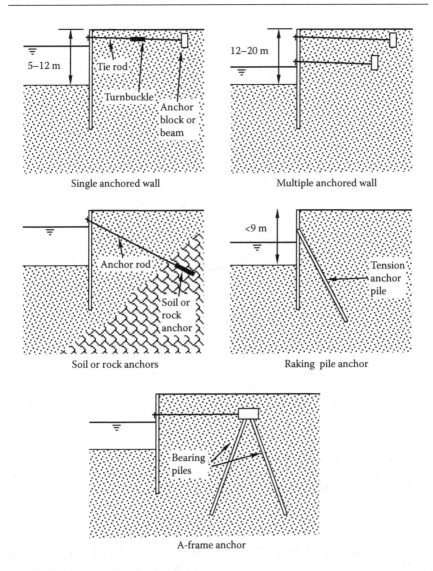

Figure 6.17 Anchored sheet-pile wall schemes.

6.3.2.1 *Steel*

Steel (Figure 6.18) is the most commonly used material in construction, due to a number of advantages such as

- Variety of cross section with a wide range of strength
- Economy

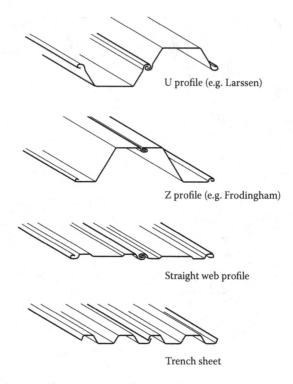

U profile (e.g. Larssen)

Z profile (e.g. Frodingham)

Straight web profile

Trench sheet

Figure 6.18 Examples of steel sheet-pile sections.

- Lack of buckling under heavy driving
- Availability in different combinations to increase wall section modulus
- Reusability for temporary works
- Relatively light weight
- The possibility of increasing the pile length by welding or bolting

Larssen and Frodingham sections are typically used for retaining walls such as those shown in Figure 6.16, where significant bending moments need to be resisted. The straight web profile is used where there is significant tension to be resisted in the plane of the wall or section, for example, for caissons and cofferdams (see Section 6.4.1). Trench sheet, as its name implies, is used for shallow braced excavations (see Section 6.3.1).

Durability can be an issue in aggressive groundwater or marine environments and/or where the wall is intended to serve as a permanent structure (e.g. a quayside wall rather than a temporary cofferdam). The appearance of steel sheets is usually considered rather unattractive, even when

Figure 6.19 Deep anchored sheet-pile excavation, Port Headland Western Australia. (Image courtesy of Marc Woodward, Perth.)

painted to inhibit corrosion. An example of a deep excavation supported by anchored steel sheet piling is shown in Figure 6.19.

6.3.2.2 Wood

Wood is usually used for temporary works, short spans, up to 2 m-high cantilevered walls, or braced sheeting. If used in permanent structures above water level, life expectancy is short, even with special preservative treatment. Figure 6.20 shows some sections used in practice.

6.3.2.3 Concrete

Reinforced concrete can be used for permanent structures, with a variety of cross sections. The most commonly used is straight web piling bar, provided with a tongue and groove, similar to the ones used on timber sheet-piles (Tsinker 1983). Sometimes grouting is used to make the resultant

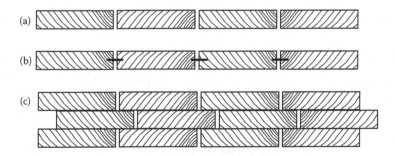

Figure 6.20 Examples of wooden sheet-pile sections. (a) Butt-end. (b) Splint-fastened. (c) Tongue and groove.

wall watertight. Prestressed concrete may be considered, since cracking of the concrete in the tension zone is thereby largely eliminated, with the corresponding advantage of reducing the possibility of corrosion of the reinforcement.

Although more durable than steel, concrete sheet-piles need a thicker section, which increases the displacement of the soil during driving, and hence the driving resistance. This may make the use of such elements uncompetitive with steel piles. Also, if concrete sheet-piles are more difficult to install, they will be more difficult to extract, affecting their reusability. Appearance-wise, concrete may be preferable to steel, but unless the sheet-piles are exposed to view permanently, this may not be a significant factor.

Figure 6.21 shows some details of reinforced-concrete sheet-piles. Sometimes grouting is used to make the wall watertight. Prestressed

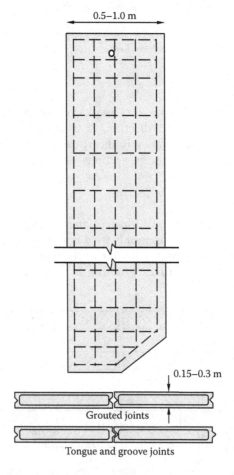

Figure 6.21 Example of concrete sheet-pile section.

concrete may be considered, since cracking of the concrete in the tension zone is thereby largely eliminated, with the corresponding advantage of reducing the possibility of corrosion of the reinforcement. The high weight of each element and the displacement of the soil during driving, however, may make the use of such elements uncompetitive with steel piles.

6.3.3 Bored pile walls

Bored pile walls have the advantage that they can be constructed in almost any ground conditions. Construction noise and vibration are relatively low, allowing installation close to existing structures. Bored piles can be included in the structural design of a building and can support high vertical loads in addition to lateral earth pressures. Bored pile walls may be

- Intermittent (spacing exceeds diameter). Large gaps between adjacent piles (spacing s > diameter d) are only feasible in overconsolidated soils, or soils with some natural cementing, and where groundwater is below excavation level.
- Contiguous (piles touching). Because of construction tolerances, contiguous piles may be just in contact along their length or have a small gap between piles. This is not easy to guarantee in practice, so watertightness cannot be ensured. In very low permeability soils, however, seepage will be negligible.
- Secant (piles interlocking). Here spacing is less than the diameter. Primary concrete piles typically have no reinforcement, and in a 'hard-soft' configuration may be constructed of cement bentonite to control groundwater flow. The secondary piles are installed whilst primary pile concrete is still 'green' and not hardened. With this form of construction, seepage will be negligible, and the overall bending stiffness of the wall will be considerably increased.

(See Figure 6.22.) The top of the piles will often be capped by a reinforced concrete beam to distribute loads.

The poor aesthetics caused by the irregularities in shape of exposed bored pile walls is often plain to see—bored pile walls are therefore typically covered (e.g. using a non-load-bearing block facing, or by casting the concrete peripheral walls of the permanent structure walls using the pile wall as a back shutter). In addition, it is not normally feasible to extract bored piles from the ground if reusability demands the recycling of materials for use elsewhere.

Plant requirements can range from relatively modest (tripod rig with shell and auger) to very substantial (large diameter rotary). Bored pile walls should be of low maintenance unless the workmanship is of low quality.

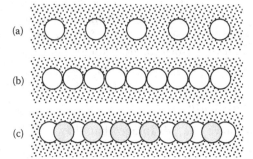

Figure 6.22 Plan view through typical bored pile wall configurations: (a) intermittent, (b) contiguous, (c) secant.

Figure 6.23 Anchored and propped bored pile retaining walls. Castle Mall, Norwich.

Bored pile walls can be constructed in a wide range of diameters and to almost any geometric layout (see Figure 6.23). Secant pile walls can have a high degree of watertightness. Horizontal deformations can be restricted to 1%–2% of the retained height when tied back with anchors. As with diaphragm walls (see later section), bored pile walls can support high vertical loads in addition to lateral loads and can provide very efficient load-bearing basement walls. Unlike diaphragm walls, bentonite may not be required and casing can be used to provide borehole stability under favourable ground and groundwater conditions. Bored pile walls are more expensive than sheet-pile or soldier pile walls, but tend to be cheaper to construct than diaphragm walls.

6.3.4 Diaphragm walls

Diaphragm walls can be used as retaining structures and as load-bearing elements (barrettes) for deep basements of buildings, traffic underpasses, underground mass-transit stations, cut-and-cover tunnels, car parks, underground industrial facilities, docks and waterfront installations, and waterworks.

Three factors have contributed to the expansion in the use of this type of construction:

1. The commercial availability of bentonite (used as a trench supporting slurry)
2. Experience of construction in urban areas, which suggests that the method can satisfactorily deal with difficult conditions, including problematic soils
3. The resolution of certain practical problems, such as improvement in excavation techniques, and the development of on-site plants for processing slurries

Diaphragm wall construction in general provides maximum economy either where it can provide both temporary and permanent ground support, or where the diaphragm walling can help to avoid underpinning of adjacent (existing) structures, or the need for groundwater control. Diaphragm walling will normally allow maximum use of a plot of ground in crowded inner-city areas. Walls can be constructed to considerable depths ahead of the main excavation, so acting as support for adjacent structures.

The cost of installation and its ability to produce a low-cost structure will depend on a number of factors, such as the configuration and physical dimensions of the wall. Its cost may be influenced by

1. The required embedment below excavation level, either for stability or seepage control
2. The nature of the ground to be excavated, and the presence of boulders or other obstructions
3. Associated stability requirements, such as the need for anchorage or propping
4. Site construction factors, such as slurry treatment or disposal, the availability of services, and restrictions on time and working space

The construction sequence for a continuous slurry trench diaphragm wall is shown in Figure 6.24.

At the start of construction, a shallow reinforced-concrete lined guide wall is constructed (e.g. Figure 6.24 [1] down to a depth of about 1 m, to help alignment and to prevent collapse of the soil close to ground level). At the surface, the movement of the slurry against the side of an unlined trench could cause a collapse. Alternate 'primary' diaphragm wall panels are then constructed as shown in Figure 6.24 (2). Each is excavated under a full head of bentonite, provided by a special plant (Figure 6.25b). The end of each section, or panel, must be blocked after excavation and before concreting, either by a steel tube stop end, or the

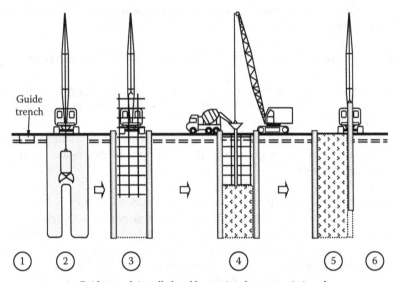

Figure 6.24 Diaphragm wall construction. (Based on a drawing from Balfour Beatty Ground Engineering Ltd.)

1. Guide trench installed and bentonite plant commissioned
2. Excavation of primary panels, under full head of bentonite
3. Reinforcement and stop end tubes lowered into excavated panels
4. Concrete tremied to the base of the panel
5. After concrete has gone off, stop ends removed
6. Secondary panels excavated, reinforced and concreted

end of a previously cut panel. Excavation is made under bentonite slurry using a purpose-built grab or 'clamshell' in soft ground (Figure 6.25a), or a drilling tool (hydromill or hydrofraise) in harder ground. The bentonite provides wall support, and thus avoids the need to introduce expensive and restrictive mechanical temporary support systems. As excavation proceeds, the trench must be kept full of bentonite slurry. After the placing of stop ends and the reinforcing cage (Figure 6.24 [3] and Figure 6.25b), the bentonite slurry must be pumped either to waste or storage, as concrete is tremied into the base of the panel (Figure 6.24 [4]). Finally, the stop ends are removed (Figure 6.24 [5]), and the process is repeated for the secondary panels. The shape of the stop ends means that adjacent panels interlock.

Prefabricated panel walls are less common. They are constructed using prefabricated reinforced concrete. A practical limit in the use of prefabricated walls is imposed by the weight of the individual units. A practicable maximum length is of the order of 15 m for a panel width of 2 m, although available headroom and other site restrictions may set other limits. There

(a)   (b)

Figure 6.25 Diaphragm wall construction. (a) Clamshell excavating slurry trench panels. (b) Reinforcement cage being lowered into slurry trench (bentonite plant in background).

are two principal types of prefabricated walls; those with identical panels, and those with beam and slab components (which act as soldier piles) with lagging. A prefabricated wall with identical panels is suitable for ground consisting of stiff or dense formations, so that the depth of embedment is restricted and most of the length of the heavy panels is used effectively in the support of soil. If great depths are required, beam and slab systems are lighter to handle.

Because of the modular nature of diaphragm walling, the method can be used to construct cellular or polygonal enclosures. Large-diameter circular enclosures can also be most effectively constructed in this way, as can buttressed walls and structures arched in plan.

Diaphragm walls offer high lateral load and moment capacity, are potentially watertight, and can carry significant vertical loads (leading to their use as deep foundation elements, termed 'barrettes'). With tieback anchors, horizontal deformations can be restricted to 1%–2% of the retained height. They can be constructed in conditions of low headroom, have a high tolerance potential, and (if using a hydraulic grab) cause relatively low noise and vibration. However, they are uneconomic in small developments, are associated with large plant and labour demands, and the length of panel excavation makes it difficult to follow irregular plan shapes. Gaps for utilities, etc., crossing the wall are not easy to accommodate. Diaphragm walls also carry a risk of bentonite spillage, and conventional plant can cause considerable noise and vibration.

6.3.5 King post ('soldier pile' or 'Berlin') walls

King post walls are typically constructed using vertical steel H-piles, driven at regular spacings, with either precast or *in situ* concrete panels placed horizontally between them. The concrete panels transfer the earth pressure horizontally to the king posts, which transmit the load vertically, and support the retained height through bending. The king posts may be supported by

- Props (e.g. inside a 'braced' excavation)
- Ground anchors (for example, placed as excavation of the retained height takes place)
- Soil beneath the retained soil (where they are driven below the base of the supported soil)

Figure 6.26a shows a plan of propped king post wall.

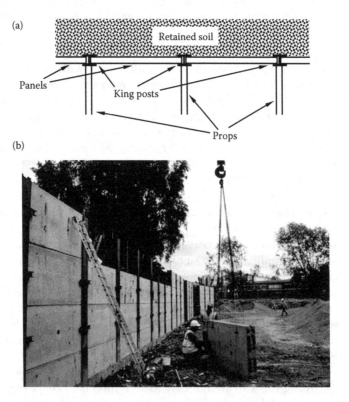

Figure 6.26 King post wall. (a) Plan showing steel H-piles and concrete panels. (b) Wall construction, using pre-cast panels. (Image courtesy of ElecoPrecast.)

Figure 6.27 Top-down construction using anchored steel H piles in-filled with cast *in situ* concrete planking (Hong Kong).

Figure 6.26b shows a system using precast concrete wall panels. This type of wall can also be used to create above-ground bulk storage, by driving steel H-piles from ground level, and infilling with precast concrete panels

Steel piles are often used as part of a braced or anchored wall system during temporary works excavations for inner-city basement construction. In many cases, H-piles can be driven from the surface to the required level in a soil, and these can be braced across an excavation as it is dug. As excavation proceeds, the soil is cut back between the H-piles, and either timber or precast or *in situ* reinforced concrete planking is placed between the H-pile webs. If the ground is too hard to allow the H-piles to be driven, then their locations can be prebored. *In situ* concrete planking can also be used with anchored walls, as Figure 6.27 shows.

6.3.6 Jet-grouted walls

Jet grouting creates a column of soil-cement by rotating a horizontal jet of grout in the ground over a specified depth of treatment, mixing grout with the *in situ* soil. Creating columns at relatively close centres will produce a wall which is similar to a secant pile wall, although with much lower strength and thus lesser ability to withstand lateral pressure.

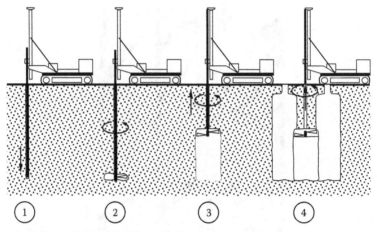

1. Drill open holing (70–120 mm dia.) to maximum wall depth
2. Cutting of soil using compressed air or grout
3. Rotation and gradual withdrawal of rods and injection nozzle, to form column
4. Infill columns formed to complete the wall

Figure 6.28 Creating a jet-grouted wall. (Based on a sketch from cee.engr.ucdavis. edu/faculty/boulanger/geo_photo_album/Ground%20improvement/Jet%20Grouting/ Jetgrouting%20Stockton%20P0.html).

The process is illustrated in Figure 6.28. An open hole is first drilled to the required wall depth. The columns are constructed from the bottom upward, with alternate columns being formed before the infill columns. In favourable ground conditions, the process is both cheaper and faster than bored piling, but the end product has poorer aesthetics and scores very low on the sustainability and reusability scale—due to the large mass of soil-cement created in the ground, and the need to remove cement-contaminated material displaced by the grout and ejected at the ground surface.

6.4 COMPOSITE WALLS AND OTHER SUPPORT SYSTEMS

According to *Eurocode 7* (EN 1997-1 Geotechnical design part 1: General rules para. 9.1.2.1, a composite wall is '... composed of elements from the [other] ... two types of wall [gravity and embedded])'.

Examples of composite walls include double sheet-pile wall cofferdams (see Section 6.4.1); earth structures reinforced by tendons, geotextiles, or grouting (Sections 6.4.2 and 6.4.3); and structures with multiple rows of ground anchorages or soil nails (Sections 6.4.4 and 6.4.5).

The distinction between these last two groups is subtle. It depends upon

a. Whether the soil or rock is *in situ*, or has been placed as backfill
b. Whether the soil or rock is reinforced throughout its volume, or is improved by the action of anchors acting on facing units

Figure 6.29 shows some of the terms used to describe retaining systems which use anchors and reinforcement. The terms 'ReinforcedEarth®' (a trademark of the Reinforced Earth Company, widely recognised in the UK) or 'mechanically stabilised earth' (MSE—common generic terminology in the USA and elsewhere) are used to describe the situation where significant load (e.g. due to the self weight of selected backfill) is transferred all along reinforcement, thus producing a self-supporting composite mass. Because of this, the facings for reinforced or MSE systems are not required to support significant load. In contrast, the term 'anchored earth' is used to describe systems where wall facing is joined to anchors, either at the back of selected backfill or at some distance inside *in situ* ground, with little load transfer along the length of the bars connecting the anchor to the facing.

	In-situ ground	Compacted backfill
Reinforced	Soil nailing	Mechanically stabilised earth
Anchored	Anchored shotcrete or pre-cast facings	Anchored earth

Figure 6.29 Classification of reinforced and anchored ground.

Generally, soil nailing and anchored shotcrete are used when the ground to be supported is relatively strong. MSE systems are suited to a wide range of ground conditions, working well on soft ground where wall flexibility is required.

6.4.1 Cofferdams

The term 'cofferdam' or 'caisson' is used for any structure built to facilitate construction or repair in areas that are normally submerged, allowing work to be carried out in the dry. For example, a cofferdam might be used to construct spread foundations in the middle of a river. A circular sheet-pile wall could be driven from a barge, around the proposed site of construction and, after stiffening using wales and bracing, pumped dry. After the construction of the bridge foundation and pier, the cofferdam would be flooded before sheet-pile extraction.

Double-skin cofferdams can be categorised into two groups: double-wall cofferdams, and cellular cofferdams (Puller 2003). Both are essentially gravity structures, made up by placing granular backfill between a series of sheet-pile retaining structures.

A double-wall cofferdam consists of two parallel walls of sheet piling, connected at one or more levels by steel rods, bearing on external wall-ings. The space between the sheet-piles is filled with granular soil, rock, or hardcore.

Cellular cofferdams are constructed from interlinked smaller circular cofferdams known as cells, made from straight-web steel sheet piling. A considerable variation in geometry is possible (Figure 6.30). These are used in large projects such as dam construction, where rivers need to be diverted, or for major dock construction. Cellular cofferdams are more economical than double-wall cofferdams for greater water depths, larger retained heights, longer structures, and where bracing and anchoring is not possible.

6.4.2 Reinforced soil (MSE) structures

Reinforced soil (MSE) consists of a number of components (Figure 6.31a):

- A strip foundation
- Facing units
- Reinforcement
- Capping beam

Essentially, reinforced soil structures are gravity walls, where the wall is made from a combination of compacted soil and a relatively large number

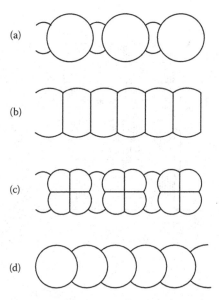

Figure 6.30 Some forms of cellular cofferdam. (a) Circular cells connected by arcs. (b) Semi-circular cells with straight diaphragm cross walls. (c) Clover leaf cells—four circular arcs of sheeting fixed on two transverse walls, connected by smaller circular arcs. (d) Repeated circular arcs. (From Puller, M., *Deep Excavations: A Practical Manual*, Thomas Telford, London, 2003.)

of closely spaced reinforcing elements. The wall can be built up from a prepared base and strip foundation using a lightweight plant. The reinforcement is initially unstressed, taking load only as the soil mass tries to deform under its self weight, and any applied loads. As the reinforcement interacts with the surrounding soil, it develops bond stresses along its whole length. Because of this, reinforced soil generally requires only lightweight facing, which acts to support the material adjacent to the face, also providing a tidy appearance.

Modern concepts of fill reinforcement can be traced to the late 1920s with systems patented by Coyne in France and Munster in the United States. By the 1960s, Vidal developed a system using concrete facings and steel strips, termed 'La Terre Armée' or 'Reinforced Earth' (see Figure 6.31), which has become one of the success stories of modern civil engineering; thousands of 'Reinforced Earth' structures have been built around the world. Figure 6.31b shows an example of the facing and reinforcement components of this system. Cruciform precast concrete facing units are attached to galvanised steel reinforcing by means

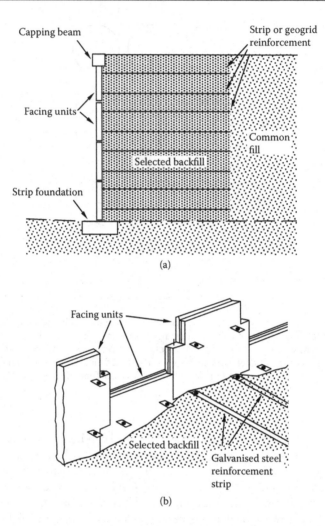

Figure 6.31 Components of reinforced fill. (a) General layout. (b) Detail showing Reinforced Earth facing units and strip reinforcement. (b, Redrawn from Ingold, T.S., Reinforced Earth. ICE Publishing, London, 1982a.)

of bolts and galvanised connectors. Figure 6.32 shows its application on a highway in Venezuela.

Over the past 50 years, a number of other materials have been used for facing, and for reinforcement. Polymer grid reinforcement (Figure 6.33) and geotextile reinforced walls (Figure 6.34) are now common.

Figure 6.32 Reinforced earth walls (Caracas, Venezuela).

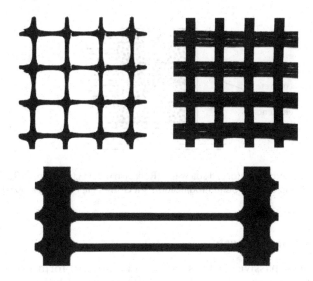

Figure 6.33 Types of polymer grid reinforcement.

A major advantage of using reinforced fill for retaining structures is its flexibility, and also the fact that the retaining structure is made simultaneously with the filling. The method is suited for the construction of highway embankments on steep sidelong ground and also in the construction of abutments and wing walls of bridges. Large settlements and differential rotations can be tolerated by reinforced soil without damage. Savings of up to 20%–30% can be obtained in comparison with conventional reinforced concrete walls, especially for heights over 5 m.

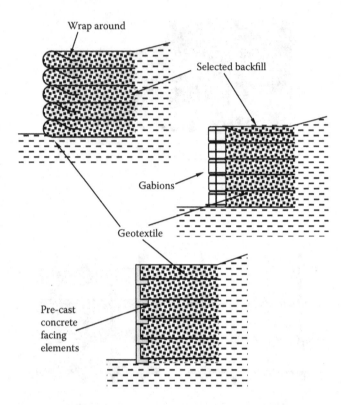

Figure 6.34 Geotextile reinforced soil walls. (Redrawn from Hausmann, M.R., Engineering Principles of Ground Modification. McGraw-Hill, New York, 1990.)

6.4.3 Anchored earth

Closely spaced passive anchors can also be used to form gravity retaining structures from engineered fill. Such a system is known as anchored earth (Figure 6.35).

Soil reinforcement (MSE) and anchored earth provide two different methods of binding a soil mass together. Anchored earth generally uses a similar number of elements as MSE, but is composed of a bar or strip with a relatively small surface area, terminating at a passive block or hoop at the rear of the backfill. Whilst MSE interacts with the surrounding backfill and develops bond stresses along the whole length of each reinforcing element, anchored earth transmits most load directly from the wall facing to the remote block or hoop. Both systems use relatively lightweight facings.

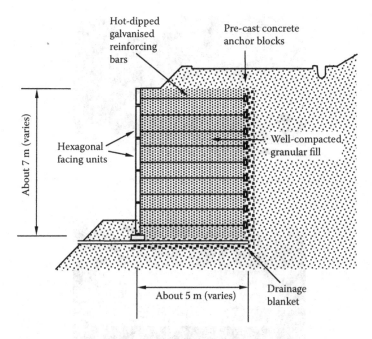

Figure 6.35 An example of anchored earth from the KL–Karak Highway, Malaysia. (Available at http://www.nehemiah.com.my/app_case_karak.htm).

6.4.4 Support using ground anchors

Modern ground anchors were introduced in the mid-1930s (by Coyne) when they were used to provide a vertical prestress through a masonry dam into underlying rock foundations. Many thousands of rock anchors have since been successfully installed (mostly in dams and tunnels), and in more recent years, anchors have been used to a great advantage in retaining walls. The use of anchors has already been noted in connection with deep excavations. They are also widely used in other ways, for example, to support protective skins such as chunam and shotcrete, which are placed on excavation faces, built top down, to protect the ground from tropical weathering (Figure 6.36).

Anchoring of *in situ* ground can also be used with precast facing units. Figure 6.37 shows the use of individual precast concrete 'slabs' used to provide a retaining structure at the toe of a preexisting area of slope instability, which, if unsupported, might threaten the highway below. Each slab is held back by two anchors, which presumably pass back into stable ground. Such a system could potentially provide a rapid means of restabilising a slope if unexpected problems arose during construction.

Figure 6.36 Anchored shotcrete wall (Caracas, Venezuela).

Figure 6.37 Precast slabs used with anchors to provide a retaining structure at the toe of a failing slope (South Wales).

The use of ground anchors to hold back retaining structures is distinctly different from reinforced soil, which aims to bind a soil mass together to form a gravity structure. Ground anchors comprise a smaller number of more widely spaced elements that are highly stressed before the retaining structure is commissioned. Reinforced soil interacts with the surrounding backfill and develops bond stresses along its whole length, whereas ground anchors transmit load directly from the wall to a remote 'fixed length' with little stress transfer along the 'free length' in between. Soil reinforcement generally requires only lightweight facing whereas ground anchors must be used with a substantial structural wall (precast concrete, sheet-pile, diaphragm, etc.) capable of distributing load from the anchor head.

6.4.5 Soil nailing

Soil nailing is a method of reinforcing the ground *in situ*, in which steel bars are either driven (system Hurpinoise), drilled and grouted, or fired ballistically into the excavated face. Nail installation proceeds in parallel with staged top-down excavation, usually with some form of shotcrete and steel mesh facing being applied in panels (Figure 6.38).

As with ground anchoring, the technique is best suited to near vertical faces in relatively good ground. An example, forming a bridge abutment, is shown in Figure 6.39. The technique has also proved quite successful in

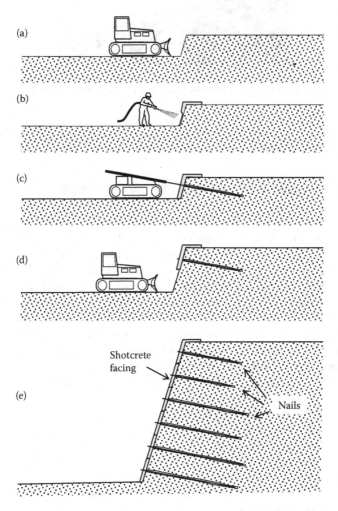

Figure 6.38 Soil nailing. (a) Excavation; (b) shotcreting; (c) nail installation; (d) excavation; (e) completed wall. (After Gassler, G., In situ techniques of reinforced soil. Proc Int Reinforced Soil Conf. organised by the British Geotech. Society, Glasgow, A McGown, K Yeo and KZ Andrawes, eds., 10–12 September, pp. 185–197, 1991.)

Figure 6.39 Soil-nailed wall along Midland Expressway, UK. (Image courtesy of Phi Group.)

stabilising old masonry walls showing excessive deformation. A summary is given by Bruce and Jewell (1986) and by Gassler (1991).

6.5 PRELIMINARY SELECTION OF WALL TYPE

The available options and constraints on a retaining structure will often make preliminary design a complex proposal. Initially, there will be a need to support soil, or a structural load or any adjacent structure. There will be a desired geometry for the completed structure, but, in addition, constraints due to subsoil and groundwater conditions, available construction methods, and local experience of those methods will also play their part in the choice of the retaining structure. Often, the designer will choose a particular type of structure because he has used it successfully before, has confidence in his ability to design and build it, and feels (using 'engineering judgement') that it will work within the given situation.

The factors which may influence the choice of structure are

- Height of the ground to be supported
- Type of retained soil
- Type of foundation soil
- Groundwater regime
- Adjacent structures
  - Magnitude of external loads
  - Allowable movements
- Available space for construction and construction plant
- Experience and local practice
- Available standards and codes of practice

- Available construction techniques and equipment
- Cost

Many of these factors are interlinked. For example, a quay wall would normally be constructed of interlocking steel sheet piling because of ease of driving in soft sediments, difficulty of access for *in situ* construction, experience with similar structures, foundation soil, and cost.

Chapter 7

Avoiding failure

Chapter 6 described the different types of retaining structure that may be used, and Chapters 8 to 11 will consider the calculation of earth pressure and the resistance of these walls. This chapter describes how walls can fail to perform satisfactorily and ways in which the designer can ensure that such failure does not occur.

At the time of writing, there are two broad approaches in use across the world in the avoidance of failure of geotechnical structures:

- A traditional approach based upon recommendations from piece-meal research reported in academic and professional journals. In this approach, unsatisfactory retaining wall performance has been avoided for recognised failure (ultimate) states by applying a single, lumped, factor of safety to resisting forces for each failure mode. The semi-empirical nature of design calculations is justified on the basis of previous satisfactory behaviour. Design practices vary from structure to structure and, of course, from country to country.
- A unified approach, proposed recently in Eurocode 7, which attempts a more rational avoidance of unsatisfactory behaviour (termed 'limit states') in soil-structure interaction and combines this with the so-called 'partial factors' that are applied both to driving and resisting forces, as well as to soil parameters.

This chapter considers both approaches.

7.1 DEFINING FAILURE

Retaining structures must be designed, constructed and maintained in such a way that they are fit for use throughout their entire working life [ISO 2394]. In particular, they should perform satisfactorily under both expected and extreme conditions. They should not be damaged by accidental events (e.g. impact, overdredging, explosion, fire, etc.) to an extent disproportionate to the likelihood and magnitude of such events. Design is carried out to ensure that the reliability of a structure is appropriate for the consequences of failure, which might include risks to life, economic performance, society and the environment.

According to the Eurocodes, structures must be designed, built, and maintained so that they meet the following requirements:

- Serviceability. Over its intended ('design') life, the structure should meet specified service requirements, with sufficient reliability and at reasonable maintenance cost.
- Safety. The structure should survive all events (for example, accidental impacts) likely to occur during its construction and use.
- Fire. Structural performance should remain satisfactory for a required period of time (e.g. for building evacuation to take place).
- Robustness. The structure should not be damaged by extreme events (e.g. explosions and human error) to an extent disproportionate to the severity and likelihood of the event.

Limit state design separates desired states of the structure from undesired states. Serviceability limit states (EN 1990 Clause 1.5.2.14) are 'states that correspond to conditions beyond which specified service requirements for a structure or structural member are no longer met'. An example would be the excessive settlement of a gravity wall to the extent that it appeared to the public or to users of a facility to be unsafe. Ultimate limit states are much more serious as according to EN 1990 (Clause 1.5.2.13); they are 'states associated with collapse or with other similar forms of structural failure'. Traditionally, the design of earth retaining structures has been based largely on avoidance of ultimate limit states with serviceability limit states being avoided by applying high lumped factors of safety in ultimate limit state calculations.

Eurocode 7 lists five ultimate limit states to consider:

- Verification of static equilibrium (abbreviated in the Eurocodes as EQU)
- Verification of resistance to uplift (abbreviated as UPL)
- Verification of resistance to hydraulic failure due to large hydraulic gradients (abbreviated as HYD)
- Verification of (ground) strength (abbreviated as GEO)
- Verification of (structural) strength (abbreviated as STR)

Figure 7.1 gives examples of each of these.

Loss of EQU is avoided when, for cases where soil strength does not have much impact on stability, the sum of the driving (destabilising) forces

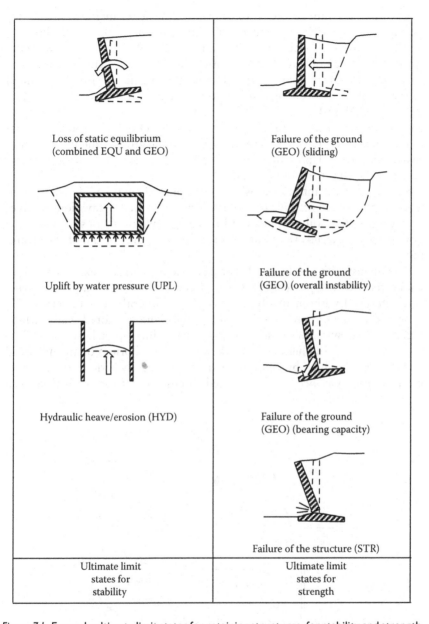

Loss of static equilibrium
(combined EQU and GEO)

Failure of the ground
(GEO) (sliding)

Uplift by water pressure (UPL)

Failure of the ground
(GEO) (overall instability)

Hydraulic heave/erosion (HYD)

Failure of the ground
(GEO) (bearing capacity)

Failure of the structure (STR)

Ultimate limit states for stability	Ultimate limit states for strength

Figure 7.1 Example ultimate limit states for retaining structures, for stability and strength (brackets give Eurocode 7 abbreviation).

is less than the sum of the stabilising forces. Checks might be made, for example, considering moment equilibrium (toppling) of a wall about its toe for a gravity retaining wall. If insufficient, the width of the wall base is increased. In practice, loss of static equilibrium is mainly relevant to structural design. In geotechnical design, EQU verification will be limited to rare cases, such as a rigid foundation bearing on rock (EN 1997-1, cl. 2.4.7.2(2) P Note 1).

UPL can occur as a result of the Archimedes effect (i.e. when the mass of the new structure is less than the weight of the ground and groundwater that it displaces). Uplift may be resisted to some extent by shear stresses between the structure and the soil.

Seepage forces, internal erosion and piping (HYD) can be significant issues in the design of waterfront structures, or braced excavations, in fine-grained granular soils. Piping and hydraulic heave occur in soils when the pore pressure exceeds the total stress at, or near, the bottom of the retaining structure.

Overall instability, sliding and bearing capacity failure are examples of situations (termed GEO) where failure can occur in the soil (or on the soil/structure interface) without involvement of the structural strength. Figure 7.2 shows some possible mechanisms for gravity and anchored walls, whilst Figure 7.3 shows how these may vary, depending upon how the wall is supported. For a further example, the vertical equilibrium of an embedded retaining wall (which by definition can have relatively little bearing capacity at its toe) may need to be checked to ensure that the implied vertical

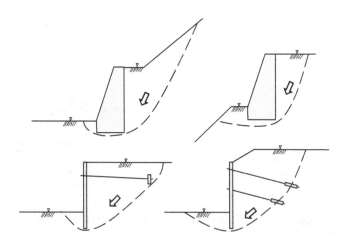

Figure 7.2 Examples of overall instability for gravity and anchored retaining walls. (Redrawn from Clayton, C. et al., *Earth Pressure and Earth-Retaining Structures*, Second Edition, Taylor & Francis, Jan 7, 1993.)

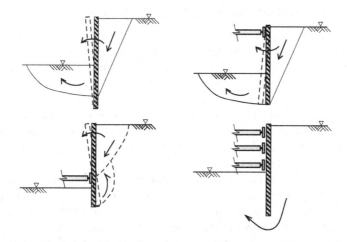

Figure 7.3 Examples of failure modes for a range of embedded walls.

component of anchor forces can be met by the resistance due to wall/soil friction.

Finally, the ability of the structure to support the loads imposed by the ground (termed STR) must also be checked. This includes not only the strength of the retaining structure, but also structural elements such as anchorages, wales and struts, and failure of the connection between such elements.

In summary, for any retaining wall/soil system, the engineer must identify all possible failure mechanisms (serviceability and ultimate limit states) and then design to prevent them. The critical mechanisms of potential failure may change during construction (e.g. as an excavation is deepened and as struts are placed).

7.2 UNCERTAINTIES IN DESIGN

The designer typically has to deal with a number of uncertainties. For example, the lateral and vertical extent of different types of ground (sands, clays, rock, etc.) will typically not be completely defined, as a result of limited site investigation. Different methods of calculation can lead to different results, for example, in terms of predicted earth pressures. And there will often be uncertainty about the precise values of geotechnical parameters, such as undrained shear strength, which can typically be determined in a number of different ways.

7.2.1 Uncertainties in the ground model

Figure 7.4 shows the results of a survey carried out to assess the sources of ground-related problems during construction. Almost all of the problems encountered on 28 construction projects could be attributed to just seven causes:

- Soil boundaries. The position or thickness of unfavourable types of ground (for example, soft materials, rock, etc.) was not as envisaged from the information given in the tender. In other words, the sub-soil geometry was not as expected.
- Soil properties. Parts of the ground were weaker or stronger than expected, for example, causing problems during excavation or piling.
- Groundwater. The groundwater level was higher than expected, or groundwater existed where none was expected.
- Contamination. Contaminated land was encountered, for example, in man-made ground, where none was expected.
- Obstructions. Excavation was made difficult by the presence of man-made obstructions, such as old foundations.

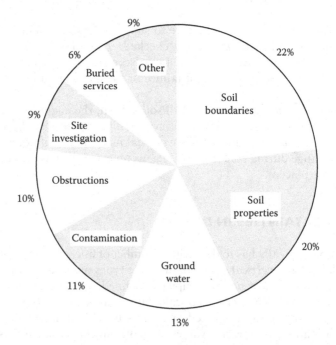

Figure 7.4 Causes of problems during construction, from a survey of 28 projects. (Redrawn from Clayton, C.R.I., *Managing Geotechnical Risk. Improving productivity in UK building and construction*, Thomas Telford, London, 2001.)

- Site investigation. The site investigation was inadequate, giving an incorrect expectation of ground conditions, as a result of poor planning or interpretation.
- Services. Pipes and cables were encountered where none were expected and caused delays.

A good design will explicitly recognise and take account of uncertainties and unknowns in these factors.

Some form of ground model should be developed by the designer. This will ideally use an expert knowledge of local geology in combination with the results of a desk study, field observation and testing, trial pitting, and boreholes in order to produce sections which show the expected geometry of different types of ground, the groundwater regimes, and any expected man-made deposits (for example, fill, which may be contaminated). As an example, Figure 7.5a reproduces one of many geological models described by Fookes (1997). A number of features should be noted.

First, the geometry in this geological setting (and in most near-surface ground conditions) is complex. In this case, cambering, valley bulging, dip and fault structures, and solifluction occur as a result of glacial action. As a result, there is little prospect that any reasonable number of boreholes will be able to define the complete sub-soil geometry. This conclusion is supported by a survey carried out on cost overruns on UK highway projects (Figure 7.6). The data show that construction cost overruns are significantly reduced as expenditure on site investigation is increased. But expenditure would have to reach an unrealistic 7% or 8% of total construction cost to bring additional costs down to less than 10% of tender price. At present, considerably less than 1% of total construction tender price is typically spent on site investigation, and the data show that at this level of expenditure cost overruns in excess of 100% are then possible, even when (as in these projects) high levels of engineering skill and care are used.

In the specific case illustrated in Figure 7.5a, boreholes have failed to locate the cavities and deposits associated with the gulls, allowing a much simpler interpretation of stratification than in reality exists. This situation certainly occurs in practice. A geotechnical model interpreted solely from the borehole records, such as in Figure 7.5b, portrays a much simpler ground geometry than in fact exists. In doing so, it misses some important features of the site.

In preparing a geotechnical ground model, some simplification is always required. It can be seen from the geological model in Figure 7.5a that although some boundaries between different ground types (e.g. the mudstone and the limestone) are quite distinct, others (e.g. between mudstone and clay and between terraces and gravel and silty clay) are gradational.

(a)

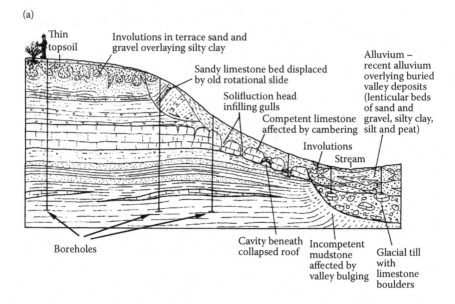

(b)

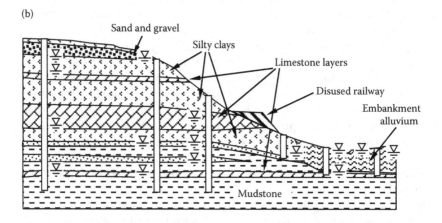

Figure 7.5 Uncertainty in interpretation of borehole records. (a) Geological ground model, from Fookes (1997). (b) Inferred geotechnical model, from borehole records. (From Fookes, P.G., *Q.J. Engineering Geology and Hydrogeology* 30, 293–424, 1997.)

In preparing an engineering borehole record, the engineering descriptions must be simplified, potentially hiding the natural variability in each layer and placing hard boundaries where they do not exist. Thus, each of the layers in Figure 7.5b contains material that is spatially variable, in terms of strength, stiffness and permeability.

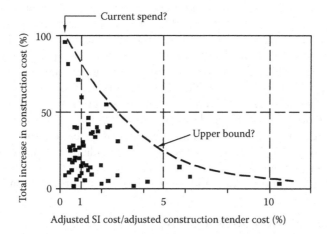

Figure 7.6 Impact of ground investigation expenditure on increase in ground conditions. (Redrawn from Clayton, C.R.I., *Managing Geotechnical Risk. Improving productivity in UK building and construction*, Thomas Telford, London, 2001.)

Also shown in Figure 7.5b are some examples of the sort of groundwater information that might be recorded on borehole records. For simplicity, only the level of groundwater strikes is shown. In practice, this may well be all that the design engineer gets, because

- Although site investigation specifications call for drilling to cease when groundwater is encountered, allowing time for the water level in the borehole to rise and stabilise, this is not always done.
- Casing is typically used to support boreholes. This reduces or eliminates the flow of water into the borehole, perhaps preventing a water strike from being noticed.
- Water is often added to boreholes to allow progress to be made. This occurs both during light percussion boring in sands and gravels and when rotary coring. Groundwater strikes may not be detected under these conditions.

Despite all of this complexity, the design engineer must produce a model of the expected groundwater regime(s) in the areas of any proposed retaining structures.

Further, there is a need to identify man-made ground and old foundations, because this can contain obstructions (which may affect piling, for example), weak and highly variable materials, and contamination. Figure 7.5b includes, as a simple example, an embankment for a disused railway line. The likely range of contaminants can be estimated, given this previous land use, but receptors and pathways will not be easy to define at this stage.

In summary, near-surface ground conditions are geometrically complex, and for the limited resources available for ground investigation, there will always be geometric uncertainty. Given the geological setting of a site, certain features (such as, in this case, solifluction and gulls) may be suspected, but the precise three-dimensional geometry of the ground will always be difficult to determine with a high level of certainty. Estimating ground geometry from borehole data is a subjective business, given the limited number of investigation points, the gradational boundaries between some materials, and the need to idealise the ground as a series of zones containing material with similar geotechnical properties. Design is made more uncertain by the difficulty in determining the groundwater regime, and the need to identify made ground, contamination and obstructions left by previous land use.

7.2.2 Calculation uncertainties

Acoording to EC7, serviceability and ultimate limit states can be checked in a number of ways:

- By adopting 'prescriptive methods' or 'rules of thumb' (Section 8.1)
- Using calculations, which may vary from simple hand calculations to sophisticated numerical (finite element or finite difference) modelling (Sections 8.2 through 8.7)
- On the basis of the results of physical modelling and load tests
- Using an observational method

Calculations, for example, those described in the following chapters, are most commonly used in the design of earth retaining structures.

Calculations are used to compare available resistance (of a structure or pile, for example) with the expected loads that needs to be designed for. Some loads (e.g. vertical loads resulting from dead weight) can be expected to be more-or-less constant during the life of a structure. Others, such as wind loads, snow loads, earthquake loads and accidental impacts, are temporary. Eurocode 7 identifies four 'design situations' that must be satisfied:

- Permanent loads resulting from normal use. These may be caused by self weight, uneven settlement and structural component shrinkage, for example.
- Transient loads from temporary conditions. These may be caused by 'live' floor loads due to building usage, temperature changes, wind load and snow load.
- Accidental loads, from exceptional conditions such as vehicle or ship impact, explosion, or fire.
- Seismic loads from earthquakes.

Calculations are generally carried out using commercial computer software, in effect a 'black box' which the user cannot easily check for performance. Hopefully, the user manual will give the results of benchmark testing to give some confidence in the product. However, calculation methods involve simplifications. For example, hand calculation methods based on closed-form analytical solutions (equations) can generally only be derived by simplifying the problem, so that mathematical solutions can be obtained. Methods based on different simplifications will give different solutions, as the extensive geotechnical literature on limit equilibrium analysis for slope stability and bearing capacity demonstrate.

Forrest and Orr (2011) suggest that modelling uncertainty is not significant in the design of spread foundations since the uncertainties in soil strength parameters or the loads in the case of an eccentrically loaded foundation control the reliability of the design. But this situation is not likely to be universally true. For example, Whittle and Davis (2006) suggest that the choice of an inappropriate constitutive model in a finite element analysis may have been a significant factor in the Singapore Nicoll Highway collapse in 2004. Benchmarking exercises using various finite element programs have demonstrated that differences in boundary locations and conditions, and mesh discretisation, can have a very significant influence on the displacements predicted around deep basement retaining structures.

7.2.3 Parameter uncertainties

The geotechnical ground model (see Section 7.2) provides a geometric simplification that is used in calculating the stability or displacements of the retaining structure. As has been described above, the ground is divided into geometric zones which are thought to contain material with similar (but variable) geotechnical properties. Groundwater regimes are interpreted from site investigation data. Parameters are then required for each soil or rock zone that significantly affects the stability or displacement of the structure.

The parameters used in calculations cannot be known with certainty, for the following reasons:

- During simplification to obtain the ground model, geometric errors will occur because of the limited number and depth of ground investigation holes and in situ tests. Compare the geometries of the limestone layers shown in Figures 7.5a and 7.5b.
- As noted, the ground model is a simplification of the complex reality, where some boundaries are gradational. Material properties therefore vary within each zone (see Figure 7.5).
- Measured properties, whether determined from in situ or laboratory tests, will be affected by a range of factors, not all relating to the parameter to be determined. For example, the scatter of measured

undrained shear strength in Figure 7.7a results from the small size of the laboratory test specimens relative to the fissure spacing in the London Clay.

- Directly measured ground properties, such as undrained shear strength, depend upon the method used to determine them. An example of the magnitude of this can be seen from data given by Marsland

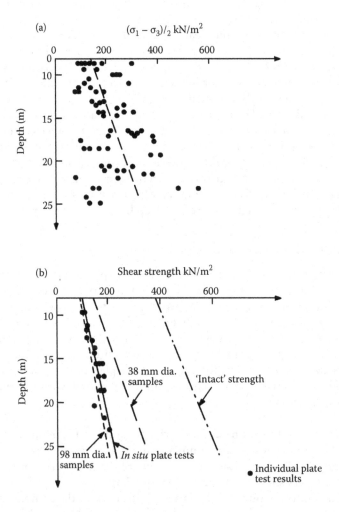

Figure 7.7 Comparison of different methods of measuring the undrained shear strength of fissured London Clay. (a) Triaxial test results for 38 mm diameter specimens. (b) Effect of test size on measured undrained shear strength. (Redrawn from Marsland, A., *Ground Engineering* 5 (6), 24–31, 1972.)

for the London Clay (Figure 7.7b) where results from 38 mm and 98 mm test specimens are compared with the results of 865 mm diameter plate bearing tests. The undrained shear strengths obtained from larger tests, which include representative fissuring, are lower on average than those from smaller tests.

- Properties determined from in situ tests will be affected by borehole disturbance, caused by forming the exploratory hole (e.g. for the SPT), or the in situ test pocket (e.g. for the self-boring pressuremeter). This will vary from test to test.
- Sample disturbance will affect the measured strength properties, such as undrained shear strength and effective strength parameters obtained from laboratory testing. In stiff clays, undrained shear strength may be increased (Hight 1986), whilst in softer clays it may be decreased (Clayton et al. 1998).
- Indirectly deduced properties, such as the effective angle of friction of sand or gravel, depend not only upon the method of in situ testing (e.g. standard penetration test [SPT] or cone penetration test [CPT]), but also on the particular correlation used to determine the value of the required parameter from the test result.
- The excess pore pressures set up in slow-draining materials such as clays dissipate during and after construction. Both short-term and long-term calculations need to be carried out (BS EN 1997 Clause 2.1(1)) because the exact state of drainage will not be known during, or at the end of, construction. These use different parameter sets. For example, in a clay, undrained shear strength (c_u) may be used for short-term stability calculations, with the effective angle of friction (φ') being used for long-term stability calculations.
- When calculating ground movements (e.g. behind a retaining structure for a deep basement in an inner city site) many different constitutive models are incorporated in the available numerical modelling (e.g. finite element and finite difference) codes. In almost all cases, it is not possible or practical to obtain values of all the parameters required for an analysis. To give a simple example, a drained transversely isotropic constitutive model requires the measurement of five independent stiffness parameters, for example: E_v, E_h, G_v, G_h, v_{vh}. Only two of these are readily measurable. The remainder must be estimated.

The above suggests that geotechnical calculations will always, to some extent, involve uncertainty and empiricism. A margin of safety is required to allow for the above unknowns, which can be broadly classified as

- Geometric uncertainties
- Calculation uncertainties
- Property uncertainties

7.3 PROVIDING FOR UNCERTAINTY—
INTRODUCING SAFETY AND RELIABILITY

Figure 7.8 shows a simple example of a gravity retaining wall, with smooth vertical faces, founded in, and retaining, dry sand. Most of the forces shown in Figure 7.8 are unknown, except under particular circumstances. The active force (P_a), for example, is applied only when the wall moves forward sufficiently to develop the failure wedge behind it. The passive force (P_p) requires proportionately greater movement, relative to the height of the soil/wall surface, so that even if active conditions occur, the full passive force may not be fully mobilised. In reality, the designer of such a wall will have checked for horizontal sliding, overturning about the toe and bearing capacity failure, so that the base width of the wall will be greater than required for stability, and therefore the driving and resisting forces will not be at their limiting (active or passive, for example) values under working conditions.

Limit equilibrium analysis is commonly used in geotechnical engineering (for example in slope stability and simple earth pressure problems, Section 8.5). It involves simplification of the analysis to the point that it is relatively easy to carry out, by relaxing the requirements for full equilibrium (e.g. by making assumptions about interslice forces in calculating slope instability, by considering separately the various requirements for force and moment equilibrium in retaining structure analysis, or by abandoning the need for kinematic admissibility). Traditionally, the system is brought to equilibrium by applying a 'Factor of Safety' to reduce one or more of the stabilising forces involved in the problem.

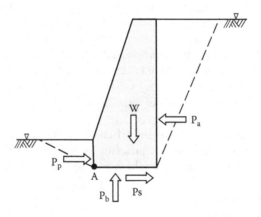

Figure 7.8 Forces on a gravity wall.

For the example shown in Figure 7.8, wall stability can be checked by calculating

- The factor of safety against horizontal sliding. This is found by dividing the failure value of P_s (the force developed between the base of the wall and the underlying soil) by a factor of safety F_s, to achieve horizontal force equilibrium. (In practice, passive pressure in front of the wall may also be neglected because of the possibility that excavation, e.g. for service trenches, may take place close to the toe, during the life of the structure.)
- The factor of safety against overturning. This is found by dividing the sum of the restoring moments caused by W and P_p by the overturning moment (about point A) caused by P_a.
- The factor of safety against bearing capacity failure. This is found by dividing the ultimate bearing capacity by the resultant of the applied forces on the base of the wall.

In this example of (traditional) design, three so-called 'lumped' factors of safety are calculated and compared with values specified in national codes. Other values of factor of safety could be calculated, for example, by reducing the density of the wall until failure occurred. Other mechanisms should be checked, for example, internal overstressing of the wall itself.

7.3.1 Lumped 'factors of safety'

The first design standard for earth retaining structures in the UK ('Civil engineering code of practice no. 2', known as 'CP2') was published by the Institution of Structural Engineers in 1951. A margin of safety was introduced into earth pressure calculations by applying a single factor of safety F_p of at least 2.0 on gross passive earth pressures (see Figure 7.9). In this figure, the thrusts from active and passive earth pressures are represented by the symbols P_a' and P_p' respectively; and those from pore water pressures on either side of the wall by U_a and U_p. The lever arms of these forces about the wall's point of rotation (at the top of the wall) are also shown, leading to the following equation for F_p:

$$F_p = \frac{P_p' L_{pp}}{P_a' L_{pa} + U_a L_{ua} - U_p L_{up}} \tag{7.1}$$

In addition, values of active wall adhesion were limited to 1000 lbf/ft² (47.9 kPa), with active wall friction taken as 15° against uncoated steel piling and 20° against concrete and brick. Values of passive wall adhesion and friction were specified as half their equivalent active values. In stiff clays,

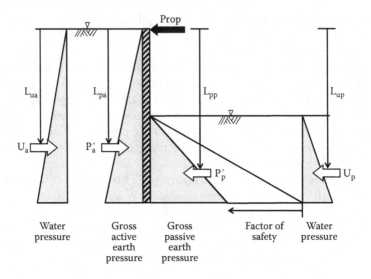

Figure 7.9 Factor of safety on gross passive earth pressures.

the minimum total active pressure acting on a wall was recommended to be no less than that caused by a fluid with a density of 30 lbf/ft³ (480 kg/m³).

Burland et al. (1981) criticized the factors of safety on gross pressures when applied to undrained conditions since it can lead to two possible required depths of embedment. Rowe and Peaker (1965) indicated that a safety factor of 1.5 is necessary to limit deformations in loose granular deposits and to allow for progressive failure in dense granular deposits.

The successful completion of the Bell Common Tunnel (Hubbard et al. 1979), a cut and cover excavation supported by bored pile retaining walls, showed that the factor of safety of 2.0 required by CP2 was unnecessarily conservative for the plastic London Clay, which has effective angles of shearing resistance in the range 20–25°. At Bell Common, a factor of safety of 1.5–1.6 was used for long-term effective stress analysis with satisfactory results.

British Steel Corporation published the first edition of its Piling Handbook—the so-called 'blue book'—in 1997 (British Steel Corporation 1997). Subsequent editions appeared regularly, latterly in 1988 and 1997, and in 2005 under the Arcelor name. Instead of applying a safety factor to gross passive pressures as required by CP2, the example calculations in the Piling Handbook applied a factor to net passive earth pressures, as illustrated in Figure 7.10. The use of net pressures has the advantage of reducing the amount of calculation needed to check wall stability.

In this diagram, earth and pore water pressures are reduced by an equal amount on both sides of the wall (by ignoring the hatched areas), so that at

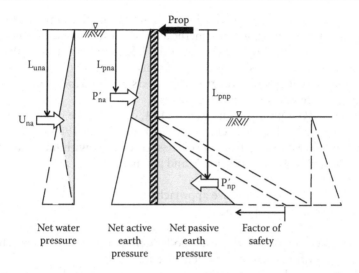

Figure 7.10 Factor of safety on net passive earth pressures *(British Steel Piling Handbook).*

any level pressures appear on one side of the wall only. The thrusts from the net active and net passive earth pressures are represented by the symbols P'_{na} and P'_{np}, respectively; those due to net pore water pressures by U_{na}. The lever arms are therefore reduced for active earth pressures and increased for passive earth pressures, relative to the gross pressure method illustrated in Figure 7.9. The equation for F_{np} is

$$F_{np} = \frac{P'_{np}L_{pnp}}{P'_{na}L_{pna} + U_{na}L_{una}}$$ (7.2)

A margin of safety was introduced into the Piling Handbook calculations for anchored retaining walls (using the free earth support method) by applying a single factor $F_{np} = 2.0$ to net passive earth pressures. However, no factor of safety was employed in the calculations for cantilever retaining walls (using the fixed earth support method, see Chapter 10).

In addition, values of active wall adhesion were limited to 50 kPa and active wall friction taken as 15° against uncoated steel piling and 20° against concrete and brick. Values of passive wall adhesion and friction were specified as half their equivalent active values. In stiff clays, the minimum total active pressure acting on a wall was recommended to be no less than that caused by a fluid with bulk unit weight 5 kN/m³.

Burland et al. (1981) and Potts and Burland (1983) found that the factor of safety on net passive pressures increased rapidly with increasing depths of embedment and was numerically much higher for the same wall length than that calculated using other methods. For these reasons, the net pressure method has fallen out of favour in the UK in modern times, although there are many instances of its successful use in practice. One reason for this may be that the soil strength parameters recommended in the Piling Handbook were particularly conservative.

In 1984, CIRIA published its Report 104, covering the design of retaining walls embedded in stiff clays (Padfield and Mair 1984). This report recommended two alternative approaches for dealing with uncertainties in the selection of soil strengths, loads and geometry:

- A moderately conservative approach
- A worst credible approach

The report commented that the moderately conservative approach was the one 'most often used by experienced engineers'.

In the moderately conservative approach, conservative best estimates of soil parameters, loads and geometry are used together with generous factors of safety. Moderately conservative soil parameters have since been likened to Eurocode 7's characteristic values (see below), which are based on a 'cautious estimate of the value relevant to the occurrence of the limit state' (Gaba et al. 2003). The worst credible approach employs pessimistic soil parameters and loads and geometry that are very unlikely to be exceeded, together with less conservative factors of safety (Padfield and Mair 1984). Worst credible parameters are very unlikely values, 'the worst that the designer reasonably believes might occur' (Gaba et al. 2003). Simpson et al. (1979) regard worst credible parameters as values that have a 99.9% chance of being exceeded.

CIRIA 104 described three distinct methods of introducing a margin of safety into the design of retaining walls, by applying lumped factors on

- Embedment length
- Soil strength
- Passive resistance

The factor on embedment, F_d, is applied to the wall's embedded length, d_o (see Figure 7.11):

$$d = F_d \times d_o \tag{7.3}$$

This is one of the simplest methods to employ, since it requires no modification to soil parameters or earth pressures, merely an extension of the pile's embedded length, once the necessary depth of embedment for equilibrium has been found. CIRIA 104 suggested values of $F_d = 1.1–1.2$ for

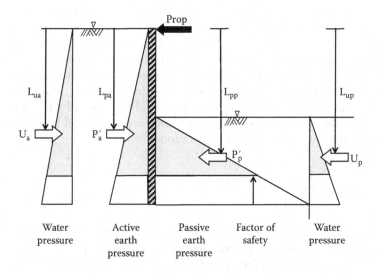

Figure 7.11 Factor of safety on wall embedment (CIRIA Report 104).

temporary and 1.2–1.6 for permanent works, when using moderately con-servative parameters based on effective stress calculations. For the design of temporary works based on total stresses, F_d = 1.1–1.2. Values of F_d recommended by Teng (1962), Tschebotarioff (1973) and the USSI (1984) vary between 1.2 and 2.0. CIRIA 104 recommended that the results of this method should be checked against one of the other methods described below. In recent years, this method has fallen into disuse.

A margin of safety can also be introduced by applying a factor, F_s, to the soil's strength parameters:

$$c'_m = \frac{c'}{F_s}$$

$$\phi'_m = \tan^{-1}\left[\frac{\tan \phi'}{F_s}\right] \tag{7.4}$$

$$c_{um} = \frac{c_u}{F_s}$$

In this method, soil strength parameters are reduced in value and all subsequent calculations (for example, of the earth pressure coefficients K_a and K_p) are based on these 'mobilised' values (hence the subscripts 'm' above). As a consequence, active earth pressures are increased and passive pressures decreased (see Figure 7.12). CIRIA 104 suggested values of F_s = 1.1–1.2 for temporary and 1.2–1.5 for permanent works, when using moderately conservative parameters based on effective stress calculations. For the design of temporary works based on total stresses, F_s = 1.5.

In Figure 7.12, the mobilised force due to active earth pressures (P'_{ma}) is enhanced by the increase in K_a caused by the decrease in φ'_m, and the thrust from the mobilised passive earth pressure (P'_{mp}) is reduced by the decrease in K_p. The thrusts from pore water pressures are unaffected by the factors of safety, leading to the following equation:

$$\frac{P'_{mp}L_{pmp}}{P'_{ma}L_{pma}+U_aL_{ua}-U_pL_{up}}\geq 1 \tag{7.5}$$

In the 'Factor on passive resistance' method, a factor F_p, F_{np}, or F_r is applied either to the gross passive, net passive, or revised passive resistance (and hence to the restoring moment), respectively, e.g.

$$M_{O,gross}\leq \frac{M_{R,gross}}{F_p} \text{ or } M_{O,nett}\leq \frac{M_{R,nett}}{F_{np}} \text{ or } M_{O,revised}\leq \frac{M_{R,revised}}{F_r} \tag{7.6}$$

In these methods, the resisting moment based on gross, net, or revised passive earth pressures is reduced by a single lumped factor F_p, F_{np}, or

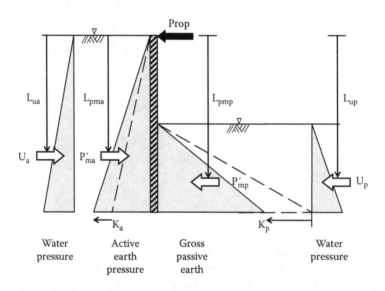

Figure 7.12 Factor of safety on strength (CIRIA Report 104).

F_r respectively. The gross and net pressure methods have been discussed earlier and the revised method is discussed below.

CIRIA 104 suggested values of F_p = 1.2–1.5 for temporary and 1.5–2.0 for permanent works, when using moderately conservative parameters based on effective stress calculations. For the design of temporary works based on total stresses, F_p = 2.0 is used. Because the net pressure method requires large values of F_{np} to achieve the same reliability as the other methods, CIRIA 104 did not recommend its use.

The 'revised' or Potts and Burland method (1983) was developed in an attempt to counteract the undesirable features of the gross and net pressure methods. The revised method is based on an analogy with bearing capacity theory which, in effect, treats the pressures coming from the ground above formation level as unfavourable and those from below formation level as favourable. Reliability was introduced into the calculations by applying a single factor of safety F_r = 1.3–2.0 on revised passive earth pressures, as shown in Figure 7.13.

In this diagram, effective earth pressures on both sides of the wall are reduced by an equal amount (by ignoring the hatched areas) so that the active earth pressures below formation level remain constant. The thrusts from the revised active and passive earth pressures are represented by the symbols P'_{ra} and P'_{rp}, respectively; and thrusts from pore water pressures by U_a and U_p. The lever arm of the active earth thrust is therefore greater than

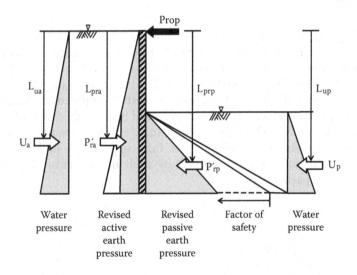

Figure 7.13 Factor of safety on revised earth pressures. (From Potts, D.M. and Burland, J.B., A parametric study of the stability of embedded earth retaining structures. Transport and Road Research Laboratory, Crowthorne, UK. Supplementary Report 813, 1983.)

in the net pressure method but less than in the gross pressure method, leading to the following equation for F_r:

$$F_r = \frac{P'_{rp}L_{prp}}{P'_{ra} + U_aL_{ua} - U_pL_{up}} \tag{7.7}$$

The revised method is more complicated than the other methods and has not been used widely in UK practice (and to the author's knowledge hardly at all outside the UK). With the publication of BS 8002 and Eurocode 7, use of the revised method has waned.

In 1994, British Standards Institution published a new code of practice for earth retaining structures to replace CP2, which was then over 40 years old. BS 8002 introduced the idea of a 'mobilisation factor' M which reduced the soil's representative peak strength (drained or undrained) to a 'design value' that was intended to keep wall movement below 0.5% of the wall height.

$$\text{design } c' = \frac{c'_r}{M}$$

$$\text{design } \phi' = \tan^{-1}\left[\frac{\tan\phi'_{r,max}}{M}\right] \tag{7.8}$$

$$\text{design } c_u = \frac{c_{u,r}}{M}$$

In addition, the code required the mobilised soil strength using effective stress parameters to be no greater than the representative critical state strength of the soil. BS 8002 recommended values of $M = 1.2$ for effective stress and 1.5 for total stress parameters. The effect of the mobilisation factors is identical to the factor of safety on strength F_s proposed by CIRIA 104.

In 2003, CIRIA Report 104 was superseded by the publication of CIRIA C580 (Gaba et al. 2003). This new guidance, for the 'economic design' of embedded retaining walls, adopted the limit state design philosophy and introduces reliability into calculations by applying factors of safety to the soil's strength parameters:

$$c'_d = \frac{c'}{F_{sc'}}$$

$$\phi'_d = \tan^{-1}\left[\frac{\tan\phi'}{F_{s\phi}}\right] \tag{7.9}$$

$$s_{ud} = \frac{s_u}{F_{ssu}}$$

where the symbol s_u is preferred to c_u for undrained strength. For ultimate limit states, CIRIA C580 recommends the use of factors $F_{sc}' = F_{s\phi} = 1.2$ and $F_{ssu} = 1.5$ when moderately conservative or most probable values of c', ϕ', and s_u are selected; and $F_{sc}' = F_{s\phi} = F_{ssu} = 1.0$ when worst credible parameters are selected. For serviceability limit states, the report recommends $F_{sc}' = F_{s\phi} = F_{ssu} = 1.0$ when moderately conservative or most probable parameters are selected (worst credible parameters are not appropriate for serviceability limit state calculations).

7.3.2 Partial factors of safety

For many years, structural design has been carried out using a range of partial factors, applied to different types of adverse and beneficial load. For example, Table 7.1 gives the load combinations and partial factors for ultimate limit state design according to British Standard BS EN 1992-1-1:2004, the code of practice for the structural use of concrete in buildings and structures. In a partial factor approach, safety can be designed into a structure by factoring both adverse and beneficial loads.

The use of lumped factors of safety by geotechnical engineers has caused a certain amount of confusion, requiring care when assessing loadings being transferred across the structural/geotechnical design interface.

The application of partial factors is undoubtedly more complex than the application of lumped factors. As Table 7.1 shows,

- Different values of partial factors are typically used for adverse and beneficial loads (in earth pressure calculations, this might involve, as a hypothetical example, multiplying active forces by a partial factor greater than unity, and passive forces by a factor less than unity).
- Different numerical values of factors may be used depending upon whether the load is permanent or transient (in this case, 'dead' or 'live').
- Different numerical values of a factor may be used for the same load combination in different limit state calculations, depending upon whether its effect is adverse or beneficial.

Table 7.1 Load combinations and partial factors for Ultimate Limit State design according to BS EN 1992-1-1:2004

| | Load type | | | | |
| | Dead | | Live | | |
Load combination	Adverse	Beneficial	Adverse	Beneficial	Wind
Dead and live	1.4	1.0	1.6	0	–
Dead and wind	1.4	1.0	–	–	1.4
Dead, live and wind	1.2	1.2	1.2	1.2	1.2

But in Europe, the use of partial factors is now becoming embedded in geotechnical design, as a result of the development and implementation of Eurocodes, principally Eurocode 7 (BS EN 1997).

7.3.3 Geotechnical limit state design using partial factors

This section provides a brief introduction to limit state design using partial factors, which is the basis of Eurocode 7—Geotechnical Design (BS EN 1997), published in December 2004. For an introductory text, the reader is referred to Anon (2006), and for a more in-depth treatment, the reader is referred to CIRIA Report C641 (Driscoll et al. 2008) and Bond and Harris (2008).

In the Eurocode 7 approach,

- Limit states separate desired states of the structure from undesired states. An example limit state might be the point at which failure of a gravity retaining wall occurs by forward sliding, for example, under reducing soil strength or increasing surcharge behind the wall.
- Partial factors are applied not only to take account of soil variability, but also to introduce a measure of safety and reliability in the face of other factors, such as parameter and model uncertainties.
- Partial factors are prescribed, but the method of calculation in which they are used is, broadly, left to the designer.

The aim is to ensure reliability. In this context, reliability is strictly defined as the probability (a numerical value between 0 and 1) that a structure will perform a required function under prescribed conditions (e.g. of loading) without failure or unacceptable deformation, for its required design life.

In geotechnical design, reliability is a more loosely used term, equivalent to the effect achieved by designing

- Using moderately conservative soil parameters
- Obtained using accepted, and preferably standardised drilling and testing methods, and good practice
- In calculations appropriate for the ground and groundwater conditions
- With adequate partial or lumped factors on applied loads and resistances

7.3.3.1 Actions, effects of actions and resistances

EC7 uses the terms 'actions', denoted by the symbol 'F', 'effects of actions', denoted by the symbol 'E' and 'resistances', denoted by the symbol 'R'.

Actions may be

- 'Direct'—sets of forces (loads) applied to a foundation by the structure, for example, as a result of self weight
- 'Indirect'—sets of imposed movements, deformations or imposed accelerations caused, for example, by shrinkage, temperature changes, earthquakes, or moisture variation

Actions may be permanent (G) or variable (Q). Characteristic actions (F_k) are the main representative values of actions used in design calculations. They are normally specified in codes as a statistical value (mean, upper or lower value).

EC7 states that the designer should consider the effects of the following geotechnical actions. This should be in addition to other recognised structural effects:

Weight of soil, rock and water	Traffic loads
Stresses in the ground	Movements caused by mining or other caving or tunnelling activities
Earth pressures and groundwater pressure	Swelling and shrinkage caused by vegetation, climate of moisture changes
Free water pressures, including wave pressures	Movements due to creeping or sliding or settling ground masses
Groundwater pressures	Movements due to degradation, dispersion, decomposition, self-compaction and solution
Seepage forces	Movements and accelerations caused by earthquakes, explosions, vibrations and dynamic loads
Dead and imposed loads from structures	Temperature effects, including frost action
Surcharges	Ice loading
Mooring forces	Imposed prestress in ground anchors or struts
Removal of load or excavation of ground	Downdrag

'Design values' of actions (F_d in EC7) are calculated by multiplying partial factors (γ_F) by 'representative' values of actions (Figure 7.14). The 'representative' value of actions are obtained from the characteristic values (F_k) by multiplying them by a correlation factor, ψ, which is always less than or equal to 1.0. For permanent actions $\psi = 1.0$.

The effect of actions (denoted by the symbol 'E' in EC7) may be the deflection or rotation of the whole structure, or the creation of internal forces, moments, stresses and strains within it. In EC7, permanent actions are considered to be either 'favourable' or 'unfavourable', but variable actions are always treated as 'unfavourable' (favourable variable actions are ignored). 'Representative actions' are obtained from characteristic action values by applying a correlation factor (ψ) selected to ensure that the design situations are catered for.

In geotechnical design, partial factors can be applied to three different parts of a design calculation; adverse loads, beneficial loads and soil properties.

7.3.3.2 Design approaches

Because of the diversity of practice in Europe, EC7 offers three Design Approaches, as shown in Table 7.2.

In the UK, the National Annex to Eurocode 7 states that Design Approach 1 shall be used. Partial factors for GEO and STR limit states are given in Table 7.3.

Design Approach 1 requires two separate checks to be made with different partial factors used for each check.

- In Combination 1, partial factors
  $$\gamma_G = 1.35$$
  $$\gamma_Q = 1.5$$
  $$\gamma_c = \gamma'_\varphi = \gamma_{cu} = 1.0$$
  $$\gamma_{Re} = 1.0$$

are applied.

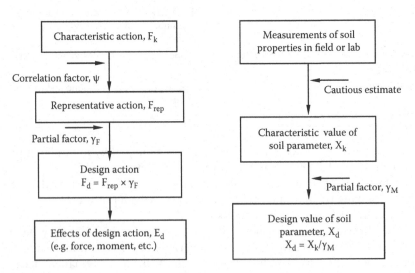

Figure 7.14 Application of design factors.

Table 7.2 Application of partial factors in different EC7 design approaches

Design approach	Partial factors applied to			
	1		*2*	*3*
combination	*1*	*2*		
Axially loaded piles and anchors	Actions	Resistances	Actions and resistances	Actions, material properties and resistances
Other structures	Actions	Material properties	Actions and resistances	Actions, material properties and resistances
Slopes	Actions	Material properties	Effects of actions and resistances	Effects of actions, material properties and resistances

Source: After Bond, A. and Harris, A., *Decoding Eurocode 7*. Taylor and Francis, 598 pp, 2008.

Table 7.3 Partial factors for EC7

Parameter		Partial factor symbol	GEO/STR partial factor sets						
			A1	*A2*	*M1*	*M2*	*R1*	*R2*	*R3*
Permanent action (G)	Unfavourable	γ_G	1.35	1.0					
	Favourable	$\gamma_{G,fav}$	1.0	1.0					
Variable action (Q)	Unfavourable	γ_Q	1.5	1.3					
	Favourable	–	–	–					
Accidental action (A)	Favourable	γ_A	1.0	1.0					
	Unfavourable	–	–	–					
Tan of effective angle of friction (tan φ')		γ'_φ			1.0	1.25			
Effective cohesion intercept (c')		γ'_c			1.0	1.25			
Undrained shear strength (c_u)		γ_{cu}			1.0	1.4			
Unconfined compressive strength (q_u)		γ_{qu}			1.0	1.4			
Weight Density (γ)		γ_γ			1.0	1.0			
Bearing capacity (R_v)		γ_{Rv}					1.0	1.4	1.0
Sliding resistance (R_h)		γ_{Rh}					1.0	1.1	1.0
Passive resistance (R_e)		γ_{Re}					1.0	1.4	1.0

- In Combination 2, the partial factors are as follows:

$\gamma_G = 1.0$

$\gamma_Q = 1.3$

$\gamma_c = \gamma'_\phi = 1.25$

$\gamma_{cu} = 1.4$

$\gamma_{Re} = 1.0$

For serviceability limit states, all partial factors are set to 1.0.

7.3.3.3 Material properties

Material properties are required for calculations to determine the effects of actions. 'Derived values' of geotechnical parameters may be obtained from test results using theory, correlation or empiricism. EC7 gives the following advice:

- Geotechnical parameters (denoted by the symbol X in EC7) should be established on the basis of 'published and well recognised information relevant to the use of each type of test in the appropriate ground conditions' with 'the value of each parameter compared with relevant published data and local and general experience...' (Clause 2.4.3 (5))
- 'The selection of characteristic values of material properties [X_k in EC7] for geotechnical parameters shall be based on results and derived values from laboratory and field tests, complemented by well-established experience.' (Clause 2.4.5.2 (1))
- 'The characteristic value of a geotechnical parameter shall be selected as a cautious estimate of the value affecting the occurrence of the limit state.' (Clause 2.4.5.2 (2))

'Design material properties' (X_d in EC7) are calculated by dividing the characteristic geotechnical parameter by the partial factor γ_M.

7.3.3.4 Geometric input

As we have seen, uncertainties in the geometry of the ground frequently give rise to problems during construction. Where changes in geometry have a significant effect on the results of design calculations, the best estimates of geometrical data (a_{nom} in EC7) should be varied to take into account possible geometric uncertainties:

$$a_d = a_{nom} \pm \Delta a \tag{7.10}$$

EC7 gives guidance relevant to retaining structures in Clauses 6.5.4(2) and 9.3.2.2.

7.3.3.5 Design calculations to EC7

Example calculations for retaining structures and foundations can be found at the end of Chapters 9 and 10, in 'A Designers Simple Guide to BS EN 1997', published by the UK Department for Communities and Local Government in 2006, in CIRIA Report C641, and in Bond and Harris (2008).

In summary, the following procedure is adopted:

1. Identify limit state for which design calculation is to be carried out.
2. Decide on partial factor set to use in this calculation (e.g. in the UK, Design Approach 1, Combination 1: A1 '+' M1 '+' R1: or Design Approach 1, Combination 2: A2 '+' M2 '+' R1) (note that both calculations are required).
3. Determine the ground geometry.
4. Determine 'moderately conservative', characteristic values of parameters for the ground, e.g. c'_k, φ'_k, c_{uk}, γ_k (bulk unit weight of soil, termed 'weight density' in EC7).
5. Calculate the design values of relevant parameters (e.g. $\tan\varphi'_d = (\tan\varphi'_k)/\gamma'_\varphi$).
6. Determine groundwater level and condition (e.g. hydrostatic, artesian, perched, under-drained), and from this, pore pressure profile.
7. Calculate geotechnical forces (e.g. mass of a gravity wall, active force on a retaining wall, passive force on a retaining wall, resistance to sliding on the base of the wall) using the design values of geotechnical parameters (e.g. φ'_d).
8. Apply partial factors to actions (e.g. for sliding, if H is the sum of the horizontal forces destabilising the wall, $E_d = H\gamma_{G:dst}$).
9. From this, determine the adequacy of the design (e.g. for sliding compare the design effect, E_d, the factored sum of the horizontal design forces, with the design resistance, R_d), e.g. $E_d \leq R_d$.

7.4 SUMMARY OF PRACTICE 1951–2011

Table 7.4 summarises various factors that have been recommended for use in the UK and elsewhere over the past 50 years or so.

Table 7.4 Factors recommended for use in the UK and elsewhere

		Factor of safety using moderately conservative parameters	
Design standard	*Applied to*	*Temporary works*	*Permanent works*
CP2	Gross passive pressures	2.0	
Piling Handbook	Nett passive pressures	1.0 for cantilever walls 2.0 for propped walls	
CIRIA 104	Embedment	1.1–1.2 2.0 on total stress	1.2–1.6 (1.5)
	Gross passive pressures	1.2–1.5 2.0 on total stress	1.5–2.0
	Revised passive pressures	1.3–1.5 2.0 on total stress	1.5–2.0

	Partial factors on			
Design standard	*tan φ'*	*c'*	*c_u*	*Condition/note*
CIRIA 104	1.1–1.2	1.1–1.2	1.5	Temporary works
	1.2–1.5	1.2–1.5	–	Permanent works
HK Geoguide 1	1.2	1.2	2.0	
BS 8002	1.2	1.2	1.5	On peak strength
	1.0	1.0	–	On critical state strength
CIRIA C580	1.2[a]	1.2[a]	1.5[a]	Moderately conservative
	1.0	1.0	1.0	Worst credible[b]
	1.2[a]	1.2[a]	1.5[a]	Most probable
EN 1997-1	1.0	1.0	1.0	Design Approach 1, Combination 1[c]
	1.25[a]	1.25[a]	1.4[a]	Design Approach 1, Combination 2[c]

[a]   1.0 for serviceability limit states.
[b]   Not applicable for serviceability limit states.
[c]   Partial factors also applied to actions.

Chapter 8

Introduction to analysis

Earth retaining structure may be analysed in many different ways and a variety of methods are available to the designer. Each has a valuable place in the 'tool box' and it is up to the engineer to appreciate the assumptions and limitations of each in order to select the most appropriate for a given task. They range from the very simple (perhaps requiring only hand calculations) to the very complex (requiring significant computational power). Demand for the latter has arisen from the need to demonstrate that displacements around excavations are within acceptable limits. Because each method is applicable to more than one type of earth retaining structure, this chapter presents an overview of the main types of analysis that are available. Subsequent chapters will focus on specific wall types and will show the application of these methods—which include

- Rules of thumb
- Evidential methods
- Closed-form solutions
- Upper and lower bound solutions
- Limiting equilibrium
- Discrete-spring models
- Continuum models

8.1 RULES OF THUMB

In general, a so-called 'rule of thumb' is an easily applied procedure for making some determination. Rules of thumb are common in geotechnical design. They permit the preliminary identification of key dimensions or proportions of foundations, excavations, tunnels and walls. The preliminary design of some types of earth retaining structure can be carried out using such rules, which are based on historical experience and do not require formal calculations. Examples include

- Tentative dimensions for mass concrete walls to ensure that the resultant (of wall weight and external soil pressure) falls within the middle third of the base
- Initial sizing of base and stem elements of semi-gravity T and L cantilever walls
- Penetration depth for embedded cantilever walls (e.g. penetration ≥ 2 × retained height)
- Preliminary proportioning for reinforced soil walls (in which, for example, the length of reinforcement is related to overall retained height)
- Driveability of sheet-piles based on cross-section area
- Piping/heave in a cofferdam using blocks and submerged unit weights

These methods originate from a variety of sources. Take, for example, the 'middle third' rule, which is based on the structural mechanics of columns. It has been known for a long time that the transverse stress distribution on a column depends not only on the magnitude of the axial force carried by the column but also its line of action—specifically, its eccentricity about the principal axes x–x and y–y. For a rectangular cross-section $b \times d$, it can be demonstrated that (e.g. Morley 1912) the transverse normal stresses are everywhere compressive if the line of action passes through the *kernel* or *core* of the section—a rhombus defined by points at $\pm b/6$ and $\pm d/6$ on either side of the centroid (see Figure 8.1). An important consequence of this in masonry columns is that tension cannot develop and all joints will thus remain closed and able to mobilise shearing resistance across the full area of contact.

The application in a mass concrete wall is very similar, the aim being to maintain compressive normal stress all along the base, thus preventing tension at the wall heel. Unlike a column, the force is inclined to the verti-

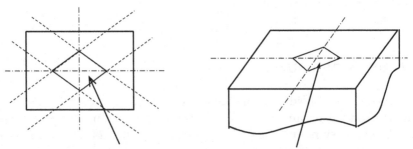

Compressive normal stress everywhere <u>only</u> if axial force falls within this zone

Figure 8.1 Middle third rule in column analysis.

cal because it is the resultant of the wall's self weight and external earth pressure, but the principal is considered.

8.2 EVIDENTIAL METHODS

Evidential methods (sometimes called *empirical* methods) are based on what has been found to work in practice—usually informed by field observations from a number of case records. Such observations have allowed a variety of design methods to be established, often presented in chart form. Evidential methods may be used for routine work, provided that the risk of damage to adjacent structures is low and the wall is relatively small. Examples include

- Estimates of maximum likely prop forces in braced excavations, based on worst case envelopes derived from field measurements
- Estimates of horizontal and vertical ground movements adjacent to retained excavations, based on observed settlement and deflection profiles in different soil types
- Guidelines for reducing bending moment in sheet-pile walls based on wall flexibility

In this section, some historical background is given for two of these examples, to provide an appreciation of the rationale and assumptions involved.

8.2.1 Envelopes for prop loads

It is common to estimate the strut forces on smaller braced excavations using the evidential methods by Peck and others. To do this rigorously requires fairly high-level finite element analyses (see Section 8.8), capable of modelling *in situ* K_0 stresses, non-linear stress-strain behaviour and possibly yielding. Analysis of this sort is complicated by the difficulty of obtaining realistic input parameters and so numerical modelling methods are only normally used in sensitive or prestigious projects, such as the underground car park excavations for the House of Commons adjacent to Big Ben in London (Burland and Hancock 1977) and for the Post Office Square in Boston (Whittle, Hashash and Whitman 1993).

Evidential methods are based on the monitoring data from a number of full-scale braced excavations, using the loads measured in the props (struts, braces) to estimate actual earth pressures on the back of the wall. By grouping the observations according to soil type and by identifying the worst-case earth pressures, it has been possible to construct envelopes of pressure that can be used for preliminary sizing of props. These envelopes represent an upper bound to the earth pressures (and hence prop forces) that are likely to be experienced, rather than indicating the *actual* pressures or forces under working conditions.

There are now a number of evidential methods for the prediction of strut loads and soil stresses on walls (e.g. Terzaghi and Peck [1967], Peck [1969], NAVFAC [1982], Tschebotarioff [1973], Goldberg et al. [1976]). The most commonly used methods are the Terzaghi and Peck (1967) distribution and Peck's (1969) version of the Terzaghi and Peck distributions. All versions are based upon the Terzaghi and Peck distributions and are somewhat similar.

Figure 8.2 shows the original Terzaghi and Peck (1967) distribution, as modified by Peck (1969). The pressures in Figure 8.2c are largely derived from envelopes of monitored strut loads in the stiff fissured clays of Chicago.

Peck introduced the idea that the behaviour of the soil-bracing system depends on how closely bottom instability is approached. Peck's stability number, $N = (\gamma H/c_u)$, gives a guide to the proximity of the soil to (negative bearing capacity) failure, which is approached when $N > 6$–7. If $N < 4$, classical earth-pressure theory leads to zero pressures on the wall. However, the behaviour of clays in braced excavations may approach an elastic deformation and the reduction in pressure from K_0 conditions will be strongly dependent on the horizontal movements allowed during construction.

The conditions for the satisfactory application of the basic Terzaghi and Peck diagrams are

a. Deep excavation (>6 m).
b. Water table assumed below the base of the excavation.
c. Sand assumed drained (i.e. diagram leads to effective stresses).
d. Clay assumed undrained (i.e. diagram leads to total stresses).
e. Bottom stability must be checked separately.

There are a number of difficulties in applying the diagrams to real soil conditions (e.g. how to classify clays and sands, how to treat groundwater, how to select the correct value of undrained shear strength for a soil profile with varying undrained shear strength, how to treat soil profiles with interbedded layers of sand and clay and how to treat silt). All of these problems must be the subject of judgement for the individual designer.

A comparison of strut loads predicted by various methods and those measured in an instrumented excavation is presented in Figure 8.3. It can be seen that the loads predicted by different methods are quite scattered and that agreement between predicted and observed strut loads is often unsatisfactory.

In the 40 years since Peck's original work, further research and field studies in the UK and elsewhere have led to a refinement of this evidential approach. CIRIA Report C517 (Twine and Roscoe 1999) offers a modern update in the form of distributed prop load (DPL) diagrams and its application to the design of braced excavations is covered in Chapter 10.

Pressure distribution

(a) Sands

$K_a = \tan^2(45 - \phi'/2)$

$= (1 - \sin\phi)/(1 + \sin\phi)$

Add groundwater pressures where groundwater is above the base of the excavation

Total force

$P_t = \text{trapezoid} = 0.65\,K_a\gamma\,H^2$

$P_a = \text{Rankine} = 0.50\,K_a\gamma\,H^2$

$P_t/P_a = 1.30$

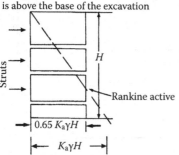

(b) Soft to medium clays* $(N > 5\text{--}6)$

$K_a = 1 - m(4c_u/\gamma\,H) = 1 - (4/N)$

$m = 1.0$ except where cut is underlain by deep soft normally consolidated clay, when $m = 0.4$

$m = 1.0$

$P_t = 0.875\gamma\,H^2\,(1 - (4/N))$

$P_a = 0.50\gamma\,H^2\,(1 - (4/N))$

$P_t/P_a = 1.75$

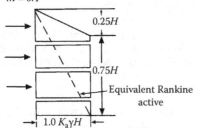

(c) Stiff clays*

For $N < 4$ (for $4 < N < 6$, use the larger of diagrams (b) and (c))

$P_t = 0.15\gamma H^2$ to $0.30\gamma H^2$

$P_a/N = 4, P_a = 0$

$N < 6, P_a < 0.$

Note: equivalent Rankine active = 0.

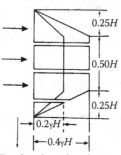

*For clays, base the selection on $N = \gamma H/c_u$.

Figure 8.2 Design earth pressure diagrams for different soil conditions on internally braced excavations. (a) Sands. (b) Soft to medium clays. (c) Stiff clays. (From Clayton, C. et al., *Earth Pressure and Earth-Retaining Structures*, Second Edition, Taylor & Francis, Jan 7, 1993.)

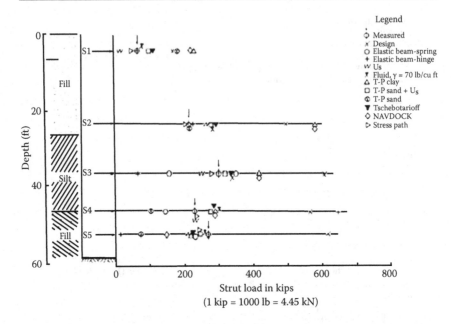

Figure 8.3 Strut loads in an instrumented braced excavation. Comparison between observed and predicted values. (From Clayton, C. et al., *Earth Pressure and Earth-Retaining Structures*, Second Edition, Taylor & Francis, Jan 7, 1993.)

8.2.2 Displacement correlations

The prediction of soil movements adjacent to a braced excavation can be made by evidential methods or using finite element methods. Major challenges exist in using computer methods for this purpose because of the need to obtain realistic input parameters for a complex soil model and the need to model detailed construction technique and sequences.

An evidential estimate of soil movements can be obtained from graphical summaries of information from previously instrumented excavations. Figure 8.4 shows an example of measured movements for an excavation in soft clay in Oslo. Points to note are that

    i. Surface settlements extend back about 1.5 times the depth of excavation from the edge of the excavation
    ii. Maximum inward wall movements occur at about the bottom of the excavation
    iii. The amount of surface settlement is largely controlled by the amount of inward yield of the support system.

An evidential estimate of soil movements can be obtained from graphical summaries of information from case records such as those in Figure 8.4. Axes are usually normalised—often by the depth (height) of excavation—to extend

(a)

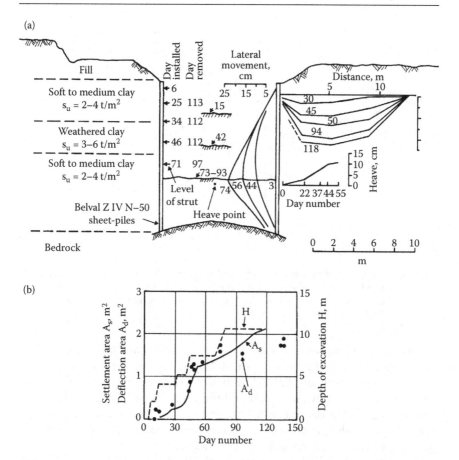

(b)

Figure 8.4 Lateral movements and surface settlements for a braced excavation at Vaterland 1, Oslo based upon data from NGI Technical Report No. 6. (a) Relationship between excavation, lateral movement and settlement adjacent to a braced excavation. (b) Relationship between volume of settlement (A_s) and volume of wall displacement (A_d) for the excavation in (a). (From Clayton, C. et al., *Earth Pressure and Earth-Retaining Structures*, Second Edition, Taylor & Francis, Jan 7, 1993.)

the applicability of the chart. An early example of this was a graph presented by Peck (1969) for the prediction of soil settlement adjacent to an excavation, taking into account soil type, depth of excavation and distance from the excavation (Figure 8.5). This graph considers the soil to be classified into one of the following groups:

a. Non-cohesive sands
b. Granular cohesive deposits
c. Saturated very soft to soft clays
d. Stiff clays

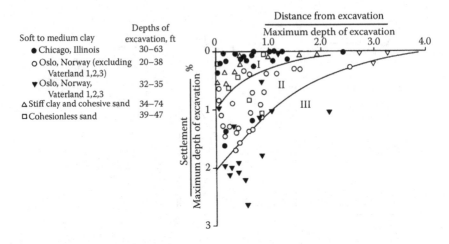

Figure 8.5 Summary of settlements adjacent to braced excavations in various soils, as a function of distance from the edge of the excavation. (From Clayton, C. et al., *Earth Pressure and Earth-Retaining Structures*, Second Edition, Taylor & Francis, Jan 7, 1993.)

Zone I: Sand and soft to hard clay, average workmanship. Zone II: Very soft to soft clay, (i) Limited depth of clay below bottom of excavation, (ii) Significant depth of clay below bottom of excavation but $N_b < N_{cb}$; settlements affected by construction difficulties. Zone III: Very soft to soft clay to a significant depth below bottom of excavation and with $N_b \geq N_{cb}$. Note: All data shown are for excavations using standard soldier piles or sheet-piles braced with cross-bracing or tiebacks.

The early work of Peck was subsequently developed by Mana and Clough (1981) using finite element analysis and later by Clough and O'Rourke (1990). Their charts for ground surface settlement and horizontal wall deflections have now largely superseded those of Peck and are presented in Chapter 10.

8.2.3 Moment reduction factors for sheet-pile walls

The work of Rowe (1952) is described in Section 3.3.8 and will not be repeated here. It is another example of an evidential method, but differs from those described above principally in that it is based on model wall tests rather than field observations.

8.3 CLOSED-FORM SOLUTIONS

Strictly speaking, in mathematics, a closed-form solution is one that can be expressed analytically in terms of a bounded number of certain elementary functions (excluding infinite series, etc.). Most engineers would understand a closed-form solution to be one that can be expressed by an explicit equation and computed without the need for iteration; for example, the collapse load of a strip footing on clay based on the solution for a rigid punch indenting a metal surface Prandtl, L. (1921). It is the mode of application rather than the mode of derivation that determines whether or not it is closed-form. In this section, the term closed-form will be used for solutions to governing equations that have been obtained by analytical means.

A rigorous closed-form solution in continuum mechanics is one that satisfies the equations of equilibrium, compatibility and constitution (i.e. stress-strain relationships) and the two main theories that have furnished solutions of practical use in retaining wall design are the theory of *elasticity* (Love 1927) and the theory of *plasticity* (Hill 1950).

8.3.1 Solutions based on elasticity theory

The essential characteristic of elastic behaviour is that strains are *reversible*—deformation is fully recovered upon the removal of load. Elastic behaviour need not be linear, although the vast majority of closed-form solutions assume linearity (probably because there is no universally accepted way of describing non-linear stiffness in a simple manner). In contrast, non-homogeneity and anisotropy are describable in fairly simple terms (e.g. $E = mz$ or $E_h/E_v = n$, respectively) and so have been incorporated into a number of available solutions. Details of elastic theory can be found in various texts (e.g. Timoshenko and Goodier 1970).

If linear elastic behaviour is assumed, two important principles can also be assumed. The first is the *principle of superposition*, wherein the results from two separate load cases may be added to or subtracted from each other, thus extending the usefulness of the solution. In a combined system of loads, it does not matter in which order the loads are applied; the resultant strains will be the same because they are independent of the loading path in an elastic material.

The second is *Saint-Venant's principle*, which may be stated as follows:

> If forces acting on a small portion of the surface of an elastic body are replaced by a statically equivalent system of forces acting on the same portion of the surface, the redistribution of loading produces substantial changes in the stresses locally but has a negligible effect on the stresses at distances which are large in comparison with the linear dimensions of the portion of the surface on which the forces are changed. (Saint-Venant 1855)

In a retaining wall context, for example, Saint-Venant's principle might permit a patch load on the crest of a wall to be replaced by a statically equivalent point force (and *vice versa*), with negligible difference in the computed increase in soil pressure on the back of the wall.

Poulos and Davis (1974) provide a comprehensive treatment of elastic solutions of particular use in geotechnical engineering. Many of the solutions assume the elastic body to be a 'semi-infinite half-space' which, in a retaining wall context, implies that the ground is of infinite depth and lateral extent. Correction factors can be applied if, for example, there is a rigid base at finite depth, but further departures from the concept of the half-space (such as multi-layer systems and vertical boundaries) will require numerical modelling.

The solutions of particular use to the retaining wall designer include

- Earth pressures due to an external point load (Boussinesq 1885) and line load (Flamant 1892)
- Elastic heave in the base of an excavation (Butler 1975)
- Internal forces in an embedded wall (Hetenyi 1946)

As it may be considered representative of the development and approach of all elastic solutions, a brief background to Butler's method will now be given.

8.3.1.1 Excavation heave

Butler (1975) has suggested that adequate estimates of excavation heave can be made using simple charts, derived with Steinbrenner's method—originally intended for estimating settlement (Steinbrenner 1934). One of Butler's charts is reproduced in Figure 8.6 for an excavation square in plan ($L/B = 1$) and for undrained loading conditions ($v = 0.5$), appropriate for estimating short-term heave. Further charts for $L/B = 2$ and 5 are also available (see Chapter 10).

In Butler's charts, Young's modulus, E, varies linearly with depth according to

$$E = E_0(1 + k.z/B) \tag{8.1}$$

where E_0 is Young's modulus at the surface and k expresses its rate of increase with respect to the depth/foundation width ratio (z/B). H (on the vertical axes of the charts) is the thickness of the layer in which the excavation is made and $I\rho$ is an influence factor for vertical movement.

The heave is estimated from the equation

$$\rho = I\rho \left[\frac{q.B}{E_0} \right] \tag{8.2}$$

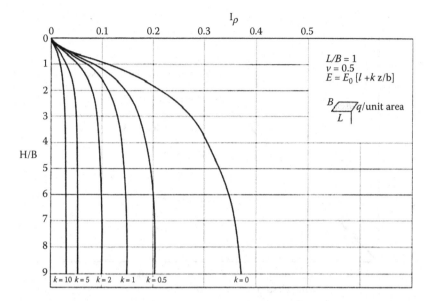

Figure 8.6 Chart for estimating undrained heave and reloading settlement. (From Clayton, C. et al., *Earth Pressure and Earth-Retaining Structures*, Second Edition, Taylor & Francis, Jan 7, 1993.)

where q is the amount of vertical stress reduction due to excavation. Using undrained Young's modulus values derived from undrained triaxial compression tests on 102 mm diameter specimens, on the basis that

$$E_u = 220c_u \tag{8.3}$$

Butler has noted good agreement between observed and predicted heaves using this method (Figure 8.7). If excavations are to be left open for long periods, or perhaps permanently, then allowance must be made by using a lower, drained, value of Young's modulus

$$E' = 140c_u \tag{8.4}$$

combined with Butler's charts for Poisson's ratio $\nu = 0.1$ (for which the same three L/B ratios are available).

8.3.1.2 Wall bending

Another important type of calculation that has its basis in elastic theory is that used for bending deformation—(i.e. the curvature and deflections of a wall under working conditions). The beam theory of Euler and Bernoulli

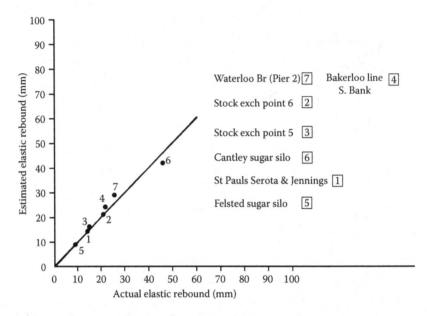

Figure 8.7 Comparison of predicted and measured heaves using Butler's method. (From Clayton, C. et al., *Earth Pressure and Earth-Retaining Structures*, Second Edition, Taylor & Francis, Jan 7, 1993.)

(nowadays referred to simply as engineering beam theory) is described in a number of standard texts (e.g. Gere and Timoshenko 1991). The governing fourth-order equation is derived from the standard result linking the principal quantities shown in Figure 8.8, namely,

$$\frac{\sigma_x}{y} = \frac{M}{I} = \frac{E}{R} \tag{8.5}$$

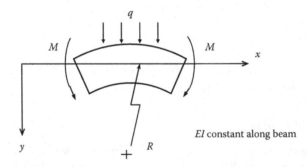

Figure 8.8 Simple beam bending.

From which

$$M = -EI\frac{d^2v}{dx^2} \tag{8.6}$$

Further considerations of moment and vertical equilibrium lead to the well-known fourth-order equation relating deflection v and loading intensity q:

$$EI\frac{d^4v}{dx^4} = q \tag{8.7}$$

For relatively simple boundary conditions (i.e. for displacement and rotation and for loading), it is possible to solve Equation 8.7 analytically to obtain distributions of deflection, moment and shear along the beam. It would be possible to adapt these solutions to a retaining wall context by considering the wall as a vertical beam and the earth pressures as the applied loading intensity. However, many earth retaining structures are statically indeterminate, rendering simple considerations of equilibrium inadequate. Furthermore, earth pressure distributions depend on wall deflection, so unless the interaction between soil and structure is incorporated, this approach is of limited usefulness.

Application of beam theory to a wall in bending implies the same deflection and curvature on any vertical section—for example, a cantilever wall propped at excavation level by a continuous concrete slab. Many walls are in fact 'two-way spanning' and exhibit curvature in the horizontal as well as the vertical direction. For this, plate bending theory is required (see, for example, Timoshenko and Woinowsky-Krieger 1959), for which the governing equation is

$$\frac{\partial^4 w}{\partial x^4} + \frac{2\partial^4 w}{\partial x^2 \partial y^2} + \frac{\partial^4 w}{\partial y^4} = \frac{q}{D} \tag{8.8}$$

where w is the deflection and $D = Et^3/12(1 - v^2)$. Direct application of Equation 8.8 to earth-retaining structures is not feasible and, as with beam theory, not particularly meaningful unless soil-structure interaction effects are included.

8.3.2 Solutions based on plasticity theory

The essential characteristic of plastic behaviour is that strains are *irreversible*—deformations are permanent and are not recovered upon the removal of load. Plastic deformation may be accompanied by elastic deformation—usually termed 'elasto-plastic behaviour'. There are three essential ingredients to a model of plastic behaviour—the yield function, flow rule and hardening law.

The *yield function* is an equation, written in terms of stresses and material parameters, that describes whether or not a material is yielding. The equation defines a locus or surface within which behaviour is purely elastic but on which yield will occur.

The *flow rule* defines the magnitudes of the components of the plastic strain increment and makes use of a plastic potential (defining a surface to which plastic strain vectors are normal). For the special case where the plastic potential is the same as the yield function, we have the case of *normality* and *associated flow* is said to apply. Where the plastic potential is different to the yield function we have the general case of *non-associated flow*.

Finally, the *hardening law* relates change in size of the yield locus/surface to the amount of plastic straining. An increase in size is usually termed 'strain hardening' and a reduction 'strain softening'. Materials that do not soften or harden after yielding are said to be *perfectly plastic*.

Unlike linear elastic behaviour, strains are not independent of loading path. This is because the direction of the principal strain increment is the same as the direction of the principal stresses, rather than the direction of the stress increment. In a combined system of loads, the order in which the loads are applied is important and so superposition is not applicable.

The theory of plasticity has provided the basis of a number of solutions of relevance to retaining walls, although most of them have been obtained using the numerical techniques described in Sections 8.5 and 8.6, rather than mathematically. An exception to this is Rankine's theory for active and passive stress states (Rankine 1857), which is of fundamental importance and in widespread use.

8.3.2.1 Active and passive stress states (Rankine)

Rankine presented a '...*mathematical theory of the frictional stability of a granular frictional mass*...' working from first principles and based only on the principle that sliding resistance was the product of normal stress and the tangent of the friction angle. This led to the now well-known expressions for the active and passive earth pressure coefficients:

$$K_a = \frac{(1 - \sin\phi')}{(1 + \sin\phi')}; \quad K_p = \frac{(1 + \sin\phi')}{(1 - \sin\phi')} \tag{8.9}$$

Rankine's approach was based on uniform states of stress and failure occurring at all points simultaneously within the retained soil mass and so was very different to Coulomb's wedge analysis nearly a century earlier. Rankine's analysis is restricted to a vertical back of wall ($\theta = 90°$) and a soil surface that is either horizontal or sloping at an angle of β to the horizontal (such that $\beta \leq \phi'$).

8.4 LIMIT ANALYSIS

It is difficult to obtain exact solutions but plasticity theorems can be used to set *bounds* for the collapse (failure) loads. Limit analysis is a way of obtaining the collapse load for a given case without having to solve the full boundary value problem. This is done by 'bracketing' the true solution with estimates that can be refined and brought closer together. By ignoring the equilibrium condition, an *unsafe* upper bound to the collapse load may be calculated; by ignoring the compatibility condition a *safe* lower bound may be calculated: the true collapse load must lie between these bounds. If the bounds are equal, the *exact* solution has been found.

The bound theorems can only be proved for materials that exhibit perfect plasticity and have an associated flow rule, which ensures that collapse loads are unique and independent of loading path.

8.4.1 Upper bound solutions

To obtain an upper bound, a plausible collapse mechanism is selected and then the plastic dissipation associated with this mechanism is equated to the work done by the applied loads. The actual collapse load is likely to be lower than this estimate—unless the true mechanism has been selected. The trial mechanism is varied in order to find a lower collapse load, until the worst case is identified.

Consider for example the undrained stability of a vertical excavation in clay, for which a possible plastic collapse mechanism is a rigid body translation of a wedge of soil along a planar slip inclined at θ to the vertical, Figure 8.9.

The internal work done ΔW is due to the undrained shear strength being overcome along the failure plane as the wedge moves through a distance of δ. The external work done ΔE is due to the self-weight of the wedge (unit weight $\gamma \times$ volume V) moving through a vertical distance of $\delta \cos \theta$. Equating

$$\Delta W = \Delta E$$

$$c_u L \delta = \gamma V \delta \cos \theta$$

$$c_u H_c \sec \theta \delta = \gamma \tfrac{1}{2} H_c^2 \tan \theta \delta \cos \theta$$

$$\therefore H_c = 2 c_u / (\gamma \sin \theta \cos \theta) \tag{8.10}$$

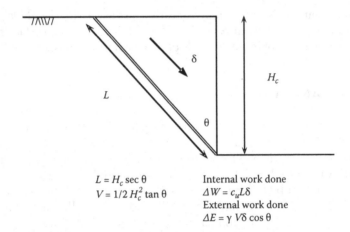

$L = H_c \sec \theta$
$V = 1/2\, H_c^2 \tan \theta$

Internal work done
$\Delta W = c_u L \delta$
External work done
$\Delta E = \gamma\, V \delta \cos \theta$

Figure 8.9 Mechanism of plastic collapse for a vertical excavation.

If $\theta = 45°$, $\sin \theta = \cos \theta = 1/\sqrt{2}$ and the upper bound collapse height $H_c = 4c_u/\gamma$. This means that a vertical excavation of height $4c_u/\gamma$ would certainly collapse although, if construction were attempted, failure would occur at a lower height—unless the true mechanism of collapse has fortuitously been found. To refine the estimate, the value of θ could be varied, although for this mechanism this would lead to bigger estimates of H_c. The mechanism of plastic collapse could be changed, for example, to that of a rigid body rotation along a circular arc centred on the top corner of the excavation. But this also leads to a bigger value of $H_c = 4.71c_u/\gamma$ and hence is not an improved estimate.

In a similar fashion, upper bound solutions can be obtained for a vertical excavation supported by a retaining wall, by taking account of the external work done by the active or passive thrust as appropriate and allowing for any wall friction. The procedures can be repeated for drained loading conditions ($\tau_f = \sigma' \tan \phi'$) although for the unsupported case, obviously $H_c = 0$. The reader is referred to Atkinson (1981) for details.

8.4.2 Lower bound solutions

To obtain a lower bound, an equilibrium stress distribution that does not exceed yield and is in equilibrium with the applied stresses is sought. Taking the same example of the undrained stability of a vertical excavation, a possible state of stress in three distinct regions separated by discontinuities is shown in Figure 8.10a. If element A were just on the horizontal discontinuity between zones I and II, then $\sigma_v = \gamma H_c$ and the maximum principal stress difference is given by the diameter of the Mohr circle at failure $= 2c_u \therefore H_c = 2c_u/\gamma$. In zone II, just on the other side of the discontinuity in element B,

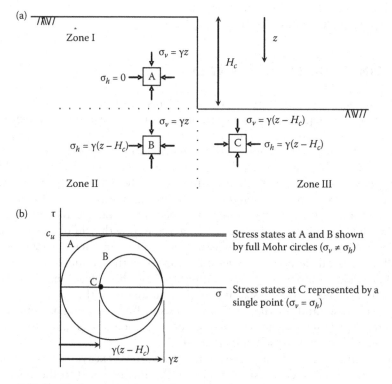

Figure 8.10 Equilibrium state of stress for a vertical excavation.

the principal stress difference is $\gamma z - \gamma(z - H_c) = \gamma(z - z + H_c) = \gamma H_c$; this can reach a maximum of $2c_u$ and so again $H_c = 2c_u/\gamma$. In zone III, the stress state is isotropic and plots as a single point, Figure 8.10b. The stress distribution is in equilibrium everywhere, whilst not exceeding the yield stress of the material anywhere—hence, a lower bound has been found.

Lower bound solutions can be obtained for a vertical excavation supported by a retaining wall, by considering equilibrium stress states relevant to the active or passive case and allowing for any wall friction. Both undrained and drained loading conditions can be accommodated.

8.4.3 Refinement

Using the upper and lower bound results obtained above, the safe height for a vertical excavation under conditions of undrained loading lies in the range $2c_u/\gamma \leq H_c \leq 4c_u/\gamma$. Further refinement can reduce the range to $2.83c_u/\gamma \leq H_c \leq 3.83c_u/\gamma$ (Atkinson 1981). It may be noted that the upper bound solution made no statements or assumptions about equilibrium, whilst the lower bound solution made no statements or assumptions about compatibility.

Each has thus ignored a requirement for a complete solution but nonetheless has provided bounds to the true collapse height.

8.5 LIMIT EQUILIBRIUM ANALYSES

Solutions to a more general range of problems can be obtained with semi-empirical limit equilibrium analyses, which combine features of upper and lower bound calculations. The limit equilibrium method is like an upper bound calculation in that it considers a mechanism of collapse and it is like a lower bound calculation in that it considers conditions of static equilibrium, but it does not satisfy the requirements of the proofs of the theorems. Although there is no proof that the limit equilibrium method leads to the correct solution, it is a very commonly used method in practice and experience shows that the solutions obtained often agree well with field observations.

An arbitrary mechanism of plastic collapse is constructed which must be 'kinematically admissible' overall. Shear strength is assumed to be fully mobilised on all slip planes. The equilibrium of each mechanism is found by resolving moments of forces acting on the boundaries of the blocks forming the mechanism and for the whole mechanism. By examining a number of different mechanisms, the critical one (for which loading is taken to be the limit equilibrium collapse load) is found.

Returning to the undrained stability of a vertical excavation, for which a possible failure mechanism is a planar slip inclined at θ to the vertical, Figure 8.11 and the shear strength is independent of the normal stress.

Considering the equilibrium of the wedge normal to slip plane and substituting for W and T

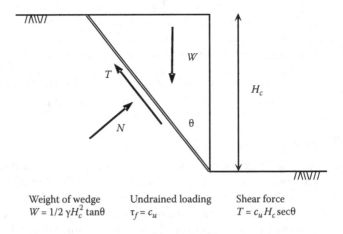

Weight of wedge	Undrained loading	Shear force
$W = 1/2\, \gamma H_c^2 \tan\theta$	$\tau_f = c_u$	$T = c_u H_c \sec\theta$

Figure 8.11 Potential failure mechanism for a vertical excavation.

$$T = W \cos\theta$$

$$\therefore c_u H_c \sec\theta = 1/2\gamma H_c^2 \tan\theta\cos\theta$$

$$\therefore H_c = (2c_u/\gamma) \sec\theta/(\tan\theta\cos\theta) = (2c_u/\gamma)/(\sin\theta\cos\theta) \qquad (8.11)$$

which is a minimum when $\theta = 45°$ and $\sin\theta\cos\theta = 1/2$ and $H_c = 4c_u/\gamma$. This is the same as the upper bound solution obtained in Section 8.4.1, although this does not of itself confirm that either solution is correct.

If the vertical excavation is supported by a retaining wall, Figure 8.11 need only be modified to show the active thrust P_a acting horizontally on the front of the wedge. Also, H_c is replaced by H, as the aim of the calculation is now to find the maximum force that the wall must provide. Equation 8.11 becomes

$$T = W \cos\theta - P_a \sin\theta$$

$$\therefore c_u H \sec\theta = 1/2\ \gamma\ H^2 \tan\theta\cos\theta - P_a \sin\theta$$

$$\therefore P_a \sin\theta = 1/2\ \gamma\ H^2 \tan\theta\cos\theta - c_u H \sec\theta$$

$$\therefore P_a = 1/2\ \gamma\ H^2 - c_u H/(\sin\theta\cos\theta) \qquad (8.12)$$

which is a maximum when $\theta = 45°$ and $P_a = 1/2\ \gamma\ H^2 - 2\ c_u H$, thus recovering the Rankine-Bell solution (Bell 1915).

Coulomb's method is, of course, a form of limit equilibrium analysis, providing for sloping back of wall, uneven ground surface and non-zero wall friction—and for both undrained and drained loading conditions.

Two very important types of limit equilibrium analysis often used in retaining wall design are the *free earth support* and *fixed earth support* methods.

8.5.1 Free earth support method

Figure 8.12 shows a typical layout for an anchored sheet-pile wall. In the free earth support method, the sheets are assumed to be rigid, rotating about point B where support is provided by an unyielding anchor. The depth of pile embedment is calculated on the basis of achieving moment equilibrium at the anchor level. The anchor force is then calculated on the basis of horizontal force equilibrium and the point of maximum bending moment is determined at zero shear force on the shear force diagram.

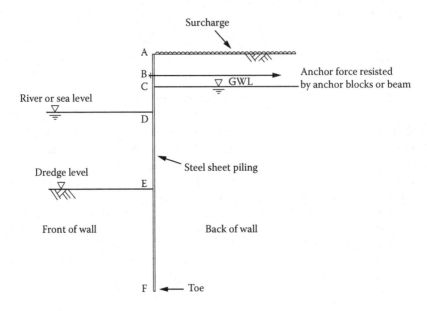

Figure 8.12 General layout for an anchored sheet-pile wall. (From Clayton, C. et al., *Earth Pressure and Earth-Retaining Structures*, Second Edition, Taylor & Francis, Jan 7, 1993.)

Assumptions in the free earth support method are that

i. Sheeting is rigid, compared with the soil.
ii. Sheets rotate about the tie level at failure, but the anchor does not yield.
iii. Despite (ii), active earth pressures occur over the full height of the retained soil; in waterfront structures anchor yield is normally sufficient to give full active pressure at the top of the wall.

8.5.2 Fixed earth support method

This method is derived from the work of Blum (1931, 1950 and 1951), in which the wall is considered flexible, but driven to sufficient depth that it may be considered fixed at its toe. Furthermore, the wall may be rigidly or flexibly anchored, or cantilevered. The stresses on the wall immediately above the toe are replaced by a single force some distance up the wall and the sheet piling is considered to be held vertical at this point. The depth of the penetration of the sheeting is found by repetitive calculation until the displacement at the anchor level is correct relative to the point of fixity (at the toe). For routine design, the anchor is assumed to be unyielding and this relative displacement must therefore be zero. Unless carried out by computer, this technique is tedious; therefore a number of simplifications are in common use.

Specific application of free and fixed earth support methods to sheet-pile and other embedded walls—including worked examples—are given in Chapter 10.

8.6 DISCRETE SPRING MODELS

If it is required to calculate wall displacements and internal forces and possibly movements in the adjacent ground, a *soil-structure interaction analysis* is required. There are several different approaches to modelling the wall and surrounding ground, as illustrated in Figure 8.13.

The physical problem in (a) is governed by fundamental equations of equilibrium, compatibility and constitution. In (b) the soil is replaced by discrete springs that are independent and have no interaction with each other. The

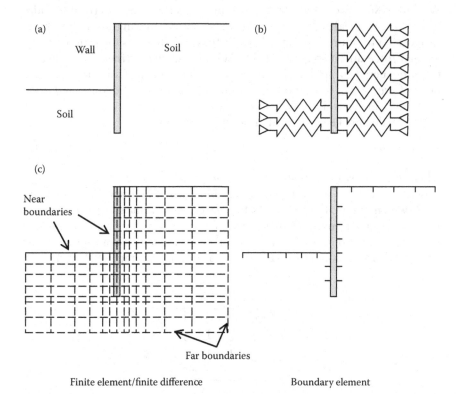

Figure 8.13 Different representations of a retained excavation. (a) Physical problem. (b) Discrete spring model. (c) Full continuum models.

wall can be modelled by engineering beam theory modified to take account of the springs, or by dividing it up into *finite elements* (described later) on spring supports. Finally, in (c) full interaction between soil and structure is represented; equilibrium and compatibility are fully satisfied and complex stress-strain behaviour is possible. It should be evident that (c) is a more faithful representation of reality than (b), but it comes with the penalty of greater modelling complexity (and hence cost). Solution of problems based on (b) can be achieved with a simple spreadsheet, whereas (c) will require specialist software either capable of formulating and solving large systems of simultaneous equations, or of iterating to an equilibrium solution. If wall displacements, moments and shears alone are sought, calculations based on (b) may be perfectly adequate. However, if surface settlements and other ground movements are required, calculations using (c) will be necessary. A brief discussion of the theories and solution procedures involved is given below.

8.6.1 Winkler spring model

Winkler (1867) proposed that the soil in contact with a structural foundation could be idealised by a bed of linear springs resting on a rigid base. Unlike a conventional spring, where stiffness is defined as the ratio between applied *force* and displacement, the Winkler spring stiffness is defined as the ratio between applied *pressure* and displacement. Consider a square plate of area $h \times h$, resting on the surface of an elastic solid and acted upon by a uniform vertical stress σ, shown in Figure 8.14.

The loaded area settles by an amount, δ, given by

$$\delta = \sigma/k \tag{8.13}$$

where k is known variously as the *Winkler spring constant, modulus of subgrade reaction*, or simply *subgrade modulus*. The dimensions of k are FL^{-3}, which can be thought of as force/area/displacement, or pressure per unit displacement (note the difference to normal spring stiffness, expressed in FL^{-1}). Despite the simplicity of the model, selecting an appropriate value for k is far from straightforward. For example, the average settlement $(\delta)_{ave}$

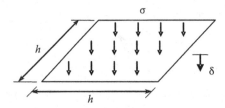

Figure 8.14 Square plate subject to uniform vertical stress.

of a square flexible footing carrying a vertical load P over an area A is (Timoshenko and Goodier 1970)

$$(\delta)_{ave} = \frac{0.95P(1-v^2)}{E\sqrt{A}} \tag{8.14}$$

Replacing P with σh^2 (see Figure 8.14) it can be deduced that

$$k = \frac{\sigma}{\delta} = \frac{E}{h}\frac{1}{0.95(1-v^2)} \tag{8.15}$$

Thus even if the soil is represented by the simplest form of continuum (i.e. a homogeneous isotropic linear elastic solid), k is not constant but depends on the size of the loaded area. This is the major deficiency of the Winkler spring model. An elastic solid needs at least two parameters to fully define it (E and v, or G and K) so attempting to describe behaviour with just one parameter can be expected to lead to difficulties—in particular, when attempting to select an appropriate value for a given practical problem. Subgrade modulus cannot, therefore, be regarded as a fundamental material parameter.

Notwithstanding this, the Winkler model has proven to be practically useful. The governing equation for beam bending (Equation 8.7) can be extended very easily to incorporate a bed of springs along the underside representing soil support:

$$EI\frac{d^4u}{dx^4} + b.k.u = b.q \tag{8.16}$$

where b is the beam width.

A general solution to the so-called *beam on elastic foundation* has been obtained by Hetenyi (1946), for beams of both finite and infinite length, subject to point loads and applied moments. Equations furnishing displacement, moment and shear are expressed in terms of periodic exponential functions, which depend on relative stiffness. This is not directly applicable to a retaining wall, but if a beam on elastic foundation is rotated through 90° so that the supporting soil is present on both sides, the analogous problem of the laterally-loaded pile is obtained, for which Equation 8.16 becomes

$$EI\frac{d^4u}{dz^4} = -bk_h u \tag{8.17}$$

where k_h is the horizontal modulus of subgrade reaction. Various solutions to Equation 8.17 have been obtained; initially in the context of the laterally-loaded pile (e.g. Matlock and Reese 1960) and subsequently for an embedded retaining wall (e.g. Haliburton 1968). Analytical solutions are possible for simplified variations of k_h with depth, with more complex variations handled conveniently by the finite difference method. Terzaghi's (1955) practical advice on the choice of k_h for both laterally-loaded piles and anchored 'bulkheads' (sheet-pile walls) is still widely followed to this day.

8.6.2 Intelligent spring models

A number of the limitations that exist with the basic Winkler spring model have overcome (at least partly) as follows:

a. *Linearity*—Equation 8.13 implies a constant value of k. This restriction may be lifted by adopting a relationship between σ and δ that is described by a continuous curve (such as a hyperbola), or by the greater complexity of the so-called *p-y curves** (McClelland and Focht 1958), which are based on empirical curve fitting and were developed for use with laterally-loaded piles.

b. *Limiting pressure*—Equation 8.13 also implies the possibility of an unbounded stress or reactive pressure, whereas the finite shear strength of the soil will impose a limit on what can be applied. In a retaining wall context, this can be overcome by imposing active and passive limiting pressures on the Winkler springs, depending on the sense of the wall movement, stress level, etc. In its simplest form, this would give a bilinear *p-y* curve and would introduce non-linearity into the solution.

c. *Single parameter*—as discussed above, at least two parameters are required for even the most idealised continuum. One approach to this has been to connect the ends of the Winkler springs to a membrane that deforms only by transverse shear (Pasternak 1954), thus providing a second parameter.

d. *Independence*—this is a variation on the single parameter issue, in that both attempt to address the lack of interaction between adjacent springs. A different approach (Pappin et al. 1986) has been to calibrate simple Winkler springs against 'prestored' finite element results, thus introducing limited continuum behaviour.

The incentive for persisting with these (and other) refinements is that the discrete spring model is generally much easier (faster, cheaper) to use in practice and so is preferred for routine design. However, as noted earlier, none of these refinements can help with the fundamental inability to model movement in the surrounding ground, away from the soil-structure interface.

---

* Relating pressure p to lateral displacement y (e.g. http://www.webcivil.com/lpile3.aspx).

8.6.3 Discretization

If using Winkler springs, a convenient representation of the wall is to use a *finite difference* approximation of Equation 8.17. This involves replacing continuous functions with discrete approximations; the governing equations, together with the boundary conditions, are thus converted to a set of linear simultaneous equations. Finite difference theory can be found in most numerical methods texts (e.g. Smith and Griffiths 1990); only the main points are summarised here.

The wall is divided (or *discretized*) into N segments of equal length, defining $N + 1$ grid points at a uniform spacing (or step size) of h. A finite difference approximation of Equation 8.17, for example, can be written as

$$(EI/h^4)[u_{i-2} - 4u_{i-1} + 6u_i - 4u_{i+1} + u_{i+2}] + b.k_{hi}.u_i = b.q_i \qquad (8.18)$$

where the subscripts $i - 1$, i, $i + 1$, $i + 2$ etc., refer to adjacent grid points and i denotes the point of concern or interest. Equation 8.18 is known as a central difference approximation and can be surprisingly accurate. The error associated with it is of the order of h^2, which means that if the step size was halved, for example, the error would be quartered. The method provides for the Winkler spring constant k and the intensity of loading q to vary down the wall by specifying appropriate values at each grid point. Variation in b requires some modification of the basic FD expression. *(NB: for a beam on elastic foundation, q is an applied load distributed along the upper surface, whereas for a laterally-loaded pile, q is not applicable and loading is handled through boundary conditions at the pile head; for an embedded wall, q is the difference between assumed pressure distributions on opposite sides—see Section 8.7.5 below.)*

Equation 8.18 is repeated at all $N + 1$ grid points and can be used to generate a matrix of coefficients and a right-hand vector, using standard stiffness method concepts. Boundary conditions at the ends of the beam (or wall) will usually be in terms of displacement and rotation, or bending moment and shear force and will require a further two 'fictitious' grid points at each end. Thus, N intervals will give rise to $N + 5$ equations in $N + 5$ unknowns, although only $N + 1$ of these are physically meaningful.

8.6.4 Procedure

The calculation typically starts off with a wall with balanced horizontal forces applied to it, arising from the *in situ* stress distribution. In principal, the analysis can be carried out with the wall being backfilled toward the top, but it is more usual to model the process of excavation from the top of the wall, downward. Spring forces are progressively set to zero, from the top of the wall toward the final depth of excavation and at the same time the spring stiffnesses are progressively halved, to model the removal of soil (and hence reduction in support) on the excavated side of the wall. As a result of the load imbalance, horizontal wall displacement and bending take place. Propping or anchoring is simulated by the addition of springs of the requisite stiffness at the appropriate level(s), as the excavation proceeds. If the long-term case is sought, the progressive softening that takes place as a result of the swelling of clays can be modelled by changing the spring stiffnesses.

8.6.5 Solution

The primary output from the analysis is wall displacements. Once these have been obtained, bending moments are found by back substitution into a finite difference approximation of Equation 8.6, *viz*,

$$M = -EI\frac{d^2u}{dz^2} = -(EI/h^2)[u_{i-1} - 2u_i + u_{i+1}] \tag{8.19}$$

Similarly for shear forces,

$$S = -EI\frac{d^3u}{dz^3} = -(EI/2h^3)[-u_{i-2} + 2u_{i-1} - 2u_{i+1} + u_{i+2}] \tag{8.20}$$

Even with a reasonably fine subdivision, the system of equations generated is not particularly big and can be solved by the matrix functions now routinely available in a spreadsheet software. This provides a powerful and convenient tool for the designer, although adjusting the number of segments is cumbersome and far better handled by a program written in a high-level language. Various commercial packages implementing a beam on Winkler spring model are available. Some use finite elements to represent the wall, but it is potentially misleading to describe this as a 'finite element analysis', as this term is normally reserved for a full continuum analysis.

8.7 CONTINUUM MODELS

Continuum models (sometimes termed 'numerical models') simplify the geometry of the soil-structure interaction problem by dividing the soil and any structural members (such as a retaining wall) into zones, or elements. Within each zone or element, the properties of the soil or structure are taken to be constant. Thus, geometry and property variations can be simplified, allowing a solution to be computed for each zone. With the rapid growth of computing power, these methods are increasingly used for retaining wall design.

8.7.1 Available methods

Three methods are available: the finite element and finite difference methods and the boundary element method. Among these, only the first two are in wide use for retaining wall analyses.

8.7.1.1 Finite element method

The finite element method (FEM) (Figure 8.13c) is a numerical technique for solving the differential equations governing a boundary value problem (Zienkiewicz 1977). The region of interest is divided into discrete areas or *elements*, often triangular or rectangular, defined by *node* points located at the vertices and sometime along the element edges (Figure 8.15).

Within each element, the behaviour is idealised, with the 'principal quantity of interest' (for example, displacement) constrained to vary in a prescribed fashion (e.g. linear or quadratic). The value of this quantity at any interior point in the element is related to its values at the nodes, through interpolation or *shape* functions, N, based on the element geometry

$$\theta = \sum_{i=1}^{i=n} N_i \theta_i \qquad (8.21)$$

where θ is the quantity and n the number of nodes. In retaining wall analyses, the displacement is the principal quantity of interest and differentiation of the shape functions yields expressions for the strain vector ε in terms of the vector of nodal displacements a:

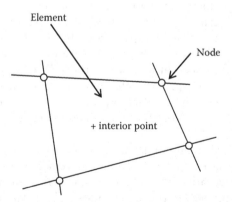

Figure 8.15 Finite element. (From Clayton, C. et al., *Earth Pressure and Earth-Retaining Structures*, Second Edition, Taylor & Francis, Jan 7, 1993.)

$$\varepsilon = Ba \tag{8.22}$$

where B depends on the element geometry. Then, an appropriate constitutive relationship can be used to relate stresses σ and strains ε within the element:

$$\sigma = D\varepsilon = DBa \tag{8.23}$$

where D depends on the properties of the material. Applying virtual work theorems, element stiffness relationships can be established between applied loads F and resulting displacements at the nodes:

$$F = Ka = \int \left(B^T DBd(vol) \right) a \tag{8.24}$$

Finally, a global stiffness matrix is obtained by assembling the contributions from each individual element. After applying boundary conditions, such as known forces and displacement 'fixities', the global system of equations is solved to yield the unknown nodal displacements. Internal strains in any element may be calculated from these displacements (Equation 8.22), followed by stresses using the constitutive relationships (Equation 8.23).

For an exhaustive treatment of the FEM in relation to geotechnical engineering in general and earth retaining structures in particular (see Potts and Zdravkovic 1999 and 2001).

8.7.1.2 Finite difference method

In the finite difference method (FDM) (Figure 8.13c), materials are represented by zones, defined between a grid of points. The user generates a grid to fit the geometry of the physical problem to be modelled. Each zone follows a prescribed pattern of stress-strain behaviour (for example, elastic or plastic) and when yielding occurs the grid distorts to update the geometry of the grid points. The simple case of one-dimensional beam bending described in Section 8.6 (where the soil-structure interaction is handled with a discrete spring model) must be generalised for a two- or three-dimensional continuum, but the substitution of continuous functions with discrete approximations is still fundamentally at the heart of the method.

The explicit FDM (Cundall 1976) uses the basic equations of motion and a time-stepping process to calculate incrementally the accelerations (and hence by integration the velocities and displacements) of the zone mass, which is lumped at the grid points. The strains obtained from this are then used in a constitutive law, to determine the corresponding stress increment for the zone. The stress increments are then summed to obtain a new out-of-balance force and the calculation cycle is repeated. The dynamic response of the system is numerically damped, so that with increasing time steps, the problem reaches equilibrium and the required solution. Note that in such an application of the finite difference method, the time-steps are used to obtain a solution, rather than to model time-dependent material behaviour.

8.7.1.3 Boundary element method

The boundary element method (BEM) is another numerical method for solving boundary value problems governed by differential equations (Banerjee and Butterfield 1981). The principal difference between this method and FEM/FDM is that the differential equations are transformed into equivalent integral equations prior to solution. Typically, the integral equations link boundary stresses to boundary displacements and so the method is particularly suited to those problems where the surface area to volume ratio is low, such as in many three-dimensional foundation problems. BEM requires only the boundary of the domain to be discretized into segments or elements (Figure 8.13c), not the interior (i.e. surface rather than volume discretization). The number of physical dimensions to be considered is effectively reduced by one, resulting in a smaller system of equations and significant savings in computing time (10 times faster than FEM for the same problem is quite typical).

This simplification is made possible by taking advantage of a so-called fundamental or singular solution, which gives the stresses and displacements at some point B due to a load or displacement acting at another point A. In geotechnical work, Mindlin's solution for a point load within a semi-infinite solid, or Boussinesq's solution for a point load acting on the surface of a half-space, are commonly used. By distributing the fundamental solution over the surface of the domain, a general solution is obtained in terms of a boundary density function. Boundary conditions are imposed by requiring the density function to satisfy an integral equation on the boundary. The solution is obtained first at the boundary and then at points within the region using the boundary solution.

There is no question as to the computational superiority of BEM in many problems, but it lacks the strong physical and intuitive appeal of FEM or FDM and is often obscured by formidable mathematics and notation. One case in which the advantage is greatly reduced is where considerable material non-homogeneity exists (each distinct zone bounded by boundary elements must be homogeneous). This partly explains why BEM has been more popular in rock mechanics applications than in soil mechanics; known applications in retaining wall and excavation analysis are very few. Consequently, discussion from hereon will be restricted to the FEM and FDM.

8.7.2 Geometric representation

This section describes some of the necessary detail in the implementation of wall and soil geometry using numerical modelling.

8.7.2.1 Discretization

Although of vital importance, the choice of mesh is entirely up to the user. While there is no single correct mesh, some will perform better than others and certain meshes may be totally incorrect (too few elements/zones) or highly inefficient (boundaries excessively remote). Experienced users will know if a mesh 'looks' right or not, but the novice may have very little idea where to start. Some element/zone boundaries will be fixed by divisions between material types, physical limits of the problem domain, the need to accommodate changing geometry, etc.—other boundaries are somewhat arbitrary. As a rule, smaller elements/zones must be used in areas where quantities of interest are likely to change rapidly and larger elements/zones should be used further away. The mesh must be graded between such areas (Figure 8.16).

Real physical problems are, of course, three-dimensional but can often be idealised as two-dimensional plane strain, plane stress, or axi-symmetry. Advantage should be taken of this where possible, provided the results are still meaningful. The additional complexity and computation time required for three-dimensional analyses is very considerable and the designer must consider carefully whether or not their use is warranted. In the context of finite element analysis of retaining walls, Simpson (1984) and Woods and Clayton (1993) present useful summaries of the issues that have to be considered; a thorough investigation of modelling assumptions and decisions is given by Woods (2003).

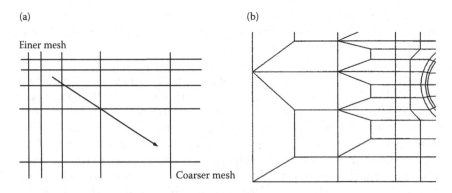

(a)

Finer mesh

Coarser mesh

(b)

Figure 8.16 Mesh grading. (a) Simple. (b) More advanced.

8.7.2.2 Boundary conditions

Near boundaries are typically defined, e.g. by the face of the retaining wall, or the base of excavation (Figure 8.13c). Far boundaries (that do not exist in reality) must be placed sufficiently remotely from the wall so as not to restrict or constrain the solution in the area of interest. An example of this would be attempting a footing settlement analysis, but placing the lower (rigid) boundary of the mesh (or grid) too close to the ground surface. In some cases, published solutions to classical problems can provide useful insights (e.g. contours of vertical stress change beneath a strip load), but sometimes the only way to check the adequacy of a mesh/grid is to investigate the effect of various alterations on some characteristic output. A simple method is to change the boundary conditions (e.g. from roller fixity to full fixity) and observe the effect on key outputs.

8.7.2.3 Types of element

For two-dimensional problems, quadrilateral elements are the most widely used, followed by triangular elements—and sometime a combination of both for particular geometries (Figure 8.17). These may be augmented by one-dimensional line elements for relatively thin material zones that have particular tensile, flexural or interfacial properties. If it is necessary to create a three-dimensional model, the equivalent 'bricks' and tetrahedra may be employed, together with two-dimensional elements for the thin zones. Special 'interface' elements may be required where it is necessary to specify the shear resistance between two parts of the mesh, or to prevent tensile stresses from developing.

The *order* of an element is defined by the interpolation functions it employs and is usually identified by the number of nodes it possesses. The 3-noded triangles and 4-noded quadrilaterals are capable of describing a linear variation in the principal quantity of interest, whereas 6-noded

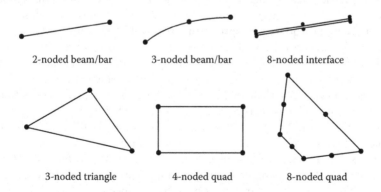

| 2-noded beam/bar | 3-noded beam/bar | 8-noded interface |

| 3-noded triangle | 4-noded quad | 8-noded quad |

Figure 8.17 Examples of two-dimensional finite elements.

triangles and 8-noded quadrilaterals can describe a quadratic variation. In a force-displacement context, the variation in strain that can be handled is one order less than that in displacement; hence the 3-noded triangle is known as a constant strain triangle and the 8-noded quadrilateral as a linear strain quadrilateral.

Another term used in connection with finite elements is *degrees of freedom* (d.o.f.), which describes the number of independent values of the principal quantity of interest associated with it. A 6-noded triangle has 12 d.o.f. because there are two components of displacement (a *vector* quantity)—in the x and y directions—at each node. Some special elements may have an additional d.o.f. at some nodes. For example, a 1D beam element, in addition to x, y displacement, d.o.f. will have additional rotation (θ) d.o.f. at each end. Another special element used in the so-called *coupled-consolidation* analysis has an additional excess head d.o.f. (a *scalar* quantity) at the vertex nodes, so that seepage can be superimposed on to a standard force-displacement analysis.

8.7.3 Constitutive models

FE and FD packages generally offer the user a number of different constitutive models. These can range from simple elastic models to highly sophisticated elasto-plastic strain-hardening/softening models and the choice is closely linked with the selection of appropriate soil parameters. The issues facing the designer are, quite simply, how much of this complexity is required in order to ensure a result that is realistic and/or fit for purpose and if it is required, can the necessary parameters be measured during site investigation, or estimated afterward with sufficient accuracy?

8.7.3.1 Linear elasticity

Sometimes referred to as homogeneous isotropic linear elastic (or *HILE*), this most basic form of constitutive behaviour requires only Young's modulus E and Poisson's ratio ν, or alternatively shear and bulk modulus G and K. Although inadequate for representing real soil behaviour, it is quite common for the structural members in an earth-retaining system to be modelled as linear elastic. There is also a strong argument for carrying out the analysis with the soil initially modelled as linear elastic and then adding increasing levels of sophistication. Not only does this help eliminate one possible source of error if the results are unexpected and/or demonstrably wrong, but it can also provide useful insights into the FE/FD model as a whole and which constitutive parameters matter most.

The extent and magnitude of ground movements in a number of field problems are often overpredicted using simple elastic models together with stiffness parameters measured in conventional laboratory tests. This is not only because conventional tests overestimate strains, but also because the strain levels in the field are typically very small (see Jardine et al. 1986). If a linear elastic analysis is to have any value, stiffnesses appropriate to the expected strain level must be selected.

8.7.3.2 Non-homogeneity

For most soils, mean effective stress (and hence stiffness) increases with depth. Non-homogeneous stiffness profile models of the form $E = E_0 + mz$ (compare to that used in Section 8.3.1.1 for Butler's [1975] charts), which usually offer sufficient versatility.

8.7.3.3 Anisotropy

Many natural soils exhibit anisotropy, i.e. different stiffness and strength properties in the horizontal and vertical directions of loading. This may be as a result of small scale layering, or of stress history. Cross anisotropic elastic models require five independent parameters (e.g. E_v, E_h, ν_{vh}, ν_{hh} and G_{vh}). Obtaining these parameters and their variation with strain level (see below), is a major challenge—see, for example, Clayton (2011)—but has been shown to be worthwhile for excavations and retaining walls in inner-city sites.

8.7.3.4 Non-linearity

The overall pattern of displacements predicted using a simple linear model may be incorrect due to the non-linear nature of most natural soils, even at relatively small strains. Non-linear models in use include simple bilinear elastic or power law models, the hyperbolic model (Duncan and Chang 1970) and the periodic-logarithmic Imperial College model (Jardine et al. 1986).

8.7.3.5 Plasticity

All soils exhibit plastic behaviour above certain levels of stress or strain. For effective stress analysis, the Mohr-Coulomb yield criterion (in terms of c', ϕ') is well established and widely available in FE and FD programs. The assumption of an associated flow rule (i.e. $\psi' = \phi'$) appears quite common, but is known to cause excessive dilation; non-associated flow is more realistic but is not generally available. For total stress formulations, the equivalent Tresca yield criterion (in terms of c_u) may be used. If the soil is non-homogeneous, c' or c_u (as appropriate) can vary with depth.

As noted in Section 8.3.2, yielding may be accompanied by hardening, softening, or no change in the size of the yield locus (perfect plasticity). The latter case is most frequently encountered in FE/FD codes, accompanying Mohr-Coulomb and Tresca yield criteria. However, reduction of strength and stiffness after yield occurs in many natural soils. Some models (e.g. the Cam-clay family) provide for strain softening (as well as hardening) although the main problem with a continuum model is reproducing the evolution of the discontinuous slip planes observed in the field, as it becomes an issue of geometry as well as constitutive behaviour. It may be noted that, in the explicit FDM, the stress change corresponding to a given strain change is directly enforced in the implementation of the stress-strain law, at each time step. This makes finite difference codes particularly useful for non-linear problems, such as those involving plastic yielding, since inaccuracies as a result of insufficient iterations are avoided.

8.7.4 Water and effective stress

The FEM is essentially a total stress method and various approaches have been adopted in modelling the (at least) two-phase nature of soil. Effective stress analysis of drained or undrained problems is usually carried out by combining the stiffness of the soil skeleton with the compressibility of the pore water (Naylor 1974). FD codes can be written in terms of effective stress, by simply subtracting the pore pressure from each of the total stress components (since water pressure is the same in all directions).

Full coupling of soil skeleton deformation and pore fluid flow is possible in both methods, allowing cases of partial drainage and long-term equalization of pore pressures following construction to be handled, though not all commercially-available programs provide this. Total stress analysis may be adequate in some cases and is performed by inputting undrained stiffness values for the soil skeleton and specifying a perfectly compressible pore fluid. (N.B. Non-geotechnical packages are unlikely to provide anything *but* total stress analysis.)

8.7.5 Construction modelling

In situations where the final distributions of movement and internal structural forces depend on the construction sequence followed, it will be necessary for the FE/FD package to be able to handle 'changing geometry'. This broadly comprises the ability to remove elements representing volumes to be excavated (e.g. a trench) and to add elements representing volumes to be placed (e.g. backfill)—sometimes in a simultaneous swop. The physical removal/ addition of material must be accompanied by the correct changes to body forces, internal stresses and boundary tractions. For the programme, it is a matter of bookkeeping (e.g. when is a particular element in the mesh) and proper accounting for the effects (e.g. has the element self weight been added/subtracted properly, etc.). If the analysis is not at the extremes of either drained or undrained behaviour, the time-dependent changes arising from consolidation or swelling must also be incorporated.

8.7.5.1 Wall installation

An FE/FD analysis can have the wall 'wished in place' (either from the start, or at a later stage through simple element swopping) or can attempt to replicate the field installation procedure. For a diaphragm wall, this involves trench excavation with drilling mud support, placing of wet concrete from the bottom up via tremie pipe and the subsequent hardening of the concrete. The principal aim of full installation modelling is to capture the reductions in horizontal stress that arise from the ground relaxing into the trench or borehole, before a rigid wall is in place (e.g. Gunn and Clayton 1992, Gourvenec and Powrie 1999). This is probably overly sophisticated for anything other than unusual and major projects—and is highly non-trivial, as it will require 3D coupled consolidation analysis in addition to any complexity in constitutive behaviour.

For a backfilled wall, it is the placement and compaction of soil fill behind the wall which is of particular importance. The principal aim of full installation modelling is to capture the locked-in horizontal stresses arising from compaction under virtually K_0 conditions (e.g. Seed 1986). This is also demanding analysis and relies heavily on the use of a bespoke constitutive model.

8.7.5.2 Excavation

Analysis of bulk excavation is straightforward enough, but relies on the program correctly handling all the appropriate removals of self-weight, etc. (which was a widespread error in some early FE programs). The mesh will need to be designed so that element boundaries correspond to the various stages of excavated geometry—including trapezoidal berms left against the wall for temporary support and intermediate levels at prop/anchor positions. Excavation is often assumed undrained, but coupled consolidation modelling removes this necessity and permits partial drainage.

8.7.5.3 Support system

Props and anchors, whether temporary or permanent, are essential element additions to the mesh, with or without prestressing as appropriate. One-dimensional structural elements (beams, bars) are often adequate for this role, except where the prop is a floor slab or road carriageway, when solid elements are more suitable. It is essential to allow for actual spacing in the out-of-plane direction when creating a plane-strain model. For example, props at a spacing S along the length of an excavation will need the axial stiffness $(E \times A)$ and any prestress force factored by $1/S$ to give appropriate values per m run. The removal of temporary props/anchors should be done with care as they can be carrying considerable force.

Chapter 9

Gravity walls

Eurocode 7 classifies retaining walls into three categories:

- Gravity walls. Walls with a base that can take significant vertical load, where the mass of the wall (and in some cases backfilled soil) are significant in maintaining stability. Examples include gabion walls, crib walls, interlocking block walls, and reinforced and unreinforced concrete walls (see Chapter 6 for examples).
- Embedded walls. Thin walls, relative to their height, where bending and shear play a significant role in supporting the ground, for example, sheet-pile walls and diaphragm walls (see Chapters 6 and 10).
- Composite retaining structures, which are composed of elements of gravity and/or embedded walls (see Chapters 6 and 11).

This chapter considers the design of gravity walls.

Gravity and reinforced-concrete cantilever walls are commonly used for relatively low retaining structures, although walls of up to 8–10 m are not unheard of. The design of this group of structures falls into five stages:

1. Selection of suitable wall type (see Chapter 6)
2. Preliminary wall sizing and layout
3. Selection of soil parameters (see Chapters 2 and 7)
4. Calculation of external stability
5. Checks on internal stability, and wall detailing

9.1 PRELIMINARY DESIGN

Preliminary design involves two stages:

- Provision of drainage, to minimise loads on the wall
- Initial sizing of the wall

293

9.1.1 Drainage and water control

Chapter 4 should be referred to for a wider discussion of the important effects of water on the pressures that need to be supported by retaining structures. Figure 9.1 illustrates some common methods of providing drainage behind a gravity retaining wall (see Teng 1962, pp. 316–317). These include using

a. Weep-holes in pervious backfill
b. Open-joint clay or perforated metal pipes in pervious backfill
c. Sub-vertical strips of filter material midway between weep-holes, linked to a horizontal strip of filter material, in semi-pervious backfill
d. Porous blocks or blanket of pervious material against the back of the wall in fine-grained backfill
e. Blanket of pervious material extending into fill in fine-grained backfill
f. Multiple blankets of pervious material in expansive clay

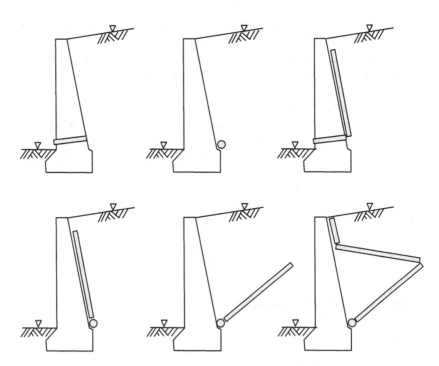

Figure 9.1 Typical ways of providing drainage to gravity retaining walls. (Redrawn from Teng, W.C., *Foundation Design*. Prentice-Hall, New Jersey, 1962.)

9.1.2 Initial sizing of gravity walls

Figure 9.2 defines some key dimensions that can be chosen during preliminary design on the basis of simple rules-of-thumb (see, for example, Cernica 1995) as summarised in Table 9.1, below.

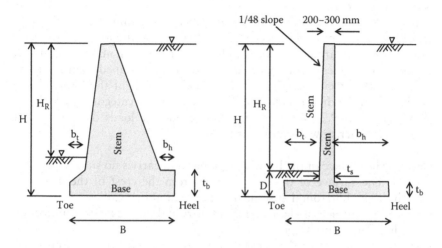

Figure 9.2 Initial sizing of gravity walls: (left) semi-gravity; (right) cantilever.

Table 9.1 Rules of thumb for initial sizing of gravity walls

Dimension		Semi-gravity wall	Cantilever wall	Counterfort wall
			Typical value for	
Base width	B	0.5H–0.7H	0.4H–0.7H $0.5H_R$–H_R[†]	0.4H–0.7H
Base thickness	t_b	H/6	H/10 $H_R/12$–$H_R/10$[†]	H/12
Stem thickness	t_s	–	H/10 $H_R/12$–$H_R/10$[†‡]	H/14
Toe extension	b_t	H/10	B/3 ≈0.13H–0.23H $H_R/10$–$H_R/8$[‡]	
Heel breadth	b_h	H/10	≈ 0.5H	
Toe embedment	D		min. 600 mm below frost line[‡]	
Counterfort spacing				0.3H–0.6H

Source: From †Coduto, D.P., *Foundation Design Principles and Practices*, Second Edition, Prentice-Hall, USA, 2001; ‡Teng, W.C., *Foundation Design*, Prentice-Hall, New Jersey, 1962; Cernica, J.H., *Foundation Design*, John Wiley & Sons Ltd., New York, USA, 1995.

9.1.3 Design charts

The limitations of classical theories have been noted in both Part I and Part II.

Bearing in mind these limitations, it has been argued that the expenditure needed to produce a detailed design for a small retaining structure is not warranted, particularly since the precise nature of the backfill material is often not known.

For gravity and L and T reinforced-concrete retaining walls, Terzaghi and Peck (1967) have produced design charts for estimating the pressure of backfill. They argue that for wall heights of less than about 6 m, it will be more economical to accept a wall that is overdesigned than to try to determine the precise properties of the backfill. Terzaghi and Peck's charts are based on the division of backfill into one of five categories, which are shown in Table 9.2. Terzaghi and Peck divide the loads on gravity and reinforced-concrete walls into four types:

Case a. The surface of the backfill is planar and carries no surcharge.
Case b. The surface of the backfill rises from the crest of the wall, but becomes horizontal at some distance behind the wall.
Case c. The surface of the backfill is horizontal, and carries a uniformly distributed surcharge.
Case d. The surface of the backfill is horizontal, and carries a uniformly distributed line load parallel to the crest of the wall.

Note 1. Very soft or soft clays will undergo large self-settlement during the life of the wall. It will not be possible to place and compact them using a normal construction plant. Their use should be avoided if at all possible.

Note 2. If the backfill cannot be protected from the ingress of water, firm to very stiff plastic clays should not be used. As the stiffness of the compacted clay increases, the increase of thrust on the wall due to wetting the clay also increases. Therefore the risk to the wall becomes greater. In practice, it is very difficult to protect backfill from the ingress of water.

Charts for cases (a) and (b) above are given in Figure 9.3 and Figure 9.4. For case (c), where the backfill is horizontal and carries a uniformly

Table 9.2 Classification of backfill for preliminary design

Soil type	Description
1	Clean, very permeable, sand or gravel.
2	Relatively impermeable silty sand or gravel.
3	Residual soil, silt, sand, gravel and cobbles with significant clay content.
4	Very soft or soft clay, organic silt or silty clay. (See Note 1)
5	Firm to stiff clay, protected from ingress of water. (See Note 2)

Source: Terzaghi, K. and Peck, R.B., *Soil Mechanics in Engineering Practice*, Second Edition, John Wiley, New York, 1967.

distributed surcharge, p, per unit area, the pressure per unit area at any depth or section ab is increased by q = C.p, where values of C are given in Table 9.3. When the surface of the backfill carries a line load q' per unit length, the load is considered to apply a horizontal line load on the section ab equal to q', where q' = C.p'. The point of application is found by drawing (Figure 9.5) a straight line (cd) at 40° to the horizontal to intersect the back of the wall (at d), and then projecting the position back by horizontal line d. If d falls below the back of the wall, the surcharge may be disregarded. The line load on ab will also produce a component of vertical force on ab. Terzaghi and Peck suggest that this force should be assessed as

p' *fb/fg*

where the force p' is assumed to spread at 60° down to the heel level, and fb and fg are distances (Figure 9.5).

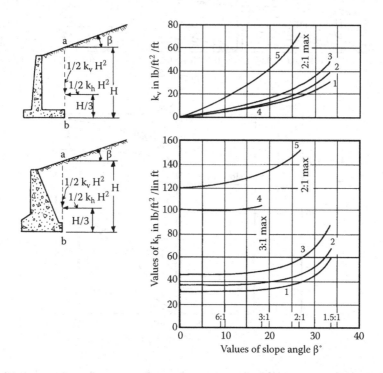

Figure 9.3 Chart for estimating the pressure of backfill against retaining walls supporting fills with a planar surface). Numerals on curves indicate soil types as described in Table 9.1. For materials of type 5, computations of pressure may be based on values of H 1.2 m less than actual value. (From Clayton, C. et al., *Earth Pressure and Earth-Retaining Structures*, Second Edition, Taylor & Francis, New York, 1993; Terzaghi, K. and Peck, R.B., *Soil Mechanics in Engineering Practice*, Second Edition, John Wiley, New York, 1967.)

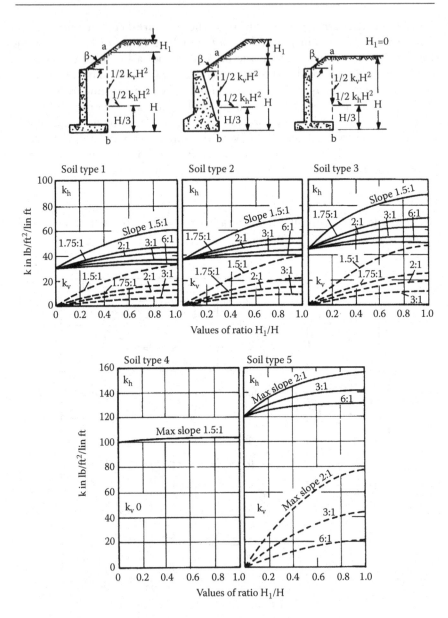

Figure 9.4 Chart for estimating the pressure of backfill against retaining walls support-ing fills with a surface which slopes upward from the crest of the wall for a limited difference. For soil type 5, computations of pressure may be based on values of H 1.2 m less than actual value. (From Clayton, C. et al., *Earth Pressure and Earth-Retaining Structures*, Second Edition, Taylor & Francis, New York, 1993; Terzaghi, K. and Peck, R.B., *Soil Mechanics in Engineering Practice*, Second Edition, John Wiley, New York, 1967.)

Table 9.3 Values of C for Terzaghi and
Peck's 'semi-empirical method'

Type of soil	C
1	0.27
2	0.30
3	0.39
4	1.00
5	1.00

Terzaghi and Peck's semi-empirical method can be used to obtain a first estimate of the backfill forces on the two types of wall it considers. These forces include the effects of seepage and 'various time-conditioned changes in the backfill', but drainage measures must be included to reduce the effect of frost action (see Section 9.1.1). The method is intended for walls on relatively incompressible foundations, where wall friction will be developed. For walls resting on soft clays, wall friction will not develop to its full extent and the values of backfill forces for material types 1, 2, 3 and 5 should be increased by 50%.

Terzaghi and Peck's charts were included in the first edition of their book, in 1948. They may, therefore, be inappropriate for some situations at the present, because of the relatively large compactive effort which is sometimes now applied to backfill. The vertical forces calculated from the charts may well be too low for horizontal backfill and too high for steeply

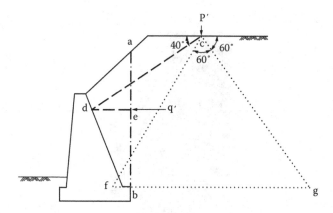

Figure 9.5 Terzaghi and Peck's construction to determine the point of application of a horizontal line load to the back of a wall, as a result of a vertical line load (P') at the soil surface. (From Clayton, C. et al., *Earth Pressure and Earth-Retaining Structures*, Second Edition, Taylor & Francis, New York, 1993.)

inclined backfill, because for the simplest case (a), the coefficients seem to incorporate Rankine's assumption (i.e. $\delta' = \beta$). In reality, the relationship between the vertical and horizontal forces must be controlled partly by the wall geometry (the length of the heel), partly by the available angle of friction in the soil, and partly by the mode of wall movement.

9.2 DETAILED DESIGN—LIMIT STATES FOR EXTERNAL STABILITY

External stability calculations treat the wall as a unit, and consider equilibrium in a number of ways:

- Overturning—the structure should have an adequate margin of safety against overturning about the toe of the wall. In practice, this is only likely to be a problem when designing semi-gravity concrete walls, reinforced brick walls, and T and L shaped reinforced concrete walls that are to bear on rock (for examples of these walls see Chapter 6).
- Toppling—the mediaeval middle third rule, where the wall is designed so that resultant force remains in the middle third of the wall section at all heights, has traditionally been used to guard against this mode of failure, which may occur in unreinforced walls composed of gabions, crib, and interlocking pre-cast blocks, for example (see Figure 9.6). The mediaeval middle third rule also provides an adequate margin of safety against overturning.

Figure 9.6 Start of toppling failure in an ancient masonry block wall in Delphi, Greece.

- Horizontal force equilibrium—neither the overall structure, nor sections of it, should slide forward under the action of the horizontal force component of the earth pressure.
- Vertical equilibrium—there should be an adequate factor of safety against bearing capacity failure.
- Overall stability—see Chapter 5—the change in ground surface geometry caused by the construction of the wall should not induce a failure around (i.e. including, but not failing) the wall itself.

9.2.1 Calculation models

A number of assumptions and simplifications are typically made when checking external stability of retaining structures. These involve, for example, assumptions regarding

- Whether the ground is *in situ* or backfilled
- If passive earth pressure at the front of the wall can be relied on
- The modes of failure in each limit state
- What constitutes the boundary between the wall and the soil

Figure 9.7 shows the geometry of a very simple gravity wall. Gravity and reinforced-concrete cantilever walls will almost always be backfilled; that is to say, an excavation will be made down to the founding level of

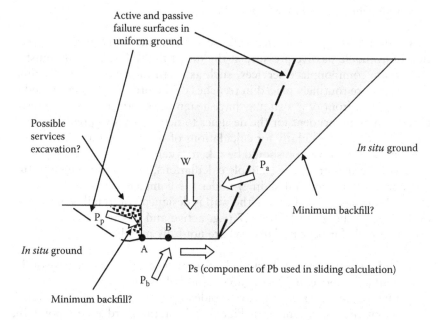

Figure 9.7 Forces used in calculations for a concrete semi-gravity wall.

the wall, the wall will be built, and the soil to be supported will then be placed behind it. The extent of the backfill will depend on local *in situ* soil conditions, and the practice of local contractors. In soft or weak soils, or uncemented granular soils, it is likely that temporary slopes will, as shown, be relatively flat (perhaps 1 on 1) and in this case, the pressure on the back of the wall will be entirely produced by the backfill. In the hard residual deposits found in many areas of the world, temporary slopes of 3 on 1 (say 70% to the horizontal) may be possible in the dry season, provided construction is carried out quickly. In this case, the loads supported by the wall will be, at least partly, a function of the *in situ* soil parameters.

For walls supporting backfill, the nature of the backfill must be known in order to carry out the design. It is suggested that coarse, relatively free-draining, granular fill is specified for the following reasons:

a. The guarantee of low pore water pressures in the retained backfill will lead to lower design forces, and a cheaper wall.
b. Coarse granular materials have high effective angles of friction, and therefore require less support than, for example, silts and clays.
c. Granular materials suffer relatively little self-settlement, even when only lightly compacted.
d. There is, at present, no theory to predict the lateral pressures from compacted clays although estimates of their magnitude may be made (see Chapter 3). Experience suggests that they are strongly time-independent, and may be very large.

A wall of this type is not generally very high, and therefore the assumption of available passive pressure at the toe of the wall may be optimistic. Even quite commonplace services, such as small diameter water distribution pipes, are routinely placed in trenches of the order of 1 m deep, and if placed just in front of a wall may have a significant effect on passive resistance. It may be prudent for the designer to neglect passive pressure at the toe of the wall, particularly for calculations of sliding stability. At the very least, passive wall friction should be taken as zero.

As will be seen from the example calculations, later in this chapter, full equilibrium (moment and sliding) under the complete set of forces is not checked during design. Because the wall is designed not to fail, some of the design forces shown in Figure 9.7 (e.g. active and passive forces, and the force available for bearing capacity) are not fully mobilised:

1. Sliding is avoided by checking horizontal resistance on the base of the wall (and above, in block, gabion and crib walls).
2. Overturning and toppling are avoided by checking moment resistance about the toe (point A in Figure 9.7) or the third point (point B), respectively.

3. Adequate bearing capacity is checked by considering the resultant of the forces acting on the base of the wall relative to the available bearing capacity.
4. Overall stability is checked using limit equilibrium calculations on trial failure surfaces.

9.2.2 Wall geometry and its effect on earth pressures

For walls with long heels (Figure 9.8 left), a full Rankine zone (Ocd) can be contained above the heel without being affected by the back of the wall. The vertical plane cd is termed the 'virtual back of the wall'. Rankine conditions apply (i.e. $\delta' = \beta$, $\beta < \phi'$ to ensure stability of the slope above the wall). Rankine's theory can be used to find the earth pressure on the virtual back of the wall. The resultant thrust, P_a, produced by this pressure is inclined to the horizontal at an angle β ($\leq \phi'$). This follows from consideration of the direction of the principal stresses in the ground (as per the Mohr's circle construction shown in Figure 9.9).

For Rankine conditions to be used, the inclination, α, of a line drawn between the back of the heel and the top of the stem (ac in Figure 9.8) must be less than the inclination of the Rankine active plane for a soil with an inclined surface, ψ, as in the case with the left hand wall. The right hand wall shown in Figure 9.8 does not meet this condition, so that the assumption of Rankine conditions on a vertical virtual back of wall cannot be

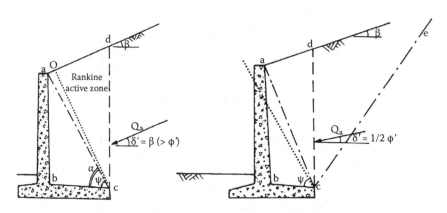

Figure 9.8 Development of a Rankine zone above the heel of a reinforced concrete cantilever wall (left) and interference with stem (right). (From Clayton, C. et al., *Earth Pressure and Earth-Retaining Structures*, Second Edition, Taylor & Francis, New York, 1993.)

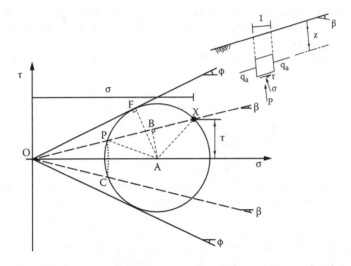

Figure 9.9 Mohr's circle of stress for gravity wall in sloping ground. X represents the stress state (t, s) on a plane at angle β to the horizontal; P is the pole of the circle; F is the stress state on the failure plane. (From Clayton, C. et al., *Earth Pressure and Earth-Retaining Structures*, Second Edition, Taylor & Francis, New York, 1993.)

made. The earth pressures on that plane are affected by the stem of the wall (albeit only slightly in this case).

Figure 9.10 gives the angle of Oc in Figure 9.8 as a function of effective angle of friction, ϕ', and angle of the slope of the retained soil, β. This graph can be used to check whether Rankine conditions can be used.

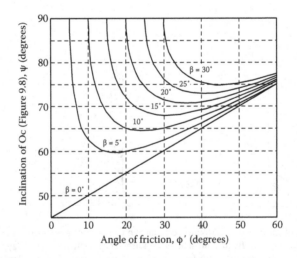

Figure 9.10 Angle at which wide heel assumption for cantilever walls no longer applies.

Conventional design uses $\delta' = \beta$ for walls with a long heel (see Figure 9.8 left) and includes the weight of soil above the heel (abcd) in the weight of the wall for calculation of overturning. For walls with shorter heels (Figure 9.8 right), the assessment of the design parameters and mechanisms is more difficult, because of the effect of the wall stem in restraining soil movement. As the length of the heel decreases, the value of δ' on the vertical plane above the heel will change from the inclination of the ground surface (β) to some value which is dependent upon the mode of wall movement (normally horizontal sliding for this type of wall), the soil properties, and the roughness of the back of the wall.

For walls with a short heel, where the stresses on the virtual back of the wall cannot be considered to be the Rankine values, it is impossible to predict the correct angle of friction on the vertical plane. Teng (1962) suggests 'using 1/2 to 2/3 ϕ' in the majority of cases', and $\delta' = \beta$ for soils with a planar ground surface.

For the sake of consistency, we suggest that the soil within abc in Figure 9.8 is taken as part of the wall regardless of heel length relative to wall height, since it should move with the wall. The virtual back of the wall can then be taken as plane ac in all cases, and since this is a soil-to-soil failure plane it seems reasonable to adopt $\delta' = \phi'$ upon it. Active stresses on plane ac can be obtained using earth pressures from the Müller-Breslau solution, and these stresses and their resultant forces can be used to assess the overall stability of the wall.

We suggest taking δ' to be

$$\frac{\delta'}{\phi'} = 1 - \left(1 - \frac{\delta'}{\phi}\right)\left(\frac{\alpha - \psi}{90° - \psi}\right) \tag{9.1}$$

(when $\psi < \alpha < 90°$), where δ is the angle of interface friction between the soil and the wall. Experimental work by Rowe and Peaker (1965) supports the use of

$$\delta' = \phi'/2 \tag{9.2}$$

Using this value, $\delta' \to \phi'$ as $\alpha \to \psi$ and $\delta' \to 1/2\ \phi'$ as $\alpha \to 90°$.

Figure 9.11 compares the orientations of the active forces produced by the two alternative calculation models that are commonly used to check the stability of reinforced concrete cantilever gravity walls. When bending moments in the stem are to be calculated, earth pressures should be derived using compaction earth pressure theory (see Chapter 3, and later).

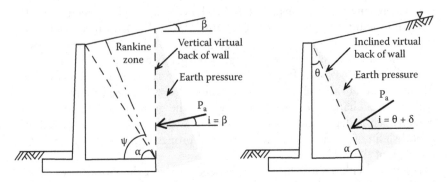

Figure 9.11 Possible calculation models for cantilever gravity wall: (left) with a wide heel; (right) with a narrow heel.

9.2.3 Earth pressures from undisturbed ground

For walls supporting undisturbed ground, undisturbed total and effective stress soil properties need to be estimated. For such soils, total and effective strength parameters can be expected to be relatively high. Groundwater conditions may well be expected to fluctuate considerably during the life of the structure, depending on local rainfall. The following information must be obtained, or estimated, for each soil type:

    i. Spatial distribution of different types of ground
    ii. Bulk density
    iii. Undrained shear strengths (c_u) of clays
    iv. Peak effective strength parameters (c', ϕ') for all soil types
    v. Worst envisaged groundwater conditions

For granular soils and normally consolidated clays, c' should be assumed zero. For overconsolidated clays, a conservative estimate of c' should be used (not more than, say, 1.5 kPa) unless a higher value can be justified from laboratory testing results.

For this type of construction, it is often difficult to compact the small wedge of backfill between the back of the wall and the *in situ* soil. If the backfill is loose tipped, and therefore compressible, it is reasonable to assume that active conditions will apply to the *in situ* soil. A suitable effective angle of wall friction should be chosen (see Chapter 2). For most cases, the detailed groundwater profile will not be known. If there is insufficient data to predict an inclined non-planar groundwater surface, then a horizontal groundwater surface should be used. This will allow the use of earth-pressure coefficients, which greatly speeds the calculation of the applied active force, provided a simple soil geometry with horizontal soil layers can be reasonably assumed. If such assumptions are not reasonable, the Coulomb active wedge analysis must be used.

9.2.4 Calculation of compaction pressures

Gravity walls normally support compacted fill, because of the need for a temporary excavation to prepare their foundation or base (Figure 9.7). This is particularly true of cantilever reinforced concrete walls, which can have a relatively long heel. Therefore, the type of fill and the type of compaction to be applied to it must be known or specified at the time of design, and compaction theory will be relevant to the design of some parts of the wall (e.g. the stem).

For granular backfill soils, lateral earth pressure may easily be estimated on the basis of the simplifications proposed by Ingold (1979) for Broms' compaction theory (see Chapter 3). An estimate of the maximum pressures to be supported by the structure, shown in Figure 9.12, can be obtained if three parameters are known, or can be estimated or specified: soil density after

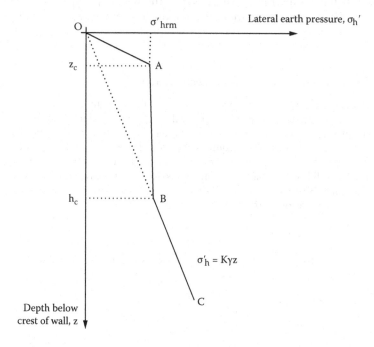

Figure 9.12 Design pressure for gravity walls retaining compacted granular backfill. (From Clayton, C. et al., *Earth Pressure and Earth-Retaining Structures*, Second Edition, Taylor & Francis, New York, 1993.)

compaction; effective angle of friction of soil; and maximum load per unit width of roller.

$$\sigma'_{hrm} = 2p\gamma/\pi \quad z_c = K\sqrt{(2p/\gamma\pi)} \quad h_c = \frac{1}{K\sqrt{2p/\gamma\pi}} \tag{9.3}$$

p = load/unit width of roller
γ = bulk unit weight of soil
K = earth pressure coefficient (K_0 for rigid walls, K_a for yielding walls)

The profile of soil pressure to be supported is given in Figure 9.12. For most walls, sufficient tilt will occur to give full active conditions below the zone of compaction pressures (i.e. deeper than h_c), and so the active pressure coefficient equals the Rankine value.

$$K_a = \frac{1 - \sin\phi'}{1 + \sin\phi'} = \tan^2(45° - \phi'/2) \tag{9.4}$$

For unusual geometries, perhaps where the wall is propped or rests on a very rigid base, earth pressure at rest conditions may be approached

$$K_0 = 1 - \sin\phi' \tag{9.5}$$

Example 9.1:   Active force on a gabion wall

A gabion wall is shown in Figure 9.13. The total height of the gabions, measured parallel to the face of the wall, is 8.0 m, of which 1 m is buried at the toe. The structure is to be battered back at 6° to the vertical. A surcharge of 18 kN/m², equivalent to a soil layer of about 1 m, is to be allowed for at the top of the wall. The effective angle of friction of the backfill is conservatively estimated as 30°, and the unit weight of the backfill is expected to be 18 kN/m³. Calculate the active force on the wall.
The inclination of the back of the wall

$\alpha = \tan^{-1}(8/2.5) + 6° = 79°$

Müller-Breslau earth-pressure coefficients (Appendix B, Table B.3) give, for

$\alpha = 80° \; \beta = 0 \; \delta = \phi' = 30°$

$K_a = 0.384$

$$\therefore P_a = \frac{1}{2}K_a\gamma H^2 = \frac{1}{2}0.384.18.\left(8.\sin(79°)\right)^2 = 213 \text{ kN/m run of wall}$$

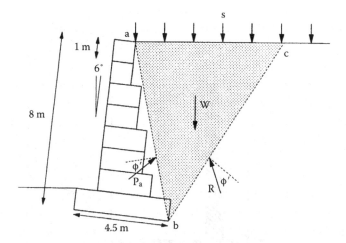

Figure 9.13 Calculation of active force on a gabion wall. (From Clayton, C. et al., *Earth Pressure and Earth-Retaining Structures*, Second Edition, Taylor & Francis, New York, 1993.)

9.2.5 Sliding—horizontal equilibrium

Gravity walls should normally be founded on reasonably good soil. If the soil on which the wall rests is different from that it supports, then undisturbed effective strength parameters are required in order to check against sliding. If groundwater is present, or can be anticipated during the wall's design life, then the pore pressure distribution across the base of the wall should be estimated.

Figure 9.14 shows the system of forces that need to be considered when checking that a cantilever gravity wall with a wide heel has sufficient resistance to sliding. W is the self-weight of the wall plus that of the backfill (block 'abcd') on its base; P_a is the inclined active thrust from the soil behind the vertical virtual back of the wall; P_p is the passive resistance from soil to the left of the wall toe (note that wall friction has been ignored), and S_{ult} is the ultimate shear resistance generated on the underside of the base.

In traditional design, a lumped factor of safety of between 1.5 and 2.0 is normally required, but depends on the assumptions made in calculation (see Chapter 7). The British Standard CP2 (1951) recommended a minimum factor of safety (F) of 2.0, when the wall is designed for active earth pressures. For the general case,

$$F = \frac{Q_{ph}/2 + S}{Q_{ah}}$$

(9.6)

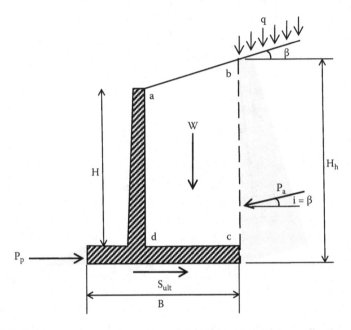

Figure 9.14 Earth pressures acting on the virtual back of a cantilever gravity retaining wall.

where
  S is the maximum shear force available from the soil/structure contact at the base of the wall
  Q_{ph} is the horizontal component of the passive resistance at the front of the wall (if used)
  Q_{ah} is the horizontal component of the active force at the base of the wall.

Because there will normally be disturbance at the surface of the soil where the wall is to be placed, it is sensible to neglect the effective adhesion (c'_w) between the base of the wall and the soil when calculating the total shear resistance of the base of the wall (S). The wall should be designed to give an adequate factor of safety against sliding under active earth pressures from the retained fill, regardless of whether or not the fill is to be compacted. We recommend a minimum lumped factor of safety of 1.5 (neglecting c'_w and Q_{ph}), and a maximum factor of safety of 2.5. If too large a factor of safety is adopted, then compaction pressures can become very large; if a modest factor of safety is used the wall will (imperceptibly) slide forward, relieving the horizontal pressures before damage can be done to the stem. When the wall is to retain compacted fill, it is recommended that the toe fill (contributing to the passive pressure in front of the wall) is not placed until the backfill to the wall is completed.

For the simple case of a masonry or gabion wall, the use of a 'key' (a downward projection from the base of the wall) is not feasible. If, therefore, the factor of safety against sliding is found to be too low, three choices present themselves:

   i. Increase the width of the wall
  ii. Back tilt the base of the wall
 iii. Reduce the pore water pressures

Pore water pressure will not normally be a significant problem. Increasing the width of the wall may be possible using a toe projection. This toe projection should not be included in calculations for the base width, factor of safety against overturning, or bearing capacity, because the projection cannot sustain bending moments. Back tilting the base of the wall will certainly produce an increase in sliding resistance, but it may be an unwelcome complexity in an otherwise very simple geometry.

For back-tilted gabion walls, Officine Maccaferri suggests that a sliding check can be made by considering a horizontal sliding plane, such as that shown in Figure 9.15. The factor of safety against sliding failure is calculated as

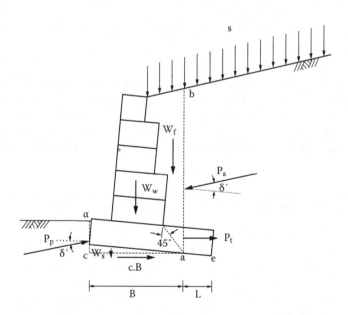

Figure 9.15 Layout and design forces for a galvanized steel gabion retaining wall. (From Clayton, C. et al., *Earth Pressure and Earth-Retaining Structures*, Second Edition, Taylor & Francis, New York, 1993.)

$$F_s = \frac{N.\tan\phi' + c'B + P_p \cos\delta' + P_t}{P_a \cos\delta'} \tag{9.7}$$

where

> N is the sum of the vertical components of forces (i.e. normal to the assumed surface of sliding)
>
> B is the assumed width of the sliding surface (taken as less than the actual width, because of the overturning moment—see below)
>
> P_t is the tensile force contributed by the anchorage heel

For the geometry shown

$$N = W_w + W_f + W_s + (P_a - P_p)\cos\delta' + S_1 \tag{9.8}$$

where

> W_w is the self-weight of the wall (typical bulk density values for a range of rock materials used in gabions are given in Table 9.4)
>
> W_f is the weight of backfill between the virtual back of wall (ab) and the actual back of the wall
>
> W_s is the (small) weight of soil between the sliding surface (ac) and the underside of the wall
>
> S_1 is the component of surcharge between the virtual back of the wall, at b, and the wall itself.

The example shown in Figure 9.15 incorporates an anchorage heel, which is introduced to increase the factor of safety against sliding. Because of its flexibility, it is assumed that the virtual back of wall originates from point a, rather than point e. The position of point a is found by assuming that the flexural rigidity of the gabion is only effective for a length equal to its

Table 9.4 Typical bulk densities for filled gabions, as a function of rock type

Type of rock used in gabion	Bulk density (kg/m³)
Basalt	1740–2030
Granite	1560–1820
Hard limestone	1560–1820
Calcareous gravel	1500–1750
Sandstone	1380–1610

Source: Officine Maccaferri S.p.A. *Flexible Gabion Structures in Earth Retaining Works.* Officine Maccaferri S.p.A. Bologna, Italy, 22 pp, 1987.

thickness, from the back of the wall. The tensile force (P_t) contributed by the anchorage heel can be estimated from

$$P_t = 2.L.\sigma'_v.\tan\phi' \tag{9.9}$$

where σ'_v is the vertical effective stress at the level of the anchorage heel.

If the gabion is placed on a concrete base, Officine Maccaferri (1987) suggests that the effective angle of friction between the concrete and the gabion should be taken as about 32°.

For the case of a reinforced-concrete cantilever wall, a key may be included if sliding resistance is a problem. In such a case, the soil resistance from the bottom of the key to the toe of the wall should be calculated, rather than the sum of the resistances of the underside of the base and the passive resistance of the key, when calculating the factor of safety against sliding.

Example 9.2: Calculate adequacy for sliding

Geometry

Consider a simple mass concrete gravity wall which is designed to retain 8 m of compacted sand and gravel with a horizontal ground surface. The proposed width of the wall is 4 m at its base and 1 m at its crest. The unit weight of concrete is expected to be 24 kN/m³.

Ground parameters

Estimated soil parameters are:

- For the sand and gravel, $\phi' = 44°$ and $\gamma = 20$ kN/m³
- For the underlying sand, $\phi' = 35°$ and $\gamma = 20$ kN/m³

The soil behind the wall is to be compacted with a vibratory roller giving an equivalent line load estimated as 50 kN/m width. Groundwater is below the base of the wall and the wall (which has a rough base) is to be founded on sand. Figure 9.16 shows the layout of the wall.

Check for resistance to sliding on the base of the wall.

If base width = 4 m

$$\alpha = 90° - \tan^{-1}\left(\frac{4-1}{9}\right) = 90° - 18.4° = 71.6°$$

For gravel, Rankine $K_a = 0.18$

For concrete or brick, CP2 gives $\delta' = 20°$

For active pressures on masonry walls, Rowe and Peaker (1965) give $\delta' \leq \phi'/2$

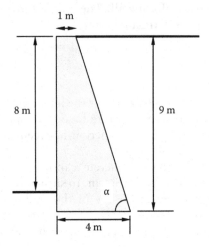

Figure 9.16 Layout of gravity wall.

∴ use $\delta' = 20°$

∴ Muller Breslau $K_a = 0.31$

Active force acts at $(20 + 18.4)°$ to the horizontal
Compaction pressures

$$\sigma' = \sqrt{\frac{2p\gamma}{\pi}} = \sqrt{\frac{2.50.20}{\pi}} = 25.2 \text{ kN/m}^2$$

$$z_c = K_a\sqrt{\frac{2p}{\gamma\pi}} = 0.23 \text{ m}$$

$$h_c = \frac{1}{K_a}\sqrt{\frac{2p}{\gamma\pi}} = 7.01 \text{ m}$$

At base of wall, $\sigma'_h = K_a\gamma.h \approx 32.4$ kN/m run.

Total thrust on wall (assuming conservatively that the active force is horizontal)

$$Q = \frac{0.23}{2}.25.2 + 25.2(6.78 + 1.99)$$
$$+ \frac{1.99}{2}7.22 = 2.9 + 221.0 + 7.2 = 231.1 \text{ kN/m run}$$

Weight of wall

$$W = 9.\frac{(4+1)}{2}.24 = 540 \text{ kN/m run of wall}$$

Base sliding resistance (recall that pore pressure u = 0)

P_s = W.tan δ′ = 540.tan (20°) = 540.0.364 = 196.5 kN/m run

∴ Sliding resistance is inadequate, as Q > Ps

9.2.6 Overturning and toppling

If the wall is reinforced and to be founded on rock or hard soil, a calculation should be made to ensure that an adequate factor of safety is achieved
against the wall rigidly rotating about the toe point 0 (see Figure 9.17).
A factor of safety of at least 2 is recommended (CP2), and may be checked
by taking moments of the various forces acting on the wall about the toe.
If unreinforced wall units (such as gabions, masonry blocks or cribs) are
to be used to form the wall, then a check on toppling failure can be made
by carrying out a similar calculation, but taking moments about the third
point of the base of the wall (Figure 9.7).

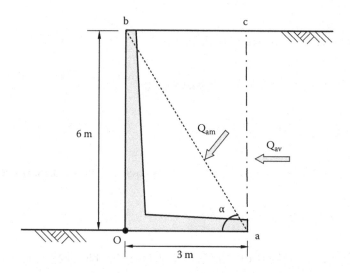

Figure 9.17 Example of two methods to calculate factor of safety on overturning.

For reinforced concrete cantilever walls, there is a need to decide where to take the 'virtual back of wall', as discussed above. Generally, for this type of wall, the most economical wall will be one in which the length of the heel is approximately twice the length of the toe. However, for the sake of illustration, an L-shaped wall is shown. The virtual back of wall could be taken as either ac or ab (Figure 9.17).

Example 9.3: Comparison of two methods of calculating a factor of safety on overturning

$$\beta = 0; \quad \gamma_{conc} = \gamma_{soil} = 20 \text{ kN/m}^3; \quad \phi' = 30°;$$
$$\delta' = 0 \text{ on ac}; \quad \delta' = \phi' \text{ on ab}; \quad \alpha = 63.4°$$

Factor of safety on overturning based on ac

$$Q_{av} = \frac{0.333}{2}.20.6^2 = 119.9 \text{ kN/m run}$$

Weight of soil = 6.3.20 = 360.0-kN/m run

$$F_{overturning} = \frac{360.0.\dfrac{3}{2}}{119.9.\dfrac{6}{3}} = 2.25$$

Factor of safety on overturning based on ab

$$Q_{am} = 0.610.\frac{1}{2}.20.6^2 = 219.6 \text{ kN/m run}$$

Vertical component,

$$Q_{amv} = 219.6 \ \cos(90 - (90 - \alpha) - \delta') = 219.6 \ \cos(63.4 - 30) = 183.3 \text{ kN/m run}$$

Horizontal component

$$Q_{amh} = 219.6 \cos((90 - \alpha) + \delta') = 219.6 \cos(26.6 + 30) = 120.9 \text{ kN/m run}$$

$$\text{Weight of soil} = \frac{6.3.20}{2} = 180.0 \text{ kN/m run}$$

$$F_{overturning} = \frac{\frac{2}{3}(3.183.3) + \frac{1}{3}(3.180)}{\frac{1}{3}(6.120.9)} = 2.26$$

Both methods give the same result, and the resistance to overturning about the toe is satisfactory

The general geometry of gravity walls was discussed in the introductory section of this chapter. Typically, base widths of 1/2 to 2/3 of the wall height are to be expected. It is unlikely that gravity walls will be constructed of mass concrete, because the additional cost of reinforcement is relatively low and allows a considerable saving in the section of the wall once tension can be allowed. Under this condition, a wall will remain stable provided that the resultant of all the forces on it (see Figure 9.17) passes within the base section. Applying a factor of safety to overturning moves the resultant back from the toe.

The base width for masonry, crib, gabion or block walls should be determined by ensuring that no tension exists in the structure. For this condition to be met, the resultant, R, of the external forces:

Q_a The 'active' force, calculated from earth pressure coefficients, or compaction theory

W The self weight of the wall

Q_p/F The factored passive force (usually ignored because of the possibility of excavation at the toe. If included, use minimum F = 2 on K_p)

must pass through the middle 1/3 of the base of the wall to avoid tensile stress (Figure 9.18). For zero vertical stress at the heel of the wall (i.e. a triangular distribution of contact pressure between the base and the soil supporting it), the resultant can be shown by simple statics to pass through the third point nearest to the toe of the wall. For this case, the factor of safety against overturning, which is found by taking moments about the toe, is between 2.5 and 3.0 for a simple rectangular wall. The required base width can be found by trial and error, using an initial width of 1/3 of the height of the wall.

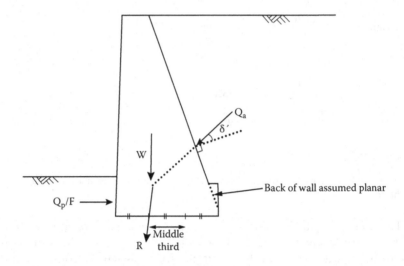

Figure 9.18 Middle third rule. (From Clayton, C. et al., *Earth Pressure and Earth-Retaining Structures*, Second Edition, Taylor & Francis, New York, 1993.)

9.2.7 Bearing capacity

The assessment of available bearing capacity is a complex problem, for two reasons. First, the base of a retaining wall, by reason of its intended use, can be expected to have both normal loads and large moments applied to it. There is a considerable disagreement in the literature about the prediction of bearing capacity for quite simple foundation loads, so that for retaining-wall foundations the precise available bearing capacity is uncertain. Second, in clays, bearing capacity is treated as a problem which is critical in the short-term. The soil load on a wall (due to active pressure) may well increase with time, however. Therefore, for a foundation supporting the loads from a retaining wall, the time at which the critical (i.e. lowest) factor of safety occurs is unknown. Since the wall loads may be applied rapidly, for example, by compaction, the short-term factor of safety against bearing-capacity failure under the maximum load should be calculated. For walls retaining *in situ* soils the long-term pressure (based on effective stress calculations) should be used. This will give a conservative design.

Short-term bearing capacity should only be used in saturated clays, because granular materials and very dry cohesive soils 'drain' as rapidly as the load can be applied. For short-term, '$\phi_u = 0$' analysis, the ultimate average bearing pressure beneath a vertically loaded foundation is

$$q_{ult} = p_0 + N_c \cdot c_u \tag{9.10}$$

where

q_{ult} is the ultimate bearing capacity

p_0 is the total vertical stress in the soil adjacent to the foundation at the founding level

N_c is a bearing capacity factor dependent on the geometry of the foundation

c_u is the average undrained shear strength of the foundation soil, from undrained triaxial compression tests on undisturbed samples.

Skempton (1951) has provided values of N_c for a strip footing, given in Table 9.5.

In the case of foundations for a gravity wall, the load will not be vertical. The components of the load can be resolved into a vertical eccentric force, and a horizontal force acting along the base (Figure 9.19).

Table 9.5 Bearing capacity factors for clay in the short-term

Ratio of depth of foundation to width D/B	Strip footing bearing capacity factor N_c
0	5.14
0.25	5.6
0.50	5.9
0.75	6.2
1.00	6.4

Source: Skempton, A.W., In *Proc. Building Research Congress*, London, 180–189, 1951.

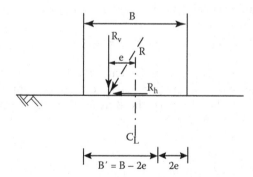

Figure 9.19 Bearing capacity for inclined eccentrically loaded strip footings (de Beer 1949; Meyerhof 1953). Effective width = $B' = B - 2e$. (From Clayton, C. et al., *Earth Pressure and Earth-Retaining Structures*, Second Edition, Taylor & Francis, New York, 1993.)

For a foundation loaded eccentrically on a cohesive soil in the short-term

$$q_{ult} = p_0 + N_c.c_u.f_{ci} \qquad (9.11)$$

where

q_{ult} is the ultimate bearing capacity

p_0 is the total vertical stress in the soil adjacent to the foundation at the founding level, which will often be small and negligible

N_c is Skempton's bearing capacity factor

c_u is the average undrained shear strength of the foundation soil, from undrained triaxial compression tests on undisturbed samples

f_{ci} is a factor to take account of eccentric and inclined loading.

According to Vesic (1975)

$$f_{ci} = 1 - \frac{2 R_h}{B' c_u N_c} \qquad (9.12)$$

where

R_h is the horizontal component of the loading

$B' = B - 2e$ (Figure 9.19)

so that

$$q_{ult} = c_u N_c - \frac{2R_h}{B'} + p_0 \qquad (9.13)$$

and the maximum vertical force per unit length of wall

$$R_{vult} = c_u N_c B' - 2R_h + p_0 B' \approx c_u N_c B' - 2 R_h \qquad (9.14)$$

For all other soils, the bearing capacity is calculated in terms of effective stress. For the basic solution of a strip foundation supporting a central vertical load, a modified form of the Buisman–Terzaghi equation is used (Buisman 1940; Terzaghi 1943):

$$q_{ult} = c' N_c + \frac{1}{2}\gamma B N_\gamma + p_0 (N_q - 1) + p_0 \qquad (9.15)$$

where

q_{ult} is the ultimate bearing capacity

c' is the effective cohesion intercept of the soil

γ   is the bulk unit weight of the soil

p_0 is the total vertical stress in the soil adjacent to the foundation at the founding level

N_c, N_g, N_q are bearing capacity factors, whose value depends on the geometry of the foundation (e.g. depth, strip, circle, square) and the effective angle of friction of the soil.

Figure 9.20 shows the classical bearing capacity solution by Prandtl (1921). Zone I is an active Rankine zone, Zone III is a passive Rankine zone, and the intermediate Zone (II) is termed the Prandtl zone. The strength of the soil above foundation level is ignored and its effect is considered solely as a surcharge, p_0. According to Vesic (1975), there is a reasonable agreement on the values of N_c and N_q but N_γ values are strongly dependent on the value of θ used to derive them, and values can be found in the literature which vary from one-third to double the values based upon θ = (45° + φ'/2) (the Rankine value—after Caquot and Kerisel, 1953), which are given in Table 9.6. The tabulated values assume no friction between the soil and the foundation, and are therefore conservative.

Once again, correction factors must be applied because the foundation load is neither vertical nor central. On the basis that

$$q_{ult} = c'N_c f_{ci} + \frac{1}{2}\gamma B N_\gamma f_{\gamma i} + p_0(N_q - 1)f_{qi} + p_0 \qquad (9.16)$$

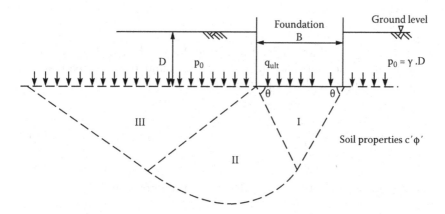

Figure 9.20 Bearing capacity of a shallow footing on drained soil. (From Clayton, C. et al., *Earth Pressure and Earth-Retaining Structures*, Second Edition, Taylor & Francis, New York, 1993.)

Table 9.6 Bearing capacity factors for long-term, effective stress calculations

Effective angle of friction of soil (degrees)	Bearing capacity factors[a]		
	N_c	N_q	N_γ
0	5.14	1.00	0.00
15	10.98	3.94	2.65
20	14.83	6.40	5.39
25	20.72	10.66	10.88
30	30.14	18.40	22.40
35	46.12	33.30	48.03
40	75.31	64.20	109.41
45	133.88	134.88	271.16

[a] Bearing capacity factors based on

$N_q = e^{\pi \tan \phi'} \tan^2(45° + \phi'/2)$

$N_c = (N_q - 1) \cot \phi'$

$N_\gamma = 2(N_q + 1) \tan \phi'$

Brinch Hansen (1961) and Sokolovski (1960) have proposed that

$$f_{qi} = \left[1 - \frac{R_h}{(R_v + B'c' \cot \phi')}\right]^2$$

$$f_{ci} = f_{qi} - \frac{(1 - f_{qi})}{(N_c \tan \phi')} \tag{9.17}$$

$$f_{\gamma i} = (f_{qi})^{3/2}$$

where R_h, R_v and B' are defined in Figure 9.19, as before. For a granular or normally consolidated soil (c' = 0), with a foundation approximately at ground level ($p_0 = 0$), these equations reduce to

$$q_{ult} = \frac{1}{2} \gamma B' N_\gamma \left(1 - \frac{R_h}{R_v}\right)^3 \tag{9.18}$$

The vertical force per unit length of wall at failure is then

$$R_{vult} = q_{ult} B' \tag{9.19}$$

where $B' = B - 2e$, as before.

The factor of safety of the wall against bearing capacity failure is calculated from

$$F = \frac{\text{Vertical force per unit length of wall } (q_{ult})}{\text{Applied vertical force per unit length of wall } (q_{app})} \qquad (9.20)$$

A minimum factor of safety of 2 has normally been required in lumped factor of safety calculations.

Example 9.4:   Check of factor of safety on bearing capacity failure

A masonry gravity wall has been sized so that the resultant of the active force and the self weight of the wall passes through the third point of the base nearest the toe. Its width is 5.8 m.

The horizontal resultant of these applied forces

R_h = 231 kN/m-run

The vertical resultant of these applied forces

R_v = 1166 kN/m-run

Passive pressure at the toe of the wall has been ignored. The effective angle of friction, ϕ', of the sand on which the wall is to be founded is 35°. Its bulk unit weight is estimated as 20 kN/m³. Calculate the factor of safety against bearing capacity failure.

$B' = 1 - 2 e = 1 - 2 \times B/6 = 2 B/3$

$R_h/R_v = 0.198$

Assuming $p_0 = 0$, and with $c' = 0$
For $\phi' = 35°$, $N_\gamma = 48.03$ (Table 9.6)

$F = q_{ult} \cdot B'/R_v = 958 \times 2/3 \times 5.8 = 3704/1166 = 3.18$

Therefore, the factor of safety against bearing capacity failure is satisfactory.

9.2.8 Overall stability

The various design stages in previous sections should have produced a design which will guarantee the stability of the wall and the soil in its immediate vicinity, but this may not be a guarantee against a failure involving a large mass of the soil surrounding the wall. Figure 9.21 gives a sketch

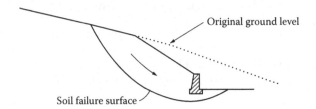

Figure 9.21 Example of loss of overall stability, as a result of regrading a slope. (From Clayton, C. et al., *Earth Pressure and Earth-Retaining Structures*, Second Edition, Taylor & Francis, New York, 1993.)

of an example of this problem, which typically involves long-term slope instability in clay soils.

The assessment of slope stability is complex, but must be tackled if the overall design is to be successful. Examples of various analyses are detailed in Chapter 5. In practice, it is important to assess the possibility that a preexisting shear surface exists in a position that will endanger the stability of the wall. The normal methods used to identify ground which is already unstable are:

- The use of air photographs—examples of unstable slopes are given in Clayton et al. (1995), and Chapter 5.
- Identification on the ground, during a site visit—unstable ground is characterised by its slope, coupled with poor drainage (i.e. 'boggy' ground), and sometimes by 'bent' trees, caused by ground movement during growth.
- Identification of smooth shear surfaces in trial pits or boreholes during ground investigation. Unless there is an engineer on site who is aware of the likelihood of unstable ground, it is easy to miss the very thin failure surfaces in a pit or borehole sample.

If a preexisting shear surface is located near to the retaining structure, it will be necessary to identify its geometry, together with the residual shear strength parameters, and the pore water pressure along the shear surface. The analysis will then probably be for a non-circular slip surface, the simplest form of which is by Janbu (see Chapter 5).

More commonly, it will be necessary to analyse a slope at the site of a proposed retaining structure where no preexisting instability can be detected. For this case, a circular slope stability analysis will normally be adequate. Peak effective strength parameters (c′, ϕ′) and estimated equilibrium pore water pressures are used in an effective strength (i.e. long-term) analysis such as Bishop's Routine Method.

Where the effective strength parameters are derived from triaxial tests on samples from the site, British Standard Code of Practice No. 2 recommended a minimum factor of safety for overall stability of 1.5. If the strength parameters are derived from a nearby failure in the same material, they are more likely to represent the strength of the soil *in situ*, CP2 then allowed a minimum factor of safety of 1.25.

BS 6031 (Code of Practice for Earthworks) is referred to by the UK retaining structures code (BS 8002 [1994]) with respect to global instability. Where the consequences of failure on neighbouring structures, railways, etc. are not particularly serious and there has been a good standard of site investigation, BS 6031 recommends the following safety factors against soil slips:

– For first-time slides: between 1.3 and 1.4
– For slides involving entirely preexisting slip surfaces, about 1.2

If the consequences of failure are serious, or if the investigation has been limited, higher safety factors are recommended.

9.2.9 Settlement and tilt

Settlement is normally only a problem because of the differential settlement and tilt which accompanies it. Forward-tilting wall faces look unsafe, and differential settlement along the length of a wall may cause concern and make the wall unsightly.

Unfortunately, it is very difficult to estimate accurately the settlement of foundations, even in well-known soil conditions and for the simple case of a foundation supporting a uniformly distributed vertical load. In the case of a retaining structure, where good information on the compressibility of the subsoil is likely to be scarce, it is unlikely that realistic estimates can be made. Nevertheless, differential settlement and tilting will occur. Tilting may be forward (away from the retained soil), as a result of the pressure from the backfill, or backward, or a result of settlement of a soft subsoil following the placement of backfill. Longitudinal differential settlement is inevitable, unless the wall is supported by piling, because of the natural variability of soil. It should be allowed for by the inclusion of construction joints at least every 6 m along the wall. Tilting should be allowed for by the use of a batter on the exposed face of the wall.

9.3 INTERNAL STABILITY

Each type of wall in this group requires special design considerations for internal stability, and must be designed in a different way. In this section, the observed mechanisms of wall failure are noted. The local actions of soil on the structure, potentially leading to damage or serviceability limit states, are noted and, where possible, quantified. Clearly, during detailed design, design of reinforced concrete, steel and other elements of the wall is required. This process is not considered here.

9.3.1 Masonry, gravity block and gabion walls

These types of wall consist of relatively strong blocks, with little or no strength at the joints between them. Their stability depends upon their self-weight. Therefore, it is necessary to check that tension is avoided on all horizontal junctions between blocks. This is done by ensuring that, at all levels, the resultant of earth pressure (including wall friction) and the self-weight of the wall above the section being considered passes through the middle-third of the wall at that level (the so-called mediaeval middle-third rule). It is clearly necessary that the factor of safety against sliding should also be adequate at all levels within the wall but, since the effective angle of friction between the bottom of the wall and the soil is normally much less than that between the wall components themselves, this is rarely a problem.

The strength of individual bricks or masonry blocks will also rarely be a problem, but for brick, a detailed design is necessary. Examples of structural design methods can be found in Haseltine and Tutt (1977).

The strength and compressibility of gabions may well be a concern, as noted by Thorburn and Smith (1985) and O'Rourke (1987). Gabion baskets may be manufactured from either polymer grid or galvanized steel. This section considers only galvanized steel gabions.

The internal stability of a gabion wall depends upon the precise type of mesh in use, and the way in which the gabions are laced together. Internal stability is therefore dependent upon the particular proprietary system in use, as well as quality of the workmanship used in its construction. Although lacing is a simple process, it is essential that it is carried out strictly to the manufacturer's recommendations, since design methods are empirical in nature, being based upon tests on full-scale gabions, and trial structures. Below, we give the design recommendations of the Italian gabion specialists, Maccaferri, as an example.

Internal stability checks are carried out to ensure that the maximum permitted stresses in the steel baskets and lacings are not exceeded. To this end, it is necessary to evaluate both the applied compressive stresses and the shear stresses on the wall, on a number of trial horizontal sections.

At any given section, the applied moment (M), horizontal resultant force (P) and vertical resultant force (N) due to all the forces acting above that level are calculated. The maximum vertical stress is evaluated on the basis of Meyerhof's expression:

$$\sigma_{max} = \frac{N}{B - 2e} \tag{9.21}$$

where
   e is the eccentricity of the point of action of the resultant of P and N
      (i.e. e = M/N)
   B is the full width of the wall at the particular height of interest.

σ_{max} must not exceed the empirically determined allowable stress for the particular gabion rock and mesh to be used, i.e.

$$\sigma_{max} \le \sigma_{am} \tag{9.22}$$

For a Maccaferi gabion

$$\sigma_{am} \approx 100 \left(\frac{5\gamma_g}{1000} \right) - 3 \tag{9.23}$$

where
   σ_{am} is the maximum allowable vertical stress, in kPa
   γ_g is the density of the gabion filling, in kg/m^3 (see Table 9.4).

The average shear stress, T/B, must not be greater than the allowable:

$$\tau_{am} = \frac{N \tan\phi *}{B} + C_g \tag{9.24}$$

$\phi*$ has been found experimentally to be related to the density of the rock in the gabion

$$\phi* = 25 \frac{\gamma_g}{1000} - 10° \tag{9.25}$$

and C_g is related to the amount of steel mesh per unit volume of gabion, P_u (in kg/m^3), by the expression

$$C_g = 3P_u - 5 \text{ (kPa)} \tag{9.26}$$

The value of C_g normally varies from about 15 to 50 kPa.
   The expected deformation of a steel mesh gabion wall can be predicted on the basis of empirical equations. Tests by Maccaferri, both on individual

gabions and on full-scale structures, have shown that the wall deforms mainly in simple shear. For each course of gabions, the outward deflection of the top edge will be approximately

$$\Delta\zeta = \Delta h \frac{T/B}{G} \tag{9.27}$$

where
  $\Delta\zeta$ is the outward displacement of the top of a single layer of gabions
  Δh is the height of the layer of gabion for which the calculation is being carried out
  T/B is the shear stress in that course
  G is the shear modulus of the gabions.

Test results have indicated that G typically varies from 250 to 400 kPa. The deflection of the top of the wall is calculated by summing the deflection, $\Delta\zeta$, of each layer of gabions. It is normal to select the backward tilt of the wall in order to ensure that after backfilling the front face remains battered backward, toward the fill.

9.3.2 Crib walls

Although, externally, crib walls behave as a simple gravity structure, their internal loading is more complex. The Hong Kong Geoguide 1 (GCO 1982) notes that 'the manufacturers of crib wall units produce design data for crib walls, but in general, care must be exercised in the interpretation and application of this data'.

Except where the wall is placed for landscaping purposes alone, the minimum front-to-back thickness of a crib should be about 1 m. As was noted in Chapter 6, the wall consists of

  i. Stretchers, which are the horizontal members parallel to the wall face; for open-faced walls the height of the stretchers should be less than one-half of their width, in order to avoid soil spilling through the face of the wall;
  ii. Headers, which connect the front and back lines of stretchers.

In addition, 'pillow blocks' or 'false headers' may be required midway between the stretchers, at the base of high walls.

Crib walling was developed in New Zealand. New Zealand Specification CD209, 'Crib walling and notes' (Ministry of Works and Development 1980), gives recommendations for the strength and testing of crib units, and the construction of crib walls. The internal stability of a crib wall is usually analysed by assuming that the wall acts as a series of silos, with the soil contained in each cell providing the vertical and horizontal forces to be resisted (both in bending and shear) by the individual crib elements. Silo theory is only strictly applicable if the silo walls do not displace significantly either in the vertical or the horizontal direction sufficiently to reduce arching action. Given that it is common to use compacted, relatively incompressible fill within the crib cells, it is doubtful if silo action can be guaranteed, and the horizontal forces on the crib components will then be relatively larger, while the vertical forces will be smaller. Schuster et al. (1975) have observed that the stresses measured in crib wall units are much higher than would be predicted on the basis of silo theory. New Zealand Code CD209 requires crib wall components to be able to withstand twice the loadings given by silo theory, following examinations of both satisfactory and unsatisfactory crib walls. Wu (1975) gives the average horizontal stress on the wall of a silo, at depth y, as

$$\sigma = \frac{ab}{2(a+b)} \frac{\gamma K_0}{\tan \delta'} \left[1 - \exp\left[-\frac{2\gamma K_0 (a+b)}{ab} \tan \delta' \right] \right] \qquad (9.28)$$

The shear stress between the soil and the wall is

$$\tau = \sigma \tan \delta' \qquad (9.29)$$

and the total vertical force transmitted by friction is

$$F = ab\gamma \left[H - \frac{ab}{2(a+b)\tan \delta' K_0} \left[1 - \exp\left(-\frac{2(a+b)}{ab} H \tan \delta' K_0 \right) \right] \right] \qquad (9.30)$$

where
    a and b are the internal length and width of the crib wall cell
    γ is the bulk unit weight of the fill
    K_0 is the coefficient of earth pressure at rest
    H is the height of the wall.

Various design methods for crib walls can be found in the literature (see, for example, Tschebotarioff 1962, 1973; Wu 1975; GCO 1982). The stretcher facings must be designed to resist both horizontal (outward) and vertical (downward) bending, and shear. The headers must be designed to hold the front and rear stretchers together, resisting the tensile force by shear on the hammer head, and to support the weight of the soil imposed on them in bending. The bearing of stretchers and headers, one on top of the other, should be checked to ensure that at the lower part of the wall, the crushing strength of the material (usually concrete or timber) is not exceeded. Pillow blocks are packed between the mid spans of the headers when such problems are foreseen, although this should only be necessary for high structures.

The failure of crib walls has been reported to occur for a number of reasons. Because drainage of the backfill is necessary for their external stability, it is essential that only free-draining coarse granular soil, or crushed rock, is used in the cells. It is normally specified that this should be heavily compacted (Hong Kong Geoguide [GCO 1982] suggests to 98% of Proctor [i.e. BS 1377 1990 Part 4, Cl. 3.3] density). Observed failures of walls have occurred because of poor detailing of reinforcement within the crib components, and due to differential settlements within the crib wall sections, leading to fracturing of headers (Tschebotarioff 1973). They might occur if the wall were erroneously assumed to be incompressible (O'Rourke 1987), but was actually moving downward relative to the retained earth.

9.3.3 Reinforced concrete cantilever walls

Structural design of a reinforced-concrete cantilever wall is relatively straightforward. The wall is considered to be composed of three structural components (i.e. the stem, the heel and the toe), each of which is designed as a cantilever (Figure 9.22).

The stem should be designed taking into account compaction pressures, and the fact that the resultant force from these cannot exceed the sum of the base sliding resistance and passive resistance above the toe. As stated previously, no reliable theory exists to predict the maximum pressures applied by cohesive materials, but since it is known that these materials can apply very high lateral pressures and have been responsible for failures of the stem due to excessive bending moment (Ingold 1979), the use of compacted clay should be avoided. Compaction theory cannot yet be modified for a sloping ground surface, and so the ground surface must be assumed to be horizontal and some arbitrary increase in the height of retained fill should be made if this case arises.

The toe and heel are designed on the basis of the simplified pressure distributions shown in Figure 9.22. For sands, these give rise to overpredictions of bending moment, but for clays, some under-prediction may occur. The exact pressure distribution cannot be computed.

Standard concrete and masonry cantilever wall details can be found in Newman (1976). These include full dimensions and steel detailing for a wide range of everyday wall applications. Alternatively, this type of wall may also be purchased as precast units.

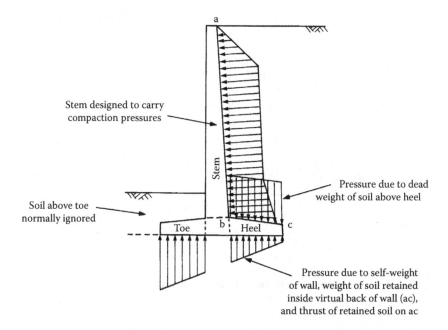

Figure 9.22 Pressure distributions for the design of stem, heel and toe of a reinforced-concrete cantilever wall. (From Clayton, C. et al., *Earth Pressure and Earth-Retaining Structures*, Second Edition, Taylor & Francis, New York, 1993.)

9.4 CALCULATIONS TO EUROCODE 7 FOR A GRAVITY WALL

As Chapter 7 has noted, Eurocode 7 requires five limit states to be considered in design. This section considers design to avoid the three following limit states:

- Overturning (EQU in EC7)
- Sliding (GEO in EC7)
- Bearing capacity (GEO in EC7)

that affect the sizing of gravity retaining walls.

Figure 9.23 shows the forces on an unreinforced masonry block wall, used in the following examples. For simplicity, it is usual to ignore the passive pressure at the toe of the wall since it generally makes only a small contribution to the overall resistance, and as noted above, this generally

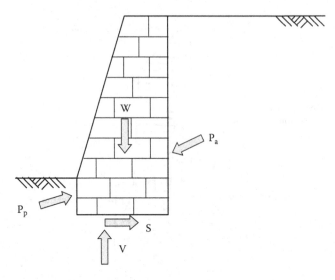

Figure 9.23 Forces on an unreinforced masonry block wall.

errs on the side of caution. However, in this example, passive pressure has been retained for illustrative purposes.

9.4.1 Partial factors

The UK National Annex states that Design Approach 1 should be used, with two combinations of partial factors:

Combination 1 (A1 + M1 + R1)
Combination 2 (A2 + M2 + R1)

as given in Table 9.7. Group A provides partial factions on actions (in this case from forces), group M provides partial factors on material properties, and group R provides partial factors on resistances. Combination 2 normally controls geotechnical design, but Combination 1, which is more critical for structural design, should routinely be checked.

Partial factor design avoids failure, and provides reliability, by using characteristic (moderately conservative) values of soil parameters (denoted with a subscript 'k'), combined with material partial factors, to calculate design values of the soil parameters (denoted with a subscript 'd'), that are then used in calculating forces. In addition, partial factors may be applied to both actions (driving forces) and to resistances. However, in Design Approach, 1 all partial factors on resistances are set to 1.0.

Table 9.7 EC7 partial factors for Design Approach I

Partial factors for Design Approach I	Combination I	Combination 2
Partial factors on actions	**A I**	**A2**
$\gamma_{G;dst}$	1.35	1.0
$\gamma_{G;stb}$	1.0	1.0
$\gamma_{Q;dst}$	1.5	1.3
$\gamma_{Q;stb}$	0	0
$\gamma_{A;dst}$	1.0	1.0
$\gamma_{A;stb}$	0	0
Partial factors on material properties	**M I**	**M2**
$\gamma_{\phi'}$	1.0	1.25
$\gamma_{c'}$	1.0	1.25
γ_{cu}	1.0	1.4
γ_{qu}	1.0	1.4
γ_{γ}	1.0	1.0
Partial factors on resistances	**R I**	**R I**
γ_{Rv}	1.0	1.0
γ_{Rh}	1.0	1.0
γ_{Re}	1.0	1.0

Notes: γ_A is a partial factor on the effect of accidental actions; γ_γ is a partial factor on self-weight, which is always unity, and is neglected below.

9.4.2 Overturning or toppling

As discussed above, overturning need only be checked where a reinforced gravity wall is founded on rock. Toppling has traditionally been checked by using similar calculations, but with moment equilibrium calculated by taking moments about the third point of the base.

To ensure sufficient margin of safety against toppling, the following inequality is checked:

$$\gamma_{G;dst}.P_a.l_a \le W.l_w + P_p.l_p \tag{9.31}$$

where

$\gamma_{G;dst}$ is the partial factor on the destabilising permanent active force P_a

P_a is the resultant force from active pressure, calculated using the design values of the soil parameters, ϕ' and δ', i.e. the characteristic (moderately conservative) values reduced by the appropriate (Material, M) partial factors

l_a is the lever arm of P_a, acting about the third point of the base of the wall

W is the self-weight of the masonry block wall section

l_w is the lever arm of W, acting about the third point

P_p is the resultant force from passive pressure acting on the toe, again calculated using the design values of the soil parameters, but in this case with wall friction at the toe (δ') conservatively set at zero

l_p is the lever arm of the passive force, P_p, acting about the toe.

9.4.3 Sliding

Sliding is checked by comparing the sum of the horizontal destabilising forces with the sum of the resistances

$$\gamma_{G;dst} \cdot P_{a,h} \le R_h + P_p \tag{9.32}$$

where

$\gamma_{G;dst}$ is a partial factor on the (permanent, unfavourable) horizontal component of the design active thrust, P_a

$P_{a,h}$ is the horizontal component of the force due to active earth pressure, which acts at δ' to the horizontal. This is calculated using the design value of the effective angle of friction of the soil. Tanϕ' for design is the characteristic value of tan ϕ' divided by the partial factor γ_ϕ (i.e. γ_ϕ is applied to tan ϕ', rather than to ϕ' itself)

γ'_ϕ is the partial factor on tan ϕ'

R_h is the (horizontal) resistance to sliding on the base, calculated using the design value of ϕ'

P_p is the (horizontal) passive resistance at the toe, again calculated using the design value of ϕ', with $\delta' = 0$.

9.4.4 Bearing capacity

The adequacy of the wall base width for bearing capacity requires the vertical component of the resultant force from self-weight and earth pressure, acting on the base of the wall, to be less than the vertical resistance. The passive pressure is omitted because with $\delta' = 0$, it has no vertical component.

$$\gamma_{G;dst} \cdot (W + P_{a,v}) \le R_V \tag{9.33}$$

where

W is the self-weight of the wall

$P_{a,v}$ is the vertical component of the force due to active earth pressure, calculated using the design value of ϕ'

R_v is the vertical bearing capacity, again calculated using the design value of ϕ', obtained by factoring tanϕ'_k by the partial factor γ_ϕ'.

In the calculations above, only two values of partial factor have been used; $\gamma_{G;dst}$ and γ'_ϕ. For Design Approach 1—Combination 1, $\gamma_{G;dst}$ = 1.35 and $\gamma'_\phi = 1.0$. For DA1—Combination 2, $\gamma_{G;dst}$ = 1.0 and $\gamma'_\phi = 1.25$.

Example 9.5: Calculation for gravity wall, using EC7 approach

Notes: The example calculations are carried out for toppling (not defined in EC7 but arguably an equilibrium limit state—EQU), for sliding on the base of the wall (a geotechnical limit state [GEO]), as well as bearing capacity failure. EC7 Section 9 requires the designer to check other limit states.

Dimensions, material properties and surcharge details are given in Figure 9.24. The forces on the wall, used in this example, are shown in Figure 9.25.

1. V_w results from the weight of the wall.
2. V_a and H_a result from the active earth pressures exerted as a result of the self-weight of the backfill.
3. V_s and H_s result from the design surcharge (characteristic value = 10 kPa) and are variable actions (i.e. forces that vary from time to time).

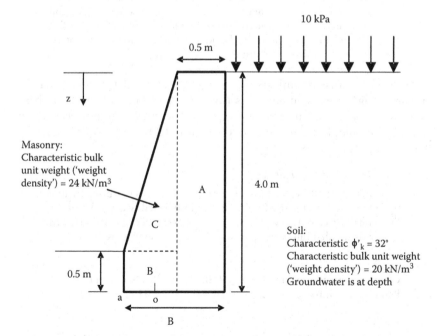

Figure 9.24 Dimensions and material properties for example calculation.

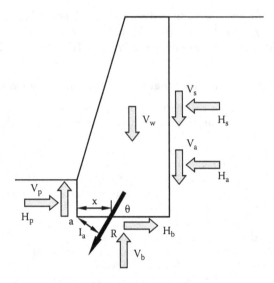

Figure 9.25 Forces on example unreinforced masonry gravity wall.

4. V_p and H_p are components of the passive resistance of the toe of the wall, mobilised as a result of forward movement on the base. Design values of geometry should take into account the possibility of excavation at the toe, according to EN 1997-1, Cl. 9.3.2.2.
5. V_b and H_b are the vertical and horizontal components of the force on the base, resulting from the above.

Since the wall is unreinforced, the line of action of the active, passive and self-weight forces should pass through the middle third of the base, to avoid toppling. In this calculation, a trial width of the base B = 2.5 m is used. The middle third starts at 'o' (Figure 9.24) which is B/3 = 2.5/3 = 0.83 m from 'a'.

Calculate weight of wall, $V_{w,k}$

The characteristic vertical load per m run of wall is obtained by multiplying the characteristic bulk unit weight (termed 'weight density' in EC7) by the area of the wall section. For convenience of calculation the section is divided into three parts: A, B and C—see Figure 9.24.

For A
Characteristic vertical load of A = 4. 0.5. 24 = 48 kN/m run of wall.
Point of action from 'O' = 2.5 – 0.83 – 0.5/2 = 1.42 m

For B
Characteristic vertical load of B = 0.5. 2. 24 = 24 kN/m run of wall.
Point of action from 'O' = 2/2 – 0.83 = 0.17 m

For C

Characteristic vertical load of C = (3.5. 2)/2. 24 = 84 kN/m run of wall.

Point of action from 'O' = (2. 2/3) – 0.83 = 0.5 m

The characteristic vertical force resulting from the characteristic self-weight of the wall

$$V_{w,k} = 48 + 24 + 84 = 156 \text{ kN/m run of wall.}$$

In calculations below, the design value $V_{w,k}$ is obtained by applying the relevant partial factor, $\gamma_{G;stb}$, which is 1.0 for both Combination 1 and Combination 2 (Table 9.7).

Check for adequacy of base width for toppling using Design Approach 1, Combination 1 (A1 '+' M1 '+' R1)

The loads on the wall are shown in Figure 9.25. Table 9.8 divides them, as required by EC7, into permanent or variable loads, and favourable or unfavourable loads for this limit state. To avoid toppling, the resultant of the self-weight of the wall, and the active and passive forces, should pass within the middle third section of the wall base.

Table 9.8 Division of loads into 'permanent/variable' and 'favourable/unfavourable' for assignment of partial factors for Design Approach I, Combination I

Load	Description	Permanent or variable (P or V)	Favourable or unfavourable (F or U)	Partial factor to apply
V_w	Self-weight of wall	P	F	$\gamma_{G;stb} = 1.0$
V_a	Vertical component of active force due to backfill	P	F	$\gamma_{G;stb} = 1.0$
H_a	Horizontal component of active force due to backfill	P	U	$\gamma_{G;dst} = 1.35$
V_s	Vertical component of active force due to surcharge	V	F	$\gamma_{Q;stb} = 0$
H_s	Horizontal component of active force due to surcharge	V	U	$\gamma_{Q;dst} = 1.5$
V_p	Vertical component of passive force	P	$V_p = 0$ as $\delta_{passive} = 0$	Not required
H_p	Horizontal component of passive force	P	F	$\gamma_{G;stb} = 1.0$
V_b	Vertical component of bearing capacity of base	Not used in this calculation		
H_b	Sliding resistance of base	Not used in this calculation		

Note: The partial factors on material properties are all set to 1.0 for calculations under Design Approach I, Combination I.

Calculate active horizontal force, due to self-weight of backfill, $H_{a;k}$

Equation C1 in Appendix C of EC7 is used to find the earth pressure normal to the wall, from which H_a can be calculated. Separating the effect of surcharge and applying the earth pressure coefficient in Appendix C of EC7.

At any depth, z

$$\sigma_a(z) = K_a \gamma z - 2c \sqrt{K_a} \tag{9.34}$$

The value of the active earth pressure can be found for the relevant value of the soil parameters, from the earth pressure coefficients in Appendix C of EC7.

$c' = 0$ (sand backfill)

$\phi'_k = 32°$

The design value of ϕ' is obtained from the characteristic value by applying the relevant partial factor:

$$\phi'_d = \tan^{-1}(\tan\phi'_k / \gamma_{\phi'}) = \tan^{-1}(\tan 32°/1.0) = 32°$$

Assuming $\delta/\phi' = 0.5$, $K_a = 0.27$ for $\phi'_d = 32°$.

Since $c' = 0$ in this case, at the top of the wall, $\sigma_a = 0$ kPa. At the base of the wall, depth of 4 m

$$\sigma_a(4) = K_a \gamma z = 0.27 \times 20 \times 4 = 21.6 \text{ kPa}$$

Therefore

$H_{a;k} = 1/2 \times 21.6 \times 4 = 43.2$ kN/m-run of wall.

Calculate active vertical force, due to self-weight of backfill, $V_{a;k}$

The shear force on the back of the wall is obtained from

$V_{a;k} = H_{a;k} \tan \delta' = 43.2 \tan (32/2) = 12.4$ kN/m run of wall.

Horizontal force due to surcharge, $H_{s;k}$

From Equation C.1 in Appendix C of EC7

$$\sigma_s(z) = K_a q$$

where q is a uniform surcharge applied behind the wall at the ground surface. Therefore, the stress normal to the back of the wall is independent of depth. As before, a partial factor of 1.0 is applied to $\tan \phi'$, and the resulting force H_{sk} is then

$$H_{s;k} = K_a \, qh = 0.27 \times 10 \times 4.0 = 10.8 \text{ kN/m-run of wall.}$$

Calculate vertical force due to surcharge, $V_{s;k}$

The shear force on the back of the wall due to the surcharge is obtained from

$$V_{s;k} = H_{s;k} \tan \delta' = 10.8 \tan (32/2) = 3.1 \text{ kN/m-run of wall.}$$

Calculate horizontal force due to passive resistance, $H_{p;k}$

The passive pressure at the toe of the wall is obtained using Equation C.2 and the attached graphs from EC7 Appendix C. For $\delta/\phi' = 0$ and $\phi' = 32°$, $K_p \approx 3.1$. Since $c' = 0$ and there is no surcharge at the toe, at the surface the passive pressure is zero. At 0.5 m,

$$\sigma_p(0.5) = K_p \, \gamma z = 3.1 \times 20 \times 0.5 = 31.0 \text{ kPa}$$

at 0.5 m depth.

The horizontal force resulting from this triangular pressure distribution is

$$H_{p,k} = \sigma_p \, (0.5) \, h/2 = 31.0 \times 0.5/2 = 7.8 \text{ kN/m run of wall.}$$

Calculate vertical force due to passive resistance, $V_{p;k}$

As $\delta = 0$, this force also equals zero.

Design check for toppling

To satisfy EC7 equilibrium check, the overturning moments of the design values of the forces about the toe should be exceeded by the restoring moments of the design values. For this calculation, however, the middle-third rule is used to ensure no tension at the base of the wall, so moments are taken about 'O' rather than the toe.

Thus, the requirement is that

$$H_{a;k} l_{ha} \gamma_{G,dst} + H_{s;k} l_{hs} \gamma_{Q,dst} \leq V_{w;k} l_w \gamma_{G,stb} + V_{a;k} l_{va} \gamma_{G,stb} + V_{s;k} l_{vs} \gamma_{Q,stb} + H_{p;k} l_{hp} \gamma_{G,stb}$$

$$43.2 \times \frac{4}{3} \times 1.35 + 10.8 \times \frac{4}{2} \times .5 \leq [48 \times 1.42 + 24 \times 0.17 + 84 \times 0.5] \times 1.0 + 12.4 \times$$

$$(2.5 - 0.83) \times 1.0 + 3.1 \times (2.5 - 0.83) \times 0.0 + 7.8 \times \frac{0.5}{3} \times 1.0$$

$$77.8 + 32.4 \leq [68.2 + 4.1 + 42] \times 1.0 + 20.7 + 0 + 1.3$$

$$110.2 \leq 135.0$$

The wall is acceptable under Design Approach 1, Combination 1.

Check for adequacy of base width against toppling using Design Approach 1, Combination 2 (A2 '+' M2 '+' R1)

The partial factors on actions (forces) shown in Table 9.9 below are applied to characteristic loads to calculate design loads.

For design Approach 1, Combination 2, the partial factors on material properties are

$\gamma_{\phi'}$	1.25
$\gamma_{c'}$	1.25
γ_{cu}	1.4
γ_{qu}	1.4
γ_{γ}	1.0

Calculate weight of wall, $V_{w,k}$

As before, the characteristic value of force resulting from the weight of the wall

$$V_{w,k} = 48 + 24 + 84 = 156 \text{ kN/m run of wall.}$$

Table 9.9 Division of loads into 'permanent/variable' and 'favourable/unfavourable' for assignment of partial factors for Design Approach I, Combination 2

Load	Description	Permanent or variable (P or V)	Favourable or unfavourable (F or U)	Partial factor to apply
V_w	Self-weight of wall	P	F	$\gamma_{G;stb} = 1.0$
V_a	Vertical component of active force due to backfill	P	F	$\gamma_{G;stb} = 1.0$
H_a	Horizontal component of active force due to backfill	P	U	$\gamma_{G;dst} = 1.0$
V_s	Vertical component of active force due to surcharge	V	F	$\gamma_{Q;stb} = 0$
H_s	Horizontal component of active force due to surcharge	V	U	$\gamma_{Q;dst} = 1.3$
V_p	Vertical component of passive force	P	$V_p = 0$ as $\delta_{passive} = 0$	Not required
H_p	Horizontal component of passive force	P	F	$\gamma_{G;stb} = 1.0$
V_b	Vertical component of bearing capacity of base	Not used in this calculation		
H_b	Sliding resistance of base	Not used in this calculation		

Calculate active horizontal force, due to self-weight of backfill, $H_{a;k}$

The design value of the effective angle of friction

$$\phi'_d = \tan^{-1}(\tan\phi'_k / \gamma_{\phi'}) = \tan^{-1}(\tan 32°/1.25) = 26.6°$$

Assuming $\delta/\phi' = 0.5$, $K_a \approx 0.36$ for $\phi'_d = 26.6°$. (Appendix C of EC7)

$$\sigma_a \text{ (at 4m)} = K_a \gamma z = 0.36 \times 20 \times 4 = 28.8 \text{ kPa}$$

Therefore

$$H_{a;k} = 1/2 \times 28.8 \times 4 = 57.6 \text{ kN/m run of wall.}$$

Calculate active vertical force, due to self-weight of backfill, $V_{a;k}$

The shear force on the back of the wall is obtained from

$$V_{a;k} = H_{a;k} \tan \delta' = 57.6 \tan (26.6/2) = 13.6 \text{ kN/m run of wall.}$$

Calculate horizontal force due to surcharge, H_s

As before, Equation C.1 in EC7 gives

$$\sigma_s(z) = K_a q \qquad\qquad (9.35)$$

where q is a uniform surcharge applied behind the wall at the ground surface. Therefore the stress normal to the back of the wall is independent of depth. A partial factor of 1.25 is now applied to $\tan \phi'$, and the resulting force H_{sk} is then

$$H_{s;k} = K_a qh = 0.36 \times 10 \times 4.0 = 14.4 \text{ kN/m run of wall.}$$

Calculate vertical force due to surcharge, V_s

The shear force on the back of the wall due to the surcharge is obtained from

$$V_{s;k} = H_{s;k} \tan \delta' = 14.4 \tan (26.6/2) = 3.4 \text{ kN/m run of wall.}$$

Calculate horizontal force due to passive resistance, H_p

The passive pressure at the toe of the wall is obtained using Equation C.2 and the attached graphs from EC7 Appendix C. For $\delta'/\phi' = 0$ and $\phi' = 26.6$, $K_p \approx 2.6$. Since $c' = 0$ and there is no surcharge at the toe, at the surface the passive pressure is zero. At 0.5 m,

$$\sigma_p(0.5) = K_p \gamma z = 2.6 \times 20 \times 0.5 = 26.0 \text{ kPa}$$

at 0.5 m depth.

The horizontal force resulting from this triangular pressure distribution is

$$H_{p,k} = \sigma_p(0.5) \, h/2 = 26.0 \times 0.5/2 = 6.5 \text{ kN/m-run of wall.}$$

Design check

As before, partial factors are now applied to characteristic forces (actions) to obtain design forces (actions).

$$H_{a;k}l_{ha}\gamma_{G,dst} + H_{s;k}l_{hs}\gamma_{Q,dst} \le V_{w;k}l_w\gamma_{G,stb} + V_{a;k}l_{va}\gamma_{G,stb} + V_{s;k}l_{vs}\gamma_{Q,stb} + H_{p;k}l_{hp}\gamma_{G;stb}$$

$$57.6 \times \frac{4}{3} \times 1.0 + 14.4 \times \frac{4}{2} \times 1.3 \le [48 \times 1.42 + 24 \times 0.17 + 84 \times 0.5] \times 1.0$$

$$+13.6 \times (2.5 - 0.83) \times 1.0 + 3.4 \times (2.5 - 0.83) \times 0.0 + 6.5 \times \frac{0.5}{3} \times 1.0$$

$$76.8 + 37.4 \le [68.2 + 4.1 + 42] \times 1.0 + 22.7 + 0 + 1.1$$

$$114.2 \le 138.3$$

The wall is acceptable under Design Approach 1, Combination 2.

Check for sliding on base (GEO) using Design Approach 1, Combination 2 (A2 '+' M2 '+' R1)

The wall is to be checked for sliding on its base by comparing the sum of the unfavourable horizontal actions, H_a and H_s, with the favourable actions, H_b and H_p.

As before, for Combination 2,

$$H_{a;k} = 57.6 \text{ kN/m run of wall}$$
$$H_{s;k} = 14.4 \text{ kN/m run of wall}$$
$$H_{p;k} = 6.5 \text{ kN/m run of wall}$$

Sliding is also resisted by the base. The shear force on the soil/wall interface is a function of the normal force resulting from the other forces on the wall, V_w, V_a and V_s (V_p is zero, as δ'_p has been assumed to be zero, conservatively), so that

$$V_b = V_w + V_a + V_s = 156.0 + 13.6 + 3.4 = 173.0 \text{ kN/m run of wall}$$

On the base, from EC7 cl. 9.5.1(6)

$$\delta_k = k.\phi'_{cv;d}$$

Assuming

— concrete cast against the soil, $k = 1.0$, and
— $\phi'_{cv;d} = \phi'_d = 26.6°$

$$H_{b;k} = V_b \tan(26.6) = 86.5 \text{ kN/m run of wall}$$

Therefore, to satisfy horizontal equilibrium

$$H_{a;k}\gamma_{G;dst} + H_{s;k}\gamma_{Q;dst} \le H_{b;k}\gamma_{G;stb} + H_{p;k}\gamma_{G;stb}$$
$$57.6 \times 1.0 + 14.4 \times 1.3 \le 86.5 \times 1.0 - 6.5 \times 1.0$$
$$76.3 \le 80.0$$

The wall satisfies the sliding requirement, under Design Approach 1, Combination 2.

Check for bearing capacity (GEO) using Design Approach 1, Combination 2 (A2 '+' M2 '+' R1)

The forces derived using characteristic values of soil parameters were found in the previous calculations. Their lever arms about the toe are as follows:

Force	Force (kN/m run of wall)	Lever arm about a (m)
W_a	48.00	2.25
W_b	24.00	1.00
W_c	84.00	1.33
$H_{a;k}$	57.60	1.33
$V_{a;k}$	13.59	2.50
$H_{s;k}$	14.40	2.00
$V_{s;k}$	3.40	2.50
$H_{p;k}$	6.50	0.16
$V_{p;k}$	0	0

Figure 9.25 shows the resultant of the applied forces. The method used to find its point of action on the base is as follows:

1. Determine horizontal resultant, R_h
2. Determine vertical resultant, R_v
3. Determine inclination (θ) of resultant of R to the vertical
4. Determine sum of moments of forces about 'a' in Figure 9.25
5. Calculate lever arm, la
6. Calculate point of application of R along base, $x = l_a/\sin\theta$.

Resolving,

$$R_h = H_a + H_s - H_p = 57.60 + 14.40 - 6.50$$
$$= 65.50 \text{ kN/m run of wall}$$

$$R_v = V_w + V_a + V_s - V_p = (48.00 + 24.00 + 84.00) + 13.59 + 3.40 - 0$$
$$= 172.99 \text{ kN/m run of wall}$$

$$R = \left(R_h^2 + R_v^2\right)0.5 = 184.98 \text{ kN/m run of wall}$$

$$\theta = \tan^{-1}(R_v/R_h) = 69.26°$$

Taking moments,

M (clockwise +ve about 'a')

$$V_w.l_w - H_a.l_{ha} + V_a.l_{va} - H_s.l_{hs} + V_s.I_{vs} + H_p.l_{hp}$$
$$= 181.83 \text{ kN m/m run of wall}$$

Lever arm, l_a = M/R = 181.83/184.98 = 0.98 m
Point of application along base = l_a/sin θ = 1.05 m along base.
For a granular soil with a foundation at approximately ground level
$(p_0 = 0)$,

$$q_{ult} = \frac{1}{2}\gamma B'N_\gamma \left[1 - \frac{R_h}{R_v}\right]^3$$

In this case

B' = B − 2 e = 2.5 − 2 × (2.5/2 − 1.05) = 2.5 − 0.4 = 2.1 m

R_h/R_v = 65.5/172.99 = 0.379

$$N_\gamma = 2(N_q + 1)\tan\phi'$$
$$N_q = e^{\pi\tan\phi'}\tan(45° + \phi'/2)$$

so that N_q = 12.59 and N_γ = 11.58 for $\phi'_d = 26.6°$

$q_{ult;d}$ = 1/2 × 20 × 2.1 × 11.58 × (1 − 0.379) = 151.01 kPa

For sufficient bearing capacity

$R_v\gamma_{G;dst} \le q_{ult}B'$
172.99 × 1.0 ≤ 177.09 × 2.1
173.00 ≤ 317.45

Therefore base width is satisfactory for bearing capacity, under
Design Approach 1, Combination 2.

Chapter 10

Embedded walls

Eurocode 7 (EN 1997-1:2004) defines embedded walls as 'relatively thin walls of steel, reinforced concrete or timber, supported by anchorages, struts and/or passive pressure'. These walls support the load applied by retained soil through bending and shear, rather than through the weight of the wall. Wall types in this group include

- Sheet-pile walls, typically used for waterfront structures, or temporary excavations
- Braced excavations, frequently used for the installation or repair of services
- Diaphragm or bored pile walls, used for deep basement excavations in inner city sites, or for cut-and-cover metro construction (see Chapter 6)

10.1 SELECTION OF SOIL PARAMETERS

In the previous chapter, we noted that many gravity walls will be designed to retain backfill. The situation is quite different when designing embedded walls, since (with the exception of backfill on part of the side of waterfront structures—see, for example, Figure 3.2), these are generally installed in natural ground. This section reviews the parameters to be used for embedded walls.

10.1.1 *In situ* soil parameters

When selecting soil parameters for use in design, it is as well to recall (from Chapter 2) that

- Whilst it is relatively easy to obtain samples of cohesive soils and determine the effective angle of friction, ϕ', from laboratory testing, it is extremely difficult to obtain reliable values of effective cohesion intercept, c'. Yet c' has a significant effect on calculated earth pressures. For normally consolidated soils or backfill, the effective cohesion intercept, c', should be taken as zero.

- It is not practical to take samples of granular soils for laboratory testing, and therefore estimates of effective angle of friction, ϕ', for granular soils will rely on correlations with the results of *in situ* tests such as the Standard Penetration Test and the Cone Penetration Test. For the reasons described previously, such estimates (although conservative) cannot be regarded as particularly accurate.
- Estimates of *in situ* pore water pressures need to be conservative, given the significant effects that they have on wall stability.

10.1.2 Wall friction

Numerous recommendations have been given for determining the angle of wall friction for use in embedded retaining wall calculations, as summarised in Table 10.1.

Table 10.1 Recommended values of wall friction for use in embedded wall calculations

Reference	Surface	Wall friction, tan δ	
		Active	Passive
Terzaghi (1954)	Steel	$\tan(\phi'/2)$	$\tan(2\phi'/3)$
CIRIA 104 (1984)	Any	$\tan(2\phi'/3)$	$\tan(\phi'/2)$
EAU	Steel	$\tan(2\phi'/3)$	$\tan(2\phi'/3)$
Piling Handbook (1997)	Steel	Usually ignored	$(2/3) \tan \phi'$
BS 6439:1984	Steel or masonry	$\tan(2\phi'/3)$	Up to $\tan(2\phi'/3)$
BS 8002	Any	$(3/4) \tan \phi'$	$(3/4) \tan \phi'$
Canadian Foundation Engineering Manual (1978)	Steel	$\tan(11-22°)$	
	Cast concrete	$\tan(17-35°)$	
	Pre-cast concrete	$\tan(14-26°)$	
CIRIA C580 Eurocode 7	Steel	$\tan(2\phi'_{cv}/3)$	
	Cast concrete	$\tan(\phi'_{cv})$	
	Pre-cast concrete	$\tan(2\phi'_{cv}/3)$	

Source: Bond, A. and Harris, A., *Decoding Eurocode 7*. Taylor and Francis, 2008.

The significant effect of these recommendations has been discussed in Chapter 3. In design, it is important to adhere to the detailed recommendations of the relevant code of practice. Puller and Lee (1996) note that BS8002 gives recommendations with regard to wall friction and adhesion without reference either to the need for relative (vertical) soil/wall movement, or to the roughness of the wall material.

10.1.3 Wall adhesion

CP2 was a total stress code, with wall adhesion values recommended (unconservatively) as a proportion of the undrained shear strength of clays. In more modern practice, c' is often taken as zero because of the remoulding that occurs as the wall is installed (by driving, augering or using a grab) into the soil. BS 8002:1994 recommends laboratory shear box testing to determine 'wall friction, base friction and undrained wall adhesion'. However, undrained analysis should not be used. In practice, laboratory tests are not carried out on clays, and BS8002 states that 'no effective adhesion c' should be taken for walls or bases in contact with soil'.

10.2 PRELIMINARY DESIGN

Preliminary design involves consideration of drainage, and making preliminary estimates of wall length and support requirements. Design charts are generally useful for making estimates for relatively low cost structures, for example, sheet-pile walls used for river protection or for quay walls, where ground conditions are variable and may never be known with great certainty.

10.2.1 Drainage and control of water

As with all earth retaining structures, the loads on embedded walls are very much affected by the water levels adjacent to the wall and in the ground. There are two mechanisms (see Figure 10.1):

1. The presence of water at the same level on both the front and back of the wall decreases the effective stress levels in the ground, relative to a wall placed above the groundwater table. Active pressures and passive pressures are decreased, but because $K_a < 1$ and $K_p > 1$, the total pressure on the active side of the wall (which the wall has to resist) is increased, and the total pressure on the passive side of the wall (which provides support) is reduced. The wall therefore needs to be longer, all other things being equal.
2. If there is an imbalance between groundwater levels on the active and passive side, then a net pressure can result in a destabilising force, for

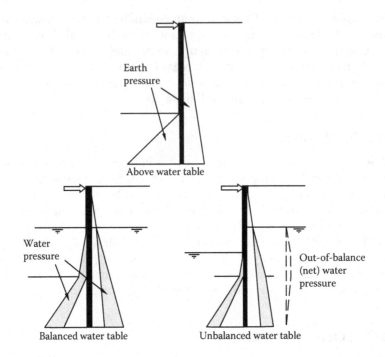

Figure 10.1 Effects of groundwater on the pressures applied to a propped embedded wall in uniform granular soil.

example, when there is a lag between the groundwater level on the active side and the water level in a dock during a falling tide. Again, all other things being equal, a longer wall will be needed than would have been the case had the imbalance not existed.

10.2.2 Design charts for waterfront structures

10.2.2.1 Cantilever sheet-pile walls

Figure 10.2 shows a design chart for estimating the depth of embedment (d) of a cantilever sheet-pile wall retaining soil of height h, under dry or fully submerged conditions. The curves on this figure are approximated by:

$$\frac{d}{h} = F_d \left(\frac{2}{3} \right) e^{\Omega} \qquad (10.1)$$

where
   $\Omega = (\phi' - 30°)/18°$ and
   F_d is the factor of safety on embedment (see Section 7.3.1).

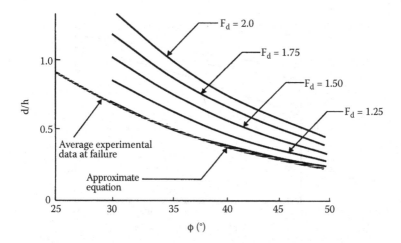

Figure 10.2 Design chart for preliminary estimation of depth of embedment of cantilever sheet-pile walls under dry or fully submerged conditions. (From Bica, A.V.D. and Clayton, C.R.I., *The Preliminary Design of Free Embedded Cantilever Walls in Granular Soil.* Proc. Int. Conf. on Retaining Structures, Cambridge, 1992.)

The exponent Ω is based on soil's plain strain effective angle of shearing resistance, which, for preliminary design purposes, can be assumed to be 10% higher than the value obtained from triaxial testing in dense soils, and equal to the triaxial value in loose soils.

The previous equation may be rearranged to give the wall's retained height h as a proportion of its total length (L = h + d) as follows:

$$h' = \frac{h}{L} = \frac{h}{h+d} = \frac{1}{(1+d/h)} = \left[1 + F_d\left(\frac{2}{3}e^{-\Omega}\right)\right]^{-1} \tag{10.2}$$

and the maximum retained height h_{max} for a given wall length L is then obtained by setting the factor of safety F_d to 1:

$$h'_{max} = \frac{h_{max}}{L} = \frac{h_{max}}{h_{max}+d} = \frac{1}{(1+d/h_{max})} = \left[1+\left(\frac{2}{3}\right)e^{-\Omega}\right]^{-1} \tag{10.3}$$

The relationship between h'_{max} and ϕ' is described by the bottom line of Figure 10.3.

Finally, the retained height h may be normalised by the maximum retained height h_{max} to give 'reduction factors' λ:

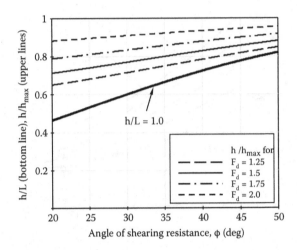

Figure 10.3 Normalised design chart for cantilever embedded retaining walls under fully drained or fully submerged conditions.

$$\lambda = \frac{h}{h_{max}} = \frac{h'}{h'_{max}} = \frac{1+(2/3)e^{-\Omega}}{1+F_d(2/3)e^{-\Omega}} \qquad (10.4)$$

The relationship between λ and ϕ' is described by the upper lines of Figure 10.3 for different values of F_d.

A preliminary estimate of the maximum retained height of a wall of length L may be obtained by reading the appropriate values of h'_{max} and λ from Figure 10.3 and then calculating

$$h = \lambda h'_{max}L \qquad (10.5)$$

Example 10.1

Determine the total length of wall needed to retain $h = 3$ m of dry soil with an effective angle of friction, $\phi' = 35°$, with a factor on embedment $F_d = 1.75$.

Figure 10.3 gives

$$h'_{max} = 0.66 \text{ and } \lambda = 0.86$$

Since $h = \lambda h_{max} L$

$$L = \frac{h}{\lambda h'_{max}} = \frac{3.00}{0.66.0.86} = 5.27 \text{ m}$$

The reason for separating the relationship between depth of embedment and angle of shearing resistance into two components is revealed when the exercise is repeated for different groundwater conditions and different definitions of the factor of safety. An analysis of the earth pressure diagram shown in Figure 10.4 leads to the following expression for d/h:

$$\frac{d}{h} = \frac{\left(1+\dfrac{\Delta h}{h}\right)(1+CC)}{\sqrt[3]{\dfrac{\dfrac{K_p}{f_p}(1-r_u)+f_w r_u}{f_a K_a (1-r_u)+f_w r_u}-1}} + \frac{\Delta h}{h} \tag{10.6}$$

where

   Δh is the height of any unplanned excavation
   K_a and K_p are the active and passive earth pressure coefficients
   r_u is a pore pressure coefficient, equal to 0 for dry conditions and γ_w/γ for saturated conditions
   γ_w is the bulk unit weight (weight density in EC7) of water
   γ is the bulk unit weight (weight density) of the soil
   f_p is the safety factor applied to passive earth pressures (via K_p)
   f_a is the safety factor applied to active earth pressures (via K_a)
   f_w is the safety factor applied to water pressures (active and passive) and
   CC is the 'cantilever correction', allowing for simplification of earth pressures below the pivot point.

The maximum bending moments to be taken by the sheet piling can, similarly, be estimated directly from published measurements. The maximum design bending moment increases with depth of embedment

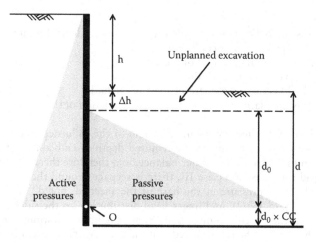

Figure 10.4 Earth pressure diagram for cantilever wall.

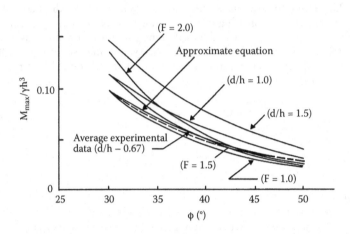

Figure 10.5 Design chart for preliminary estimation of maximum bending moment applied to free embedded cantilever walls under dry or fully submerged conditions. (From Bica, A.V.D. and Clayton, C.R.I., *The Preliminary Design of Free Embedded Cantilever Walls in Granular Soil.* Proc. Int. Conf. on Retaining Structures, Cambridge, 1992.)

(d) for any given retained height of soil (h). Therefore, when estimating the maximum bending moment from Figure 10.5, it is necessary to know either the value of d/h or of F_d. The value of the maximum bending moment can also be obtained from the approximate equation:

$$\frac{M_{max}}{\gamma h^3} = 0.095\,e^{-\left(\frac{\phi'-30°}{16°}\right)}\,e^{\left[\left(\frac{d}{2h}\right)-\left(\frac{1}{3}\right)\right]}$$

(10.7)

where
  γ is the bulk or buoyant density of the soil, depending upon the groundwater conditions
  h is the retained height of the soil
  d is the depth of embedment of the wall
  (d + h) is the total length of sheet piling and
  ϕ′ is the plane strain effective angle of friction of the soil

Figure 10.6 allows some appreciation of the influence of different groundwater conditions on the required depth of embedment. Cases I, II and IV have groundwater balance, and therefore there is no flow around the sheets. In Case III, there is seepage around the sheet piling; the pore pressure at the base of the sheet (d, not d′) has been taken as $\gamma_w(d + h/2)$, and linear head distribution up the sides of the sheet has then been assumed. In making a preliminary estimate of pile length, it should be remembered that the overturning moment is dominated by the extent of groundwater imbalance, while the restoring

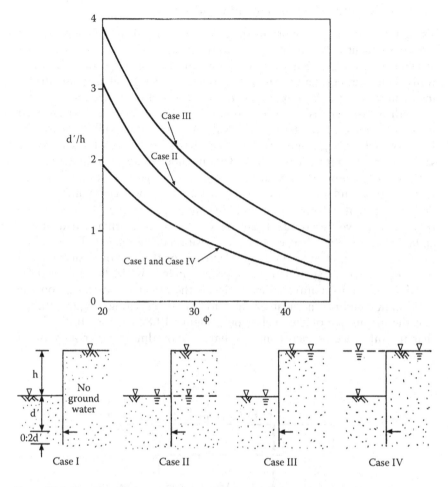

Figure 10.6 Graph to estimate preliminary value of d′ (not d) for cantilever sheet-pile wall calculations (F_p = 1.5).

moment is predominantly a function of the effective angle of friction and groundwater level below excavation level. The charts have been produced using factors of safety on passive pressure of 1.5, Mayniel's and Caquot and Kerisel's earth pressure coefficients, and a bulk density of 20 kN/m³.

From Figure 10.6, it can be seen that cantilever sheet-pile walls are best suited to low retained heights (<4.5 m) and embedment in soils with a high effective angle of friction (i.e. sands and gravels). When clay exists below dredge level and there is a large groundwater imbalance, depths of penetration rapidly become uneconomical. Under these conditions, an anchored sheet-pile wall will be preferable.

10.2.2.2 Singly anchored sheet-pile walls

Using the free earth support method (see Section 10.3.4.1), coupled with Rowe's moment reduction method (Rowe 1952, 1957), Hagerty and Nofal (1992) have produced charts for the preliminary design of anchored sheet-pile walls. The charts give all the necessary design output for sheet-piling driven into a uniform free-draining granular soil. The soil (and therefore effective strength parameters) is considered uniform, and the water level is assumed to be the same on both sides of the wall. Although not explicitly stated by Hagerty and Nofal, it appears that their charts are derived for a factor of safety on passive pressure (i.e. by the CP2 method, Chapter 7) equal to 2.

Figure 10.7 shows the geometry assumed by Hagerty and Nofal (1992). Figures 10.8a and 10.8b show the normalised depth of embedment (D/z_d) and the normalised maximum bending moment $(M_{max}/\gamma z_d^3)$, after reduction using Rowe's method. It can be seen that the normalised depth of embedment is not much affected by the position of the anchor. The bending moment to be allowed for in design decreases as the anchor position drops (i.e. z_a/z_d increases), such that it is approximately halved when z_a/z_d is 0.4.

Additional information is available on the effect of surcharge on the maximum bending moment applied to the sheet piling (Figure 10.8c), and the magnitude of the anchor pull (Figure 10.8d). Figure 10.8c shows that the influence of increasing surcharge is an almost linear increase in

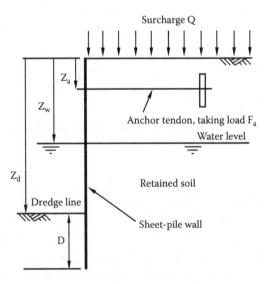

Figure 10.7 Definition of geometry for preliminary design charts for anchored sheet-pile walls. (From Hagerty, D.J. and Nofal, M.M., *Normalisation of Analytical Results for Anchored Bulkhead Design.* Proc. Int. Cont. on Retaining Structures, Cambridge, 1992.)

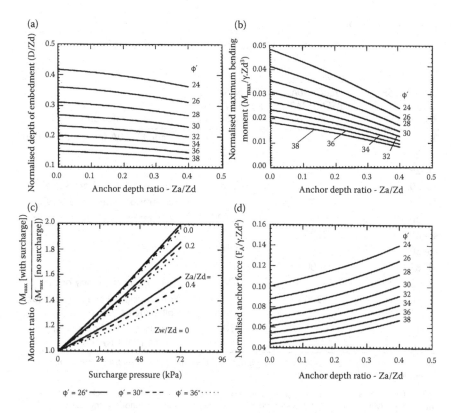

Figure 10.8 (a) Normalised depth of embedment for anchored sheet-pile walls. (b) Normalised maximum bending moment for anchored sheet-pile walls. (c) Effect of surcharge on maximum bending moment applied to sheet-pile walls. (d) Normalised anchor force for anchored sheet-pile walls. (From Hagerty, D.J. and Nofal, M.M., *Normalisation of Analytical Results for Anchored Bulkhead Design*. Proc. Int. Cont. on Retaining Structures, Cambridge, 1992.)

the maximum bending moment. Hagerty and Nofal report that similar relationships were found for all calculated output. This fact may be useful in the preliminary design stages, if the precise magnitude of surcharge remains unknown.

10.2.3 Sheet-pile drivability

The construction of sheet-pile walls requires not only that the sheet-piles are sufficiently strong to take the bending moments and shear forces that will be applied once the wall is completed; the steel section must be strong enough to resist driving into the soil without buckling and de-clutching. As the strength of the soil increases, the length of sheet-pile and the section

Table 10.2 Sheet-pile drivability

Dominant SPT (blows/ 300 mm)	Minimum wall modulus cm³/m of wall		Remarks
	BSEN 10 025 grade 430A, BSEN 4360 grade 43A	*BSEN 10 025 grade 510A, BS 4360 grade 50A*	
0–10	450		Grade FE510A for lengths >10 m
11–20		450	
21–25	850		
26–30		850	Lengths >15 m not advisable
31–35	1300		Penetration >5 m not advisable[a]
36–40		1300	Penetration >8 m not advisable[a]
41–45	2300		
46–50		2300	
51–60	3000		
61–70		3000	Some declutching may occur
71–80	4200		Some declutching may occur for pile lengths >15 m
81–140		4200	Increased risk of declutching Some piles may refuse

Source: Williams, B.P. and Waite, D., *The Design and Construction of Sheet-piled Cofferdams*. Special Publication 95, CIRIA, London, 1993.

Note: This table is based on sheet-pile sections of approximately 500-mm interlock centres, installed with panel driving techniques. Wider sections, and those installed by methods giving less control, will require greater minimum moduli.

[a] If the stratum is of greater thickness use a larger section of pile.

required to support the earth is reduced. But the size of the section needed to allow it to be driven into the ground is increased. Little is known about the necessary sections required to allow sheet-piles to be driven without damage. However, Williams and Waite (1993) suggests that in sands and gravels the minimum section can be judged on the basis of SPT N value, as given in Table 10.2.

10.3 DESIGN OF SHEET-PILE WALLS USING LIMIT EQUILIBRIUM CALCULATIONS

Simple hand calculations have traditionally been used to determine the necessary depth of embedment of walls, and the bending moments, shear forces, prop loads and anchor forces that must be resisted by these structures. Increasingly these can be carried out using commercial computer packages, such as WALLAP (by Geosolve) and ReWaRD (by Geocentrix). However, hand checking of software output remains important, and this section describes some methods that can then be used.

The hand calculation methods described below were originally developed for the design of waterfront structures:

- cantilever sheet-pile walls and
- singly-anchored sheet-pile walls

They were modified for analysis of diaphragm and bored-pile walls, which became widely used for deep basement and highway underpass excavations, and were more often used with multiple prop or anchor levels. Nowadays analysis of these types of structure is increasingly carried out using finite element, finite difference and Winkler spring methods (see Chapter 8).

10.3.1 Limit states and definitions of factor of safety

At least six ultimate limit states need to be considered during the design of embedded walls:

- Failure as a result of insufficient depth of embedment
- Failure of the wall section, because of bending or shear stress
- Failure of anchor, anchor system (e.g. anchor rod) or wall connection, or prop
- Overall instability
- Piping
- Base heave

In the past, there has been considerable debate about the most suitable way to provide a margin of safety when sizing a retaining structure. The various methods that have been proposed have been discussed in Chapter 7 (Section 7.5.1). This chapter adopts a factor of safety on effective strength parameters, which is applied to soil both on the retained and excavated side of the wall. Factors of safety suggested in CIRIA Report C580 (2003) have been superseded by those in *Eurocode 7* Design Approach 1, Combination 2.

10.3.2 Effect of groundwater

Where sheet-pile walls are used as waterfront structures they will be subjected to out-of-balance water pressures as a result of a lag between groundwater level (on the retained side of the wall) and the water in the river or dock, because of tidal or flood conditions. Fluctuations in the groundwater level on the retained (landward) side of the wall should normally be less than the tidal variation.

For routine, limit-equilibrium analyses, it is normal to consider the resultant of the unequal water pressures on either side of the wall as a separate out-of-balance pressure distribution. Effective vertical stresses in the soil, and therefore effective horizontal stresses, are calculated on the basis of a simplified, hydrostatic, groundwater pressure/depth distribution. A correction is then made, if necessary, to the active and passive forces acting on the wall, in order to allow for the effects of seepage.

Figure 10.9 shows simplified net pressure distributions for two possible out-of-balance water pressure distributions, depending on whether or not the bottom of the sheet piling penetrates impermeable ground. Table 10.3 gives the recommendations for estimating the effect of groundwater on sheet-pile walls, according to BS 6349: Part 1 (2000).

More rigorous solutions can be obtained by using flow-net sketching, or preferably finite element seepage analyses, to obtain the head distribution around the sheet piling. The pressure distribution on the face of the sheet piling can then be calculated from the head distribution via Bernoulli's equation (assuming the velocity component of head to be negligible).

Upward seepage may have a significant effect in reducing passive resistance at the toe, perhaps requiring more than the simple approach suggested

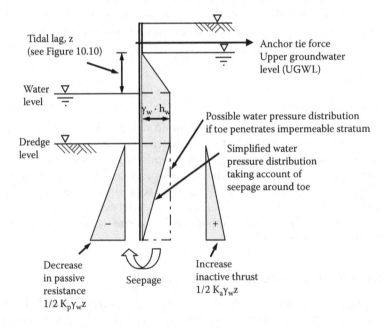

Figure 10.9 Simplified net water pressure distributions for two out-of-balance groundwater scenarios.

Table 10.3 Hydrostatic water pressure distributions for different tidal variations and drainage for sheet-piles driven into impermeable soil

Case	Description	Upper groundwater level	Differential water level, z
I	Minor non-tidal water-level variations, with weephole drainage provided.	LW + 0.5 m	0.5 m
II	High flood flows in non-tidal rivers, with weephole drainage provided	Most unfavourable between (LW + z) and HW	Max. predicted fall in 24 h
III	Large tidal variations, with no drainage provided.	Mean tide level i.e. 1/2 (MHWS + MLWS)	1/2 (MHWS − MLWS), up to an extreme of 1/2 (MHWS + MLWS) − ELW
IV	Large tidal variations, with flap valve drainage provided.	Flap valve invert level + 0.3 m	UGWL − MLWS up to an extreme of UGWL − ELW

Source: BS 6349, *Codes of Practice for Maritime Structures, Part 1*. General Criteria. British Standards Institution, London, 2000.

Note: ELW, Extreme low-water level—the lowest water level expected during the life of the structure, normally with a return period of not less than 50 years for permanent works. Reduced safety factors may be used when calculations use ELW. HW, Seasonal high-water level, in non-tidal water; LW, Seasonal low-water level, in non-tidal water; MHWS, Mean high-water spring-tide level—the average, over a long period of time, of two successive high waters at spring; MLWS, Mean low-water spring-tide level—as above, but low-water level; UGWL, Upper groundwater level, on landward side; z, Design water-level difference.

above. Solutions by Soubra and Kastner (1992), derived from the log spiral method, are shown in Figure 10.10, and can be used in place of the approximation suggested in Figure 10.9. These solutions are based upon the realistic approximation that in uniform ground conditions all head loss will occur over a length of sheet piling equal to twice the depth of penetration below dredge level. Thus, if the head loss from one side of the sheets to the other is H, and the depth of penetration is f, the maximum hydraulic gradient in the flow region will approximately equal H/(2 f). It can be seen that, for a constant effective angle of friction, K_p decreases linearly as H/f increases, becoming zero when H/f = 3. For non-zero values of K_p the reduction is a function of the effective angle of the friction (Figure 10.10b). Figures 10.10c and 10.10d give K_p values for $\delta'/\phi' = -1/3$ and for $\delta'/\phi' = -2/3$.

10.3.3 Cantilever sheet-pile walls

According to Head and Wynne (1985), the majority of retaining walls being constructed in the 1980s (up to 75%) were of the cantilever type (Figure 10.11). This is probably still true today. Little is known of their actual behaviour, but they are designed as a special case of sheet-pile walls, using

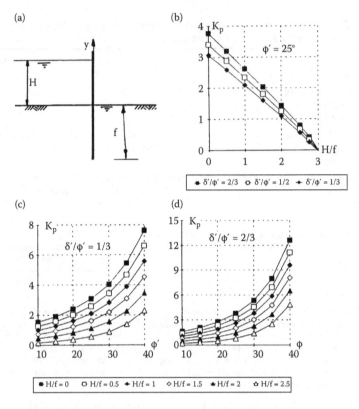

Figure 10.10 Effect of upward seepage on passive earth pressure coefficient, K_p. (a) Wall geometry. (b) K_p as a function of H/f. (c) K_p as a function of ϕ' for $\delta'/\phi' = 1/3$. (d) K_p as a function of ϕ' for $\delta'/\phi' = 2/3$. (From Soubra, A.H. and Kastner, R., *Influence of Seepage Flow on the Passive Earth Pressures*. Proc. ICE Conf. on Retaining Structures, Cambridge, 1992.)

the method of Blum (1931) which is also sometimes used for the 'fixed earth support' design of singly-anchored sheet-pile walls (Section 10.3.4). Preliminary design charts have been provided in Section 10.2.2.

The 'fixed earth support method' derives from the work of Lohmeyer (1930) and Blum (1950, 1951) and has been widely used in continental Europe since the 1930s. It is suitable for determining the rotational stability of embedded cantilever walls. The concentrated force simplification (illustrated in Figure 10.18) was also devised by Blum, although it is sometimes credited in some modern textbooks to Terzaghi. In the UK, the Institution of Structural Engineers' Code of Practice No. 2 (1951) advocated the use of the fixed earth support method for routine design of all types of embedded wall. Following the work of Rowe (1952) and Terzaghi (1954), a modified version of the 'free earth support method' (described in Section 10.3.4) has become more popular for the design of propped or anchored walls.

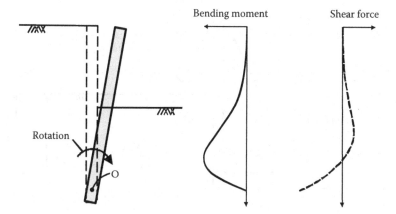

Figure 10.11 Assumed mode of movement (left) and resulting structural effects (right) under fixed earth conditions.

'Safety' has traditionally been introduced into the moment calculation by ensuring that the ratio of the restoring moment about point 'O' (M_R) to the overturning moment (M_O) does not drop below a specified factor of safety F_p (where the subscript 'p' indicates that the passive earth pressures have, in effect, been factored down). Values of F_p that have been recommended in the literature are summarized in Table 10.4.

The 'fixed earth support method' assumes (Figure 10.11) that the embedded wall is fixed at some point 'O' below formation and that the wall rotates as a rigid body about that point. The wall relies on earth pressures generated over its embedded length to maintain both horizontal and moment equilibrium. The assumed mode of wall movement is illustrated in Figure 10.11 (left) and leads to the structural effects (bending moments and shear forces in the wall) shown in Figure 10.11 (right).

Above the point of fixity 'O', ground on the retained side of the wall is assumed to go into an active state and that on the restraining side into a passive state. As a result of wall rotation the earth pressures bearing on the wall decrease from at-rest (K_0) to active (K_a) values on the retained side; and increase toward fully passive (K_p) on the restraining side, as illustrated in Figure 10.12. Limiting conditions are not mobilized at the same time on both sides—passive earth pressures normally require much greater

Table 10.4 Recommended values of F_p

Source	Recommended value of F_p
CP2 (1951)	2.0
Canadian Foundation Engineering Manual (1978)	1.5
Teng (1962)	1.5–2.0
CIRIA Report 104 (1984)	1.2–2.0

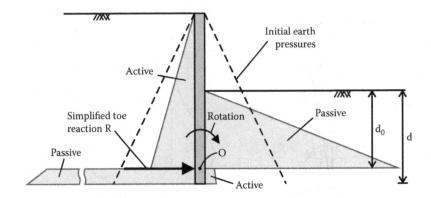

Figure 10.12 Earth pressures at limiting conditions assuming fixed earth support and rotation about point 'O'.

movement of the wall than active earth pressures to come fully into play. Hence the actual earth pressures that bear on the wall under working conditions may be very different from their limiting values.

The situation shown in Figure 10.11 is normally simplified by replacing the earth pressures below O with an equivalent reaction R (Figure 10.12) and ignoring the wall and the earth pressures applied to it below this point. This modification is known as 'concentrated force simplification' (after Blum 1931). The depth of embedment (d_O) required to ensure moment equilibrium about the point of fixity is then increased to compensate for this assumption (i.e. $d > d_O$). CIRIA Report 104 (1984) recommends using $d = 1.2 \times d_O$ for design purposes (after Tschebotarioff 1973). This is conservative since values of d/d_O are typically around 1.15.

Example 10.2:   Determination of depth of embedment of a cantilever wall using the 'fixed earth support' method

GEOMETRY

Consider a cantilever wall which is to retain h = 3 m of medium dense dry sand. A surcharge q = 10 kPa acts at the top of the wall.

Find the depth of embedment needed to provide a factor of safety $F_p = 2$ on passive earth pressures.

GROUND PARAMETERS

The sand's bulk unit weight is $\gamma = 18$ kN/m³ and its effective strength parameters are $\phi' = 35°$ and c' = 0 kPa. Assume an angle of wall friction, under active conditions, of $\delta'_a = \phi'/2 = 17.5°$ and, under passive conditions, $\delta'_p = 2\phi'/3 = 23.3°$ (see Figure 10.13).

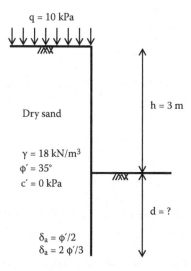

q = 10 kPa

Dry sand

γ = 18 kN/m³
φ′ = 35°
c′ = 0 kPa

h = 3 m

d = ?

$\delta_a = \phi'/2$
$\delta_a = 2\,\phi'/3$

Figure 10.13 Geometry and soil parameters for Example 10.2.

Mayniel's active earth pressure coefficient $K_a = f(\phi',\delta'_a) = 0.25$.
Caquot and Kerisel's passive earth pressure coefficient $K_p = f(\phi',\delta'_p) = 4.09$.

EARTH PRESSURE CALCULATIONS (FIRST ITERATION)

Assume a trial depth of embedment d = 2.3 m (see Example 10.1), from which we derive $d_0 = d/1.2 = 1.92$ m.

Active earth pressures on the retained side of the wall are

at a1: $z_{a1} = 0$ m

$\sigma_{a1} = K_a \times q = 0.25 \times 10 = 2.5$ kPa

at a2: $z_{a2} = h + d_0 = 3.0 + 1.92 = 4.92$ m

$\sigma_{a2} = K_a \times [q + \gamma \times (h + d_0)] = 0.25 \times [10 + 18 \times 4.92] = 24.2$ kPa

Passive earth pressures on the restraining side of the wall are

at p1: $z_{p1} = 3$ m

$\sigma_{p1} = 0$ kPa

at p2: $z_{p2} = h + d_0 = 4.92$ m

$\sigma_{p2} = K_p \times \gamma \times d_0 = 4.09 \times 18 \times 1.92 = 141.2$ kPa

Moment equilibrium calculation (taking moments about point 'O')

Overturning moment

$$M_O = \sigma_{a1} \times \frac{2}{3} \times (h + d_O) + \sigma_{a2} \times \frac{1}{3} \times (h + d_O)$$

$$= 2.5 \times \frac{2}{3} \times 4.92 + 24.2 \times \frac{1}{3} \times 4.92 = 47.8 \text{ kNm/m}$$

Restoring moment

$$M_R = \sigma_{p1} \times \frac{2}{3} \times d_O + \sigma_{p2} \times \frac{1}{3} \times d_O$$

$$= 0 \times \frac{2}{3} \times 1.92 + 141.2 \times \frac{1}{3} \times 1.92 = 90.2 \text{ kNm/m}$$

Actual factor of safety on passive earth pressures is

$$F_{actual} = \frac{M_R}{M_O} = 1.89$$

Embedment is insufficient, since $F_{actual} <$ recommended F_p

EARTH PRESSURE CALCULATIONS (FURTHER ITERATION)

By iteration, a factor of safety $F_p = 2$ is found when the depth of embedment d = 2.4 m, from which we derive d_0 = d/1.2 = 2.0 m.
    Active earth pressures on the retained side of the wall are then

at a_2: $z_{a2} = h + d_0 = 3.0 + 2.0 = 5.0$ m

$\sigma_{a2} = K_a \times [q + \gamma \times (h + d_0)] = 0.25 \times [10 + 18 \times 5.0] = 24.6$ kPa

Passive earth pressures on the restraining side of the wall are

at p_2: $z_{p2} = h + d_0 = 5.0$ m

$\sigma_{p2} = K_p \times \gamma \times d_0 = 4.09 \times 18 \times 2.0 = 147.3$ kPa

Moment equilibrium calculation (taking moments about point 'O')

Overturning moment

$$M_O = \sigma_{a1} \times \frac{2}{3} \times (h + d_O) + \sigma_{a2} \times \frac{1}{3} \times (h + d_O)$$

$$= 2.5 \times \frac{2}{3} \times 5.0 + 24.6 \times \frac{1}{3} \times 5.0 = 49.2 \text{ kNm/m}$$

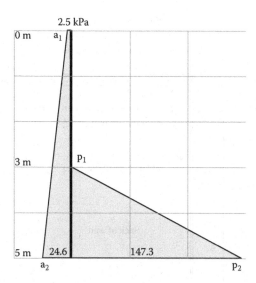

Figure 10.14 Earth pressure diagram for Example 10.2.

Restoring moment

$$M_R = \sigma_{p1} \times \frac{2}{3} \times d_O + \sigma_{p2} \times \frac{1}{3} \times d_O$$

$$= 0 \times \frac{2}{3} \times 2.0 + 147.3 \times \frac{1}{3} \times 2.0 = 98.2 \text{ kNm/m}$$

Actual factor of safety on passive earth pressures is

$$F_{actual} = \frac{M_R}{M_O} = 1.99$$

The resulting earth pressure diagram is shown in Figure 10.14, with calculation points a_1, a_2, p_1 and p_2 shown.

10.3.4 Singly-anchored sheet-pile walls

Figure 10.15 shows a typical layout for a singly-anchored sheet-pile wall, such as is often used as a waterfront structure.

Possible failure modes for singly-anchored sheet-pile walls include

i. Rotation about the point at which the anchor tendon joins the sheet piling
ii. Passive failure of the soil below dredge level, at the front of the wall, as a result of inadequate toe in
iii. Failure of the wall by bending, between a relatively rigid anchor and a deeply-embedded sheet-pile toe

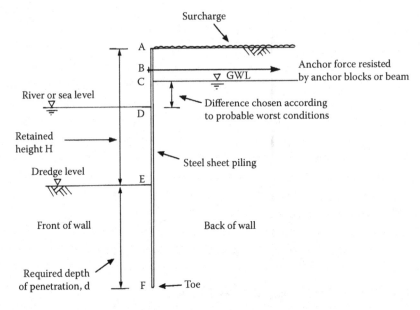

Figure 10.15 General layout for a singly-anchored sheet-pile wall.

iv. Failure of the wall in shear
v. Failure of the anchor tendon, connections to walings or of the anchor itself
vi. Piping
vii. Overall rotational failure, involving not only the mass of soil in which the sheeting is embedded, but also the soil around the anchor

A large number of methods have been proposed for the design of anchored sheet-pile walls, or 'anchored bulkheads' as they are sometimes known. Many have fallen into disuse, either because their fundamental principles have been questioned or because their complexity has made them unpopular. The two methods that have been most commonly used are the

- 'Free-earth support' method
- 'Fixed-earth support' method

In the United Kingdom, the free-earth support method is used.

10.3.4.1 Free-earth support method

According to Tschebotarioff (1973), this is the oldest and most conservative design procedure. But as the example calculations at the end of this section show, it often gives an economical design, with smaller depths of embedment but larger bending moments, than the fixed earth support method (see Section 10.4.2). Despite its age, it is in use (albeit in a modified form) in the United Kingdom, Brazil and the United States of America.

In the free earth support method, the sheets are assumed to be rigid, rotating about point B (Figure 10.15) where support is provided by a (supposedly) unyielding anchor. The depth of pile embedment is calculated on the basis of achieving moment equilibrium at the anchor level. The anchor force is then calculated on the basis of horizontal force equilibrium, and the point of maximum bending moment is determined from zero shear force on the shear force diagram. Following the work of Rowe (1952), the design bending moment used to select the sheet-pile section is obtained by reducing the maximum bending moment by a factor which depends on the relative flexibility of the sheet piling with respect to the soil.

Despite the assumption of a rigid wall rotating about an unyielding anchor, active earth pressures are assumed to act over the full height on the retained side. It might be thought that higher pressures would develop above the anchor on the retained side of the wall. However, Rowe (1952) estimated that, for typical anchored sheet-pile walls, the elastic yield of the anchor cable is of the order of H/1600, while the yield of the anchor block will be about H/800, where H is the height of the wall, and argued that all but the softest materials would be expected to achieve active conditions at these displacements. Later estimates, for example, in the South African Code of Practice 'Lateral support in surface excavations' (1989), Table 4.2, suggest that this may be optimistic.

In the free earth support method, therefore, active conditions are assumed on the entire back of the wall, and passive pressures are assumed below dredge level at the front of the wall. Typically, a surcharge is assumed to act on the ground surface on the retained side, and seepage contributes a net water pressure. Figure 10.16 shows effective earth pressures, out of balance water pressures, and bending moments for a singly-anchored sheet-pile wall.

The design process is as follows:

a. Determine soil parameters for the likely height of the sheets (see preliminary design charts).
b. Allow for unplanned excavation, by increasing the retained height (or the height below the lowest prop or anchor) by 10%.
c. For EC7 Design Approach 1, Combination 2, the 'characteristic' (moderately conservative) effective strength parameters, c' and ϕ', should be reduced by applying the partial factors $\gamma_{c'}$ and $\gamma_{\phi'}$ to obtain the design values of c' and ϕ':

$$c'_d = c'_k / \gamma_{c'} \text{ and } \phi'_d = \tan^{-1}(\tan\phi'_k / \gamma_{\phi'})$$

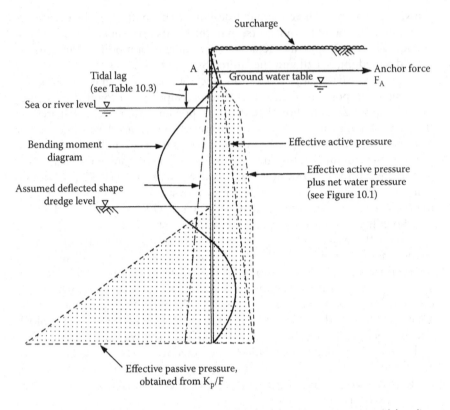

Figure 10.16 Effective earth pressures, out of balance water pressures, and bending moments for a singly-anchored sheet-pile wall.

d. The effective angle of wall friction should be taken as zero if the wall is subject to vibration. Otherwise, Terzaghi (1954) recommends

$$\delta'_a = \frac{1}{2}\phi' \quad \text{and} \quad \delta'_p = \frac{2}{3}\phi'$$

BS 6349 Part 1: (2000) recommends

$$\delta'_a = \frac{2}{3}\phi' \text{ and } \delta'_p \text{ between 0 and } \delta'_a$$

depending upon type of structure and soil density (see Table 19), and Eurocode 7 recommends $(2/3\phi'_{cv})$ for soil on steel, in both active and passive conditions.

e. Estimate tidal range, and the likely lag between the groundwater level in the retained soil and in front of the wall (Table 10.3).

f. Calculate the effective horizontal earth pressure using active earth pressure coefficients on the back of the wall. If using partial factors to EC7 Design Approach 1, Combination 2, these pressures should be calculated using design (reduced) values of c′ and φ′.

g. Calculate the effective horizontal earth pressure using passive earth pressure coefficients on the front of the wall. If using the traditional factor of safety on passive pressure approach, these pressures should be divided by a factor of safety of 2. If using EC7 Design Approach 1, Combination 2, these pressures should be calculated using design (reduced) values of c′ and φ′.

h. Calculate the out-of-balance pressure distribution on the wall due to unequal water pressure on either side. Figure 10.9 shows two possible out-of-balance water pressure distributions, depending on whether or not the bottom of the sheet piling penetrates impermeable ground. Table 10.3 gives the recommendations for groundwater on sheet-pile walls according to BS 6349: Part 1 (1984).

i. Take moments about the level at which the anchor tie is attached to the sheets, and determine the necessary depth of penetration of the sheet piling to give moment equilibrium.

j. Resolve horizontally to determine the force applied to the tie.

k. Calculate the shear force diagram for the sheets, in order to find the position of maximum bending moment. Start at the top of the wall.

l. Calculate the maximum bending moment at the point of zero shear force.

m. In sands or gravels, calculate the relative flexibility of the sheets and the soil, and reduce the bending moment as appropriate (see below).

n. Increase the depth of penetration by 20% to allow for 'the effects of unintentional excess dredging, unanticipated local scour, and the presence of pockets of exceptionally weak material' (Terzaghi 1954).

o. Increase the tie force by 10% to allow for horizontal arching (CP2).

p. Design anchors, and select tie section (see below).

Since, at the outset, the depth of penetration of the sheeting is unknown, the calculations for moment equilibrium about the anchor tie level (B in Figure 10.15 and A in Figure 10.16), can only be completed if

i. A depth is assumed or

ii. The pressure distributions at the base of Figure 10.16 are expressed in terms of the unknown depth, d (Figure 10.15).

In practice, it is normally easier to adopt the second approach. The condition of moment equilibrium then leads to a cubic equation of the form

$$Ad^3 + Bd^2 + Cd + D = 0 \qquad (10.8)$$

where A, B, C and D are known numerical coefficients. The simplest way to determine the correct value of d is by trial and error substitution, starting with a likely value (say d/H = 0.40), based upon preliminary design charts.

Chapter 3 (Section 3.3.7) explained the influence of sheet-pile flexibility on the magnitude and distribution of bending moments in a sheet-pile wall. Rowe (1952, 1957) carried out model tests and provided charts to allow the maximum bending moment calculated from the free-earth support method to be reduced in line with his experimental findings. In theory, Rowe's reduction factors can be used for any soil type, but Skempton (1953), mindful of the fact that they result from model tests, suggested that the amount of reduction should be as follows:

Sands: use 1/2 moment reduction from Rowe
Silts: use 1/4 moment reduction from Rowe
Clays: use no moment reduction

Rowe identified the stiffness of the sheet piling as follows:

$$\rho = \frac{H^4}{EI} \tag{10.9}$$

where
  H is the full length of sheet piling (i.e. retained height plus depth of embedment)
  E is Young's modulus
  I is the second moment of area of the sheet piling.

Figure 10.17 shows Rowe's moment reduction curves for sand. To use these curves, select the relevant soil condition (loose or dense) and wall height, plot a curve of bending moment v. log (p) by multiplying the maximum free-earth support bending moment for the particular wall by the values of M/M_{max} for different ρ in Figure 10.17. Next, select various possible sheet-pile sections and calculate log (ρ) and $M_{max} = fI/y$ for each, where f is the permitted maximum steel stress, and y is the distance from the neutral axis to the edge of the section. Plot the position of each of these sections on the curve. Sections giving points above the operating curve are wasteful, while those below the curve will be overstressed. Ideal sections will fall directly on the curve.

10.3.4.2 Fixed-earth support method

This method is derived from the work of Blum (1931, 1950, 1951), and has found widespread use in continental Europe and elsewhere. The sheet piling is considered flexible, but driven to sufficient depth that it may be considered fixed by the soil at its toe. Blum's general method deals with rigidly

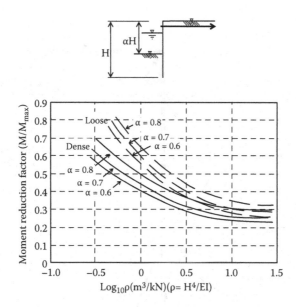

Figure 10.17 Moment reduction factors proposed by Rowe (1952) and metricated in CIRIA Report 54 (1974).

and flexibly anchored walls, and with cantilever walls. In these methods, the stresses on the wall immediately above the toe (F) are replaced by a single force some distance up the wall (Fc) and the sheet piling is considered to be held vertical at this point (Figure 10.18). The depth of penetration of the sheeting is found by repetitive calculation until the displacement at the anchor level is correct relative to the point of fixity (at the toe).

For routine design, the anchor is assumed to be unyielding, and the relative displacement between toe and anchor must therefore be zero. Unless carried out by computer, this technique is tedious; therefore a number of simplifications are in common use. These are based on Blum's 'equivalent beam method'.

The fixed earth support method was advocated by BS CP2 (1951) for routine design, but in practice, following the work of Rowe (1952) and Terzaghi (1954), geotechnical engineers in the UK now use a form of the free earth support method modified to take account of wall flexibility. This has occurred because repeated studies have shown the CP2 method to require excessive depths of penetration and steel weights compared with the requirements of other countries' national codes. Trial calculations are reported in CIRIA Report No. 54 (1974) for methods according to CP2, the German Committee for Waterfront Structures, the Danish Code, and Rowe's modification of the free earth support method. Similar comparisons, but not tied to the specific requirements of codes of practice, have been carried out by Edelman, Joustra, Koppejan, van der Veen and van

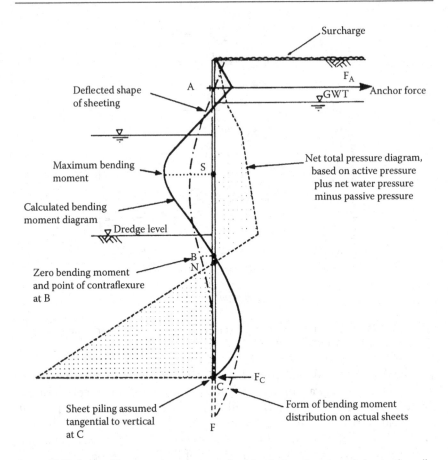

Figure 10.18 Basis of fixed earth support method for design of anchored sheet-pile walls.

Weele (1958) and Lamboj and Fang (1970). In practice, it is difficult to draw meaningful conclusions from these studies, because the detailed requirements of codes can have a more significant influence on such factors as depth of embedment, structure weight and cost, than the fundamental differences between the methods. For example, CIRIA Report 54 noted that design to the German Recommendations results in generally more economic structures compared to the other methods investigated, but that this probably results from the combination of Krey's (1963) over-optimistic passive earth-pressure coefficients (calculated on the basis of a planar failure surface) and an overconservative penetration depth calculation obtained from the fixed earth support method. Under conditions of similar assumptions (i.e. ignoring the individual requirements of the codes) Rowe's modification of the free earth support method is much more economical than the CP2 and German methods, and is simpler to use than most other methods.

On the other hand, the use of the fixed-earth support method with reasonable passive earth-pressure coefficients, such as in CP2, leads to the least economical depths and sections of sheet piling.

Methods of design in use in the UK or USA have often assumed an unyielding anchor, for simplicity (although this need not be an essential part of the method). The net total pressure diagram (Figure 10.18) is obtained by using active pressure coefficients for all soil behind the wall and full passive pressure coefficients for soil in front of the wall (i.e. below dredge level). Values of wall friction ($\delta'_a = 2/3\ \phi'$ on the active side and δ'_p between zero and δ'_a according to BS 6349 Part 1: [2000] Table 19) may be used in the absence of vibration. Groundwater pressure is included in the manner suggested by Terzaghi (1954) (Figure 10.9). At this stage, the necessary depth of sheeting is unknown, and the position of point C (which will be approximately 15% of the depth of embedment above the actual toe of the sketch) must be assumed.

The general method used for fixed earth support design is the 'elastic line method'. In this, the position of point C (Figure 10.18) is assumed, and the sheet-pile is taken as tangential to the vertical at this point. Successive integration with respect to depth of the net total pressure diagram leads in turn to the shear force diagram, the bending moment diagram, the slope diagram, and the deflection diagram. The position of point C is adjusted until the deflection of the anchor (point A) relative to point C is zero. From this, the necessary depth of sheeting may be obtained, since Blum demonstrated that the total required depth of penetration is (see Figure 10.19)

$$t \approx u + (1.05 \text{ to } 1.20)x \tag{10.10}$$

Typically, for convenience, the total required depth of penetration is taken as follows:

$$t = 1.20(u + x) \tag{10.11}$$

as in Tschebotarioff (1973), for the simplified equivalent beam method to be described below, but the actual required depth can be found from

$$t = u + x + \frac{F_c}{2\gamma' h_L (K_p \cos\delta'_p - K_a \cos\delta'_a)} \tag{10.12}$$

where
  u and x are defined in Figure 10.19.
  γ' is the average buoyant density $(\gamma - \gamma_w)$ between the top of the sheet-pile wall and point C.
  F_c is the replacement force at C.

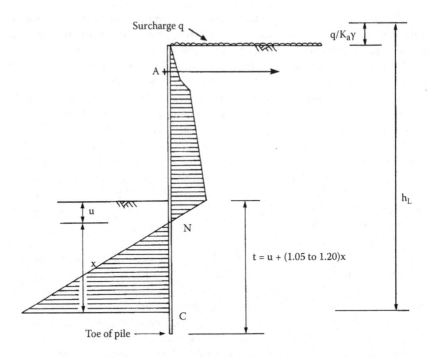

Figure 10.19 Relationship between required pile depth and the position of the substitute force at C.

K_p and represent the components of earth pressure normal to the wall.

h_L is the height of the wall (including embedment to point C), plus an allowance of $q/K_a.\gamma$ for any surcharge (q) imposed at the top of the wall.

Although in the past, the elastic line method has been solved by hand calculation, or graphically, it is now considered too time-consuming for routine use. It is, however, a relatively simple task to program a desk-top computer to provide these solutions, and commercial software is available.

In the absence of an available computer program, a number of simplifying assumptions can be made. All of these are variations of Blum's equivalent beam method. Blum's equivalent beam method (Blum 1931) uses the same simplifying assumptions with regard to the stresses at the toe of the pile as were used for the 'elastic line' method above. The stresses at the pile toe are replaced by a single force some distance above the toe. By carrying out example calculations on uniform soil profiles, Blum was able to establish the relationship between the depth to the point of sheet-pile contraflexure (y) (where the bending moment is zero—point B in Figure 10.20) and

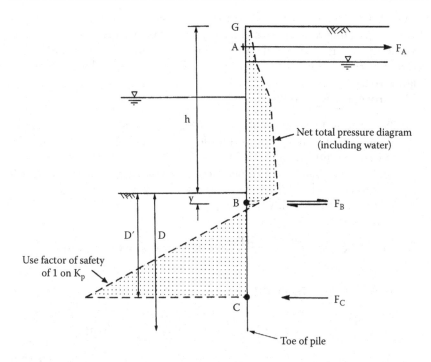

Figure 10.20 Blum's equivalent beam method.

the free height of the wall (h, from the dredge level to the top of the wall), as shown in Table 10.5.

It is reported by Tschebotarioff (1973) that these values were based on the use of

$$K_a = (1 - \sin \phi')/(1 + \sin \phi') \tag{10.13}$$

(i.e. the Rankine value, for $\delta' = 0$), and $K_p = 2/K_a$. Blum is supposed to have used this value for the passive earth-pressure coefficient, not because

Table 10.5 y/h as a function of ϕ' for Blum's equivalent beam method

Effective angle of friction of soil ϕ' (degrees)	Ratio (depth to point of contraflexure)/(free height of wall) y/h
20	0.23
25	0.15
30	0.08
35	0.03
40	−0.007

he allowed for the influence of wall friction, but because tests by Franzius (1924) (using a hinged wall in a relatively narrow box) had given similar results.

Once the point of contraflexure is known, an imaginary hinge can be inserted at that point on the sheet-pile wall, and analysis becomes trivial (Figure 10.20).

The procedure is as follows:

a. By horizontal resolution of forces on span GB, and by taking moments about B, determine the magnitude of the anchor force F_A, and the force at the hinge, F_B.

b. Take moments about C, to determine the correct length BC for which the moments about C are zero. Stresses below C are ignored.

   For a uniform soil, with $\gamma' = \gamma - \gamma_w$

$$F_B(D'-y) = \gamma' K_p y \frac{(D'-y)^2}{2} + \gamma' \frac{(D'-y)^2}{6}(K_p D' - K_p y) - \gamma' K_a (h+y)\frac{(D'-y)^2}{2}$$

$$-\gamma' \frac{(D'-y)^2}{6}(K_a D' - K_a y)$$

$$= \frac{(D'-y)^3}{6}\gamma'(K_p - K_a) + \frac{(D-y)^2}{2}\gamma'[(K_p - K_a)y - K_a h]$$

(10.14)

For moment equilibrium

$$(D'-y) \approx \sqrt{\frac{6 F_B}{\gamma'(K_p - K_a)}}$$

(10.15)

(N.B. If B is taken to be at the zero net pressure point, use F_N for F_B.)

c. Determine the final depth of embedment (which will also give a factor of safety against failure by forward movement of the piling), approximately

$$D = 1.2D'$$

(10.16)

For uniform ground conditions, the Grundbau Taschenbuch (1955) gives the more accurate equation based on the force at C, F_c (see under 'Elastic line method').

d. Determine the point of maximum bending moment from the position of zero shear force, by drawing the shear force diagram for span GB.

e. Determine the maximum bending moment.

The principal problem with this method is the determination of a correct point of contraflexure when soil conditions are non-uniform. For uniform ground conditions, B lies approximately level with the point of zero net pressure, N (see Figure 10.19). Practice in France has assumed that the point of contraflexure and the point of zero net pressure will always be approximately coincident. Given uniform ground conditions and realistic assumptions, the correlation is quite good (Figure 10.21), but the Grundbau Tachenbuch (1955) notes that one should 'be aware however of wrongly applying this method to non-uniform ground'. Here, the use of the equation (for [D′–y] based on F_b), which is no longer applicable, together with a false estimate of the position of point B, may lead to serious errors. For these conditions, the elastic line method is recommended by the Grundbau Taschenbuch, although in countries other than Germany, the zero net pressure point method has also been used.

For sheet piling penetrating clean medium-dense or dense sands, Tschebotarioff has proposed the so-called "hinge-at-the-dredge-line" procedure, which is based upon observations both in the field and the laboratory, that the point of contraflexure for this condition approximates to the dredge level. This method is only valid for limited soil types, although it can be modified slightly, but it is extremely quick to use. Figure 10.22 gives the basis of design. Horizontal force resolution on span GB, coupled with determination of moment equilibrium about B for span GB, leads to the anchor tie force, F_A. For the point of zero shear force between A and B, the maximum bending moment is calculated. The depth of penetration is fixed at 43% of the supported height αH.

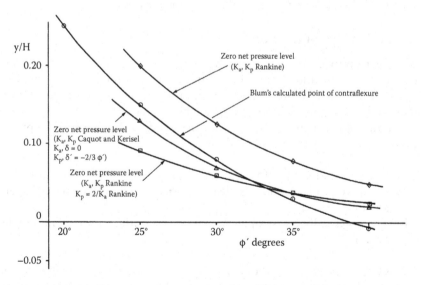

Figure 10.21 Comparison of position of zero net pressure and point of contraflexure in Blum's equivalent beam method. (From Pratt, Personal communication, 1984.)

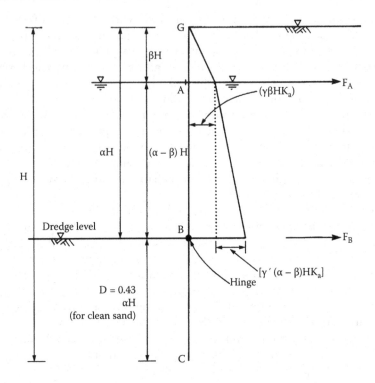

Figure 10.22 Tschebotarioff's hinge-at-the-dredge-line procedure.

10.3.5 Design of anchor systems

In order to provide support for a sheet-pile wall, it is necessary to design a tie rod, anchor and connections. Failure of sheet-pile walls commonly results either from failure of the tie rod, or of the anchor itself, but it may also result from poor detailing in the anchor system.

The tie force calculated from either the fixed or the free earth support method is normally increased to allow for

  i. The application of unforeseen surcharges
 ii. Unequal yield of anchors, leading to horizontal arching between anchors
iii. The potentially catastrophic consequences of the failure of any single anchor

While CP2 suggested that the tie force should be increased by 10%, other sources indicate much greater values. Teng (1962) and the USS Steel Sheet Piling Design Manual (USSI 1975) suggest that the tie rod force calculated by the free earth support method should be increased by 30% for the tie rod itself, while splices and other places where stress concentrations may occur should be designed for a 50%–100% increase. On this basis:

$$T_{design} = \frac{T_{fes}\ b(1 + f/100)}{\cos\alpha} \tag{10.17}$$

where

T_{fes} is the earth support value of anchor force (per metre run)
b is the horizontal spacing of the anchor tie rods
f is the % increase (30%–100%) and
α is the inclination of the tie rod to the horizontal.

At the sheet piling, the connection between the tie bar and the sheets is normally made via a wale, which often consists of two steel channel sections bolted together back to back over the tie rod. In most cases, this will be located above the high water table. Tie rods are often spaced at about 3 m centres. Figure 10.23 shows a typical layout given by the 'USS Steel Sheet Piling Design Manual' (USSI 1984), using an inside wale. Outside wales are structurally better, but are not normally used because they prevent a clear outside face. Wales are designed as single-span simply supported beams, with a maximum bending moment of $\frac{1}{8}T b^2$. Attention should be paid to the design of details such as the washer on the anchor wale.

Possible anchor systems are shown in Figure 6.17. A common form of anchor is the deadman, which usually consists of concrete blocks or continuous concrete beams. This type of anchor is most suitable for installation in relatively good ground. If the wall or block does not extend to ground level, as is commonly the case to allow the installation of services, it may still be considered for the purposes of calculation to extend to ground level, provided $H_1/H > \frac{1}{2}$ (Figure 10.24a). If this condition is not met, then it must be designed using bearing capacity theory (Smith 1957), or Krebs Ovesen's method (Krebs Ovesen 1964). For an anchor beam

$$T = \left(\frac{Q_p}{F} - Q_a\right)B \tag{10.18}$$

where F > 2, and

B centre to centre spacing of ties
Q_p and Q_a are the passive and active total forces on the front and back of the beam.

The passive force Q_p can only include for the effects of wall friction if the implied vertical forces can be resisted by the wall. Rowe and Peaker (1965) recommend that d' be taken as zero.

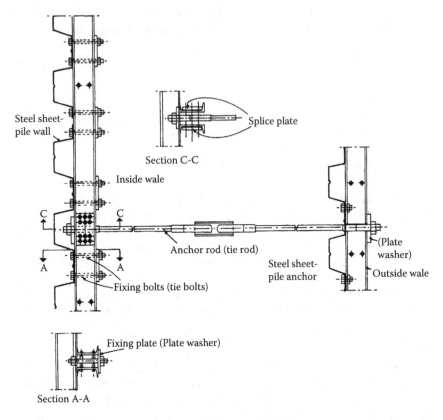

Figure 10.23 Typical wale and anchor rod details. (From USSI, *USS Steel Sheet Piling Design Manual*. United States Steel International. Updated and reprinted by US Dept. of Transportation [FHWA], 1984.)

For anchor blocks, side friction is included in the passive zone.

$$T = \left(\frac{Q_p}{F} - Q_a\right) B + 2Q_0' \frac{\tan\phi'}{F} \qquad (10.19)$$

where

$$Q_0' = \gamma H^3 / 6K_0 \tan(45° + \phi'/2) \qquad (10.20)$$

for granular soil. The use of cohesive soil is not recommended in front of anchors and it is safer to use K_a (rather than K_0) in this equation. A

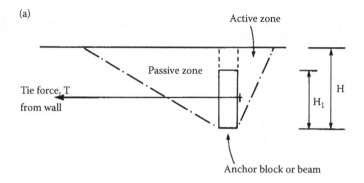

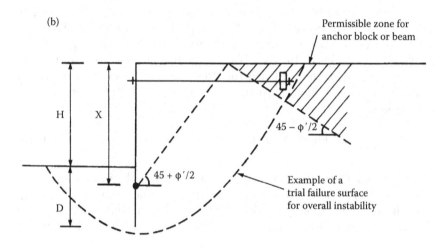

Figure 10.24 Geometry of anchor blocks or beams. (a) Strength of blocks and beams. (b) Position of anchor relative to wall.

check should be made to ensure that individual blocks do not provide more anchorage than a continuous beam of the same height.

The position of attachment of the tie to the anchor block should ideally be at the centre of net pressure (i.e. if the beam were to go to the surface, it would be at 1/3 of the beam height). If the beam does not extend to the surface, the connection should be made at between H/3 and H/2 from the base of the block. The connection should be detailed so as to avoid moment transfer from the beam to the anchor tendon should differential settlement occur. This may mean using washers on a threaded bar, in which case attention should be paid to detailing of each component in order to prevent the anchor tendon from pulling through the block.

Anchor blocks and beams must be set well back from the wall to avoid overlap of the active zone (behind the wall) with the passive zone (in front of the block). For walls driven to give end fixity,

$$X = \frac{3}{4}(H + D) \tag{10.21}$$

while for free-ended walls

$$X = (H + D) \tag{10.22}$$

If wall friction is used on either the active back of the wall or the passive front of the anchor block or beam, this assumption will be on the unsafe side since the rupture surfaces will be curved. In addition, the factor of safety of a failure surface involving the wall and the anchor should also be checked (see Figure 10.24b). If shortage of space means that anchors must be so close to the wall that the Rankine zones overlap, a reduction of available anchor force must be made. Terzaghi (1943) recommends a reduction of

$$(P'_p - P'_a)_{ab} \text{ (see Figure 10.25),} \tag{10.23}$$

where P'_a and P'_p are the passive and active forces on the vertical plane ab, obtained from the effective Rankine pressures on that plane.

If additional resistance is needed, the tie may be sloped down (away from the wall) by up to 10°. A check should be made that the wall is capable

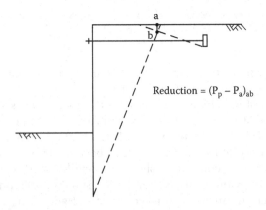

Figure 10.25 Reduction in available anchor resistance for anchors placed close to the wall.

of resisting the implied additional downward force. The soil in which the anchor is embedded should preferably be well-compacted clean granular fill. If there is soft compressible soil beneath the ties, they should either be placed in a duct or supported on piles approximately every 3 m along their length. Self settlement of the surrounding soil can otherwise lead to unexpected increases in tie forces, and subsequently to tie failure. If there is a layer of soft clay at or below dredge level, it may be necessary to use an A-frame anchor (see Figure 6.17), in order to ensure overall stability.

10.4 PROPPED AND BRACED EXCAVATIONS

Propped and braced excavations are often used for temporary support, for example for basements and service trenches. Four issues are important from a geotechnical point of view:

1. Estimation of the expected prop loads
2. Establishment of a satisfactory margin of safety against base heave, since, as with prop loads, this may lead to complete failure of the support system
3. Estimation of ground movements adjacent to the excavation, since if excessive these may cause damage to adjacent infrastructure or property
4. Overall instability

10.4.1 Calculation of prop loads

CIRIA Report C517 (1999) describes the 'distributed prop load (DPL) method' for calculating prop loads for propped temporary excavations. The DPL method is an updating of Peck's original work on this subject, discussed in Chapter 8. CIRIA Report C517 describes the back analysis of field measurements of prop loads relating to 81 case histories, of which 60 are for flexible walls (steel sheet-pile and king post) and 21 are for stiff walls (contiguous, secant and diaphragm).

The case history data covers excavations ranging in depth from 4 to 27 m to typically 5 to 15 m in soft and firm clays (soil class A, defined in Table 10.6); 10 to 15 m in stiff and very stiff clays (soil class B); and 10 to 20 m in coarse-grained soils (soil class C). The DPL method should only be used for multi-propped walls of similar dimensions and constructed in similar soil types.

Distributed prop load diagrams for soil classes A to C are provided for flexible (F) and stiff (S) walls in Figure 10.26. In this context, 'flexible' walls include timber sheet-piles and soldier pile/king post walls. 'Stiff' walls include contiguous, secant and diaphragm concrete walls.

Table 10.6 Soil classes defined in CIRIA Report C517

Soil class	Description
A	Normally and slightly overconsolidated clay soils (soft to firm clays)
B	Heavily overconsolidated clay soils (stiff and very stiff clays)
C	Coarse-grained soils
D	Mixed soils (walls retaining both fine-grained and coarse-grained soils)

Source: Twine D. and Roscoe H., *Temporary Propping of Deep Excavations—Guidance on Design (CIRIA Report C517)*. CIRIA, London, 1999.

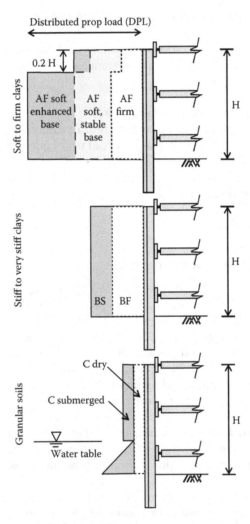

Figure 10.26 Distributed prop load diagrams for (top) Class A, (middle) Class B and (bottom) Class C soils.

Table 10.7 Magnitudes of distributed prop loads

Class	Soil	Over retained height	DPL
AS	Same as AF for firm clay		
AF	Firm clay	Top 20%	0.2 γH
		Bottom 80%	0.3 γH
	Soft clay with stable base	Top 20%	0.5 γH
		Bottom 80%	0.65 γH
	Soft clay with enhanced base stability	Top 20%	0.65 γH
		Bottom 80%	1.15 γH
BS	Stiff to very stiff clay	All	0.5 γH
BF			0.3 γH
C	Granular soil, dry	All	0.2 γH
	Granular soil, submerged	Above water	0.2 γH
		Below water	$0.2(\gamma - \gamma_w)H + \gamma_w(z - d_w)$

Source: Twine D. and Roscoe H., *Temporary Propping of Deep Excavations—Guidance on Design (CIRIA Report C517)*. CIRIA, London, 1999.

The magnitude of the 'distributed prop load' (DPL) in each case is summarised in Table 10.7.

Example 10.3

Consider a sheet-pile wall installed to retain H = 5 m of soft clay, which has a characteristic bulk unit weight ('weight density' in EC7) γ_k = 19.5 kN/m³. A separate check of the excavation's base stability has shown it to be stable.

The Distributed Prop Load (DPL) over the top 20% of the retained height (i.e. to a depth of 1m) is:

$$DPL = 0.5 \, \gamma_k \, H = 0.5 \times 19.5 \times 5 = 48.8 \text{ kPa}$$

while, over the bottom 80% of the retained height (i.e. from 1 m to 5 m depth), it is as follows:

$$DPL = 0.65 \, \gamma_k \, H = 0.65 \times 19.5 \times 5 = 63.4 \text{ kPa}$$

Consider the same wall retaining sand, with characteristic weight density γ_k = 18 kN/m³, and water, with weight density γ_w = 9.81 kN/m³, at a depth d_w = 2 m.

The DPL above the water table (i.e. to a depth of 2 m) is:

$$DPL = 0.2 \, \gamma_k \, H = 0.2 \times 18 \times 5 = 18 \text{ kPa}$$

while, below the water table (i.e. from 2 m to 5 m depth), it is:

$$DPL = 0.2(\gamma_k - \gamma_w)H + \gamma_w(z - d_w) = 0.2 \times (18 - 9.81) \times 5 + 9.81 \times (z - 2)$$
$$DPL = 8.2 \text{ kPa at } z = 2 \text{ m}$$
$$DPL = 37.6 \text{ kPa at } z = 5 \text{ m}$$

Individual prop loads are obtained by integrating the distributed prop load diagrams over the depth of influence of the prop being considered. The prop load P is given by:

$$P = s \int_{z_a}^{z_b} DPL \ dz \tag{10.24}$$

where
  s is the prop's horizontal spacing (i.e. on plan)
  z_a is the depth to a point midway between the current prop and the one above (as shown in Figure 10.27)
  z_b is the depth to a point midway between the current prop and the one below and
  DPL is the distributed prop load at depth z.

For the top prop only, the entire DPL envelope above that prop is included in the integration (i.e. $z_a = 0$ m). For the bottom prop only, the envelope is curtailed halfway toward formation level.

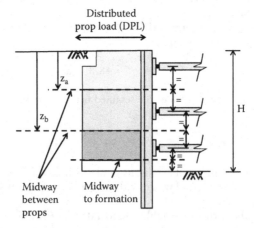

Figure 10.27 Method of assigning distributed prop load to individual props.

Example 10.4

The sheet-pile wall from Example 10.3 is supported by three levels of prop at depths $d_1 = 1.0$ m, $d_2 = 2.5$ m and $d_3 = 4.0$ m. The props are spaced at $s = 2.5$ m horizontal spacing.

For the top prop, $z_a = 0.0$ m and

$$z_b = \frac{d_1 + d_2}{2} = \frac{1.0 + 2.5}{2} = 1.75 \text{ m}$$

Hence the force carried by prop 1 is

$$P_1 = s \times \int_{0 \text{ m}}^{1.75 \text{ m}} DPL_1 \times dz$$

$$= 2.5 \times (48.8 \times 1.0 + 63.4 \times 0.75) = 241 \text{ kN}$$

For the second prop, $z_a = 1.75$ m and

$$z_b = \frac{d_2 + d_3}{2} = \frac{2.5 + 4.0}{2} = 3.25 \text{ m}$$

Hence, the force carried by prop 2 is

$$P_2 = s \times \int_{1.75 \text{ m}}^{3.25 \text{ m}} DPL_2 \times dz$$

$$= 2.5 \times 63.4 \times (3.25 - 1.75) = 238 \text{ kN}$$

For the bottom prop, $z_a = 3.25$ m and

$$z_b = \frac{d_3 + H}{2} = \frac{4.0 + 5.0}{2} = 4.5 \text{m}$$

Hence the force carried by prop 3 is

$$P_3 = s \times \int_{3.25 \text{ m}}^{4.5 \text{ m}} DPL_3 \times dz$$

$$= 2.5 \times 63.4 \times (4.5 - 3.25) = 198 \text{ kN}$$

The Distributed Prop Load method provides an estimate of characteristic prop loads. When designing according to Eurocode 7, design prop loads should be obtained by applying appropriate partial factors to increase mobilising forces (actions).

10.4.2 Base stability

There are three possible modes of instability in supported excavations in clay:

- Deep-seated failures
- Local failure adjacent to the support wall
- Bottom instability

Of these, the first two are related to the overall stability of the excavation. Local failure is of concern in soft soils where it is necessary to restrict sheeting deformations, since they occur below an excavation level adjacent to the sheeting, resulting in partial loss of support and leading to excessive inward deflections of the bottom of the wall.

In many soil types, the ground immediately below the base of an excavation can be brought close to failure. In the case of soft cohesive soils, the removal of vertical stress by excavating soil can induce a 'reverse bearing capacity' failure, driven by the loading of the soil outside of the excavation. In heavily overconsolidated cohesive soils the horizontal effective stress is already greater than the vertical, close to ground surface. Excavation leads to horizontal unloading above excavation level, and the rigidity of the wall redistributes at least part of this load onto the soil below the bottom of the excavation, bringing it to (or close to) the passive failure state. Yielding of soil in this mode is termed 'passive stress relief' (Burland and Fourie 1985) and special triaxial tests designed to follow a similar stress path have shown that under these conditions soils may develop much higher effective strength than under conventional triaxial compression testing conditions.

For braced excavations, Peck (1969) argued that the stresses applied to the struts depend on how closely base instability is approached, as measured by the stability number N:

$$N = \frac{\gamma H}{c_u} \tag{10.25}$$

where
  γ is the soil's bulk unit weight
  c_u its undrained shear strength and
  H the retained height of the excavation.

When N is greater than about 6–7, extensive plastic zones develop around the base of the excavation and settlements around the top are likely to be large. When N is between 5 and 6, earth pressures will be very large if there are great depths of soft clay. If, however, the soft clay is bounded by a stiff material near to the underside of the excavation, the plastic zone will be limited and pressures reduced.

Bjerrum and Eide (1956) defined the factor of safety against basal heave as

$$F = \frac{N_c c_u}{\gamma H + q} \tag{10.26}$$

where

γ, c_u and H are as defined above

q is the magnitude of any blanket surcharge acting on the ground surface and

N_c is a bearing capacity factor that depends on the shape and size of the excavation, given approximately by

$$N_c = 5 \times \left(1 + 0.2\frac{B}{L}\right) \times \left(1 + 0.1\frac{H}{B}\right) \tag{10.27}$$

as shown in Figure 10.28.

Base instability is a serious problem in weak clays where the strength of the soil does not increase significantly below the base of the excavation. When F is less than 2, substantial deformation may occur; when it is less than 1.5, the depth of penetration of the support system must extend below the base of the excavation (Canadian Geotechnical Society 2006).

Bjerrum and Eide (1956) developed a simple method of analysing base stability by using the chart in Figure 10.29. c_{ub} is the undrained shear strength around the base of the excavation, q is the surcharge loading, H is

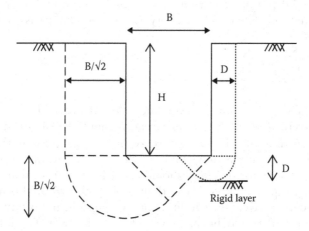

Figure 10.28 Geometry for determination of factor of safety of basal heave.

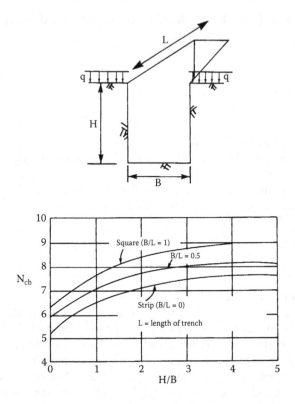

Figure 10.29 Bearing capacity factors for base instability analysis. (From Bjerrum, L. and Eide, O., *Géotechnique*, 6, 32–47, 1956.)

the proposed depth of excavation, and γ_m is the mean bulk unit weight of the excavated soil. The factor of safety against base failure is then

$$F = \frac{c_u N_{cb}}{(\gamma_m H + q)} = \frac{N_{cb}}{N} \qquad (10.28)$$

Where the soil is stratified within and below the base of excavation, the weighted average undrained shear strength should be obtained from the zone from 2.5B above the base of excavation to $B/\sqrt{2}$ below the base of excavation. To avoid plastic yielding of the soil, and to minimise ground movements, the factor of safety should exceed 2.5–3.0, so that the mobilised bearing capacity factor is less than 3.14 (Peck 1969). This method of analysis is conservative, since it does not recognise the important contribution to stability that may be made by embedment of a stiff wall below the base of the excavation.

Local failure occurs when excavations reach such a depth that lateral pressures from the base of the wall coupled with stress relief overstress the soil, leading to some loss of passive resistance and inward movement of the sheets. Figure 10.30 can be used to estimate when local failure is imminent, where flexible sheeting is used. Failure is related to the undrained shear strength of the clay and to its initial stress state in the ground. The shear stress ratio, f, is a dimensionless parameter which can vary from about +0.6 for a soft normally consolidated clay to −0.5 for a heavily overconsolidated clay. If the wall is stiff this figure will lead to conservative estimates of F (Jaworski 1973).

For walls which are continuous below excavation level, the estimation of passive stress relief stresses can be carried out using conventional calculation methods, in combination with predicted long-term (equilibrium) pore water pressures in the passive zone. The use of conventional triaxial compression effective strength parameters will yield conservative estimates of the factor of safety against failure. Because, as noted above, there is evidence that such parameters may lead to underestimates of the soil strength, if better estimates of the soil behaviour are required then special 'passive stress relief' triaxial tests might be considered (Burland and Fourie 1985).

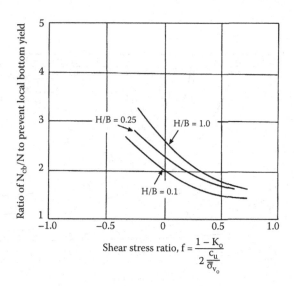

Figure 10.30 Factors of safety required to prevent local yield at the bottom of a braced excavation. Curves prepared from results of finite element analysis. (From D'Appolonia, D.J., *Effects of Foundation Construction on Nearby Structures*. In Proc. 4th Pan-Am. Conf. Soil. Mech. Found. Engin, State-of-the-Art, Vol. 1, pp. 189–236, 1971.)

10.4.3 Ground movements

The maximum lateral (i.e. horizontal) wall movement $\delta_{h,max}$ may be estimated from Figure 10.31, which relates $\delta_{h,max}/H$ (where H is the wall's retained height) to system stiffness, ρ_s, and the overall factor of safety against basal heave, F_{bh}, where the system stiffness

$$\rho_s = \frac{EI}{\gamma_w h_{avg}^4} \tag{10.29}$$

where
   E and I are the retaining wall's modulus of elasticity and second moment of area, respectively
   γ_w is the unit weight (weight density in EC7) of water and
   h_{avg} is the *average* vertical prop spacing of a multi-propped system

and the factor of safety against basal heave (Figure 10.31) is that defined by Terzaghi (1967):

$$F_{bh} = \frac{N_c c_u}{\gamma H - (H/d_b)c_u} \tag{10.30}$$

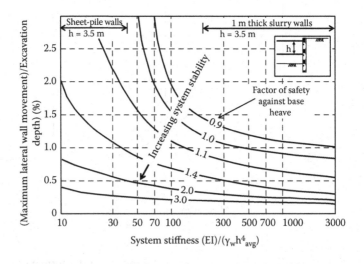

Figure 10.31 Chart for estimating maximum lateral wall movements and ground surface settlements in clays. (From Clough G.W. and O'Rourke T.D., Proc. of Design and Construction of Earth-Retaining Structures, Ithaca, NY ASCE GSP 25, pp. 430–470, 1990.)

where

N$_c$ is a stability number

γ and c$_u$ are the soil's bulk (weight) density and undrained strength, respectively

H is the wall's retained height; and the depth d$_b$ is given by

$$d_b = \frac{B}{\sqrt{2}}$$ (10.31)

in the absence of a rigid layer; and by

$$d_b = D$$ (10.32)

when one is present (see Figure 10.28).

Table 10.8 gives an alternative method of calculating h$_{avg}$ which allows Figure 10.28 to be used for cantilever and single-propped walls, which has been established for UK soils (Fernie and Suckling 1996).

Example 10.5

A PU18 sheet-pile wall is to retain H = 3.5 m of stiff clay with characteristic weight density γ_k = 20 kN/m^3 and undrained strength c$_{uk}$ = 80 kPa. The wall's Young's modulus E = 210 GPa and its second moment of area I = 38,650 cm^4/m. The breadth of the excavation is B = 25 m.

Without propping, the equivalent average prop spacing is calculated as

$$h_{avg} = 1.4 \, H = 4.9 \text{ m}$$

and the wall's system stiffness is

$$\rho_s = \frac{EI}{\gamma_w h_{avg}^4} = \frac{210 \times 10^6 \times 38650 \times 10^{-8}}{9.81 \times 4.9^4} = 14.4$$

Table 10.8 Equivalent 'average vertical prop spacing' (h$_{avg}$) for cantilever, single-propped and multi-propped walls

No. of props	Approximate value of h$_{avg}$		
	Soft soil	Medium soil	Stiff soil
None	2.4 H	1.8 H	1.4 H
Single	1.6 H	1.4 H	1.2 H
Multiple	Use maximum vertical spacing		

Note: H is the retained height of soil.

In the absence of a rigid stratum beneath the excavation, the factor of safety against basal heave is

$$F_{bh} = \frac{N_c c_u}{\gamma H - (\sqrt{2}H/B)c_u} = \frac{(\pi + 2) \times 80}{20 \times 3.5 - (\sqrt{2} \times 3.5/25) \times 80}$$
$$= 7.6$$

From the chart given above:

$$\frac{\delta_{h.max}}{H} \approx 0.37\% \Rightarrow \delta_{h.max} \approx \frac{0.37}{100} \times 3500 = 13 \text{ mm}$$

10.4.4 Overall instability

Earlier sections in Chapter 10 have considered the stability of anchored sheet-pile walls in terms of ensuring horizontal and moment equilibrium of the lateral thrusts and tie-rod forces. For walls tied back with highly stressed ground anchors, additional stability checks must be carried out as detailed below.

A check on overall equilibrium ensures that tie forces are transmitted far enough back from the wall (i.e. that the free length of anchors is adequate). The method proposed by Kranz (1953) examines the equilibrium of a wedge of soil between the back of the wall and the front of the fixed length (Figure 10.32). Considering the forces acting on the wedge, a force polygon may be constructed to determine the maximum anchor force to just maintain equilibrium. A factor of safety is defined as follows:

$$F = \frac{T_{max}}{T_{applied}} \geq 1.5 \tag{10.33}$$

If the factor of safety is too low, the free length must be increased. Kranz's method is generally thought to be conservative.

An adaptation of the Kranz method due to Broms (1968) considers a much larger wedge, including a proportion of the anchor fixed length (Figure 10.33). The method ignores the actual force in the anchor, and expresses a factor of safety in terms of passive forces:

$$F = \frac{(P_p) \text{available}}{(P_p) \text{required}} \geq 1.5 \tag{10.34}$$

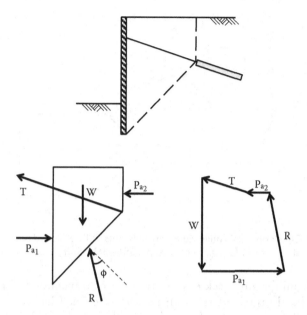

Figure 10.32 Check on overall equilibrium. (From Kranz, E., *Uber die Verankeruny von Spundwiinden*, 2. Aufl. Berlin, 1953.)

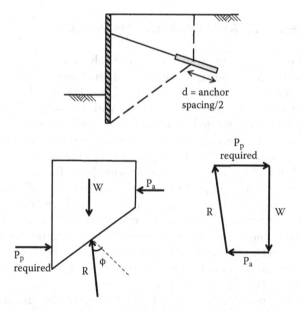

Figure 10.33 Check on overall equilibrium. (From Broms, B.B., *Proc. 3rd Budapest Conf. Soil Mech. Found. Engng.*, Budapest, pp. 391–403, 1968.)

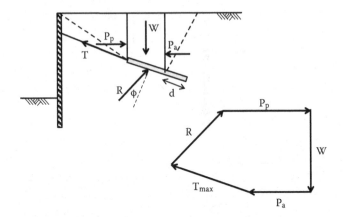

Figure 10.34 Check on local equilibrium. (From Broms, B.B., *Proc. 3rd Budapest Conf. Soil Mech. Found. Engng.*, Budapest, pp. 391–403, 1968.)

A local equilibrium check ensures that forces transmitted to the soil do not cause local failure of the soil; i.e. that the fixed length is adequate. Broms (1968) suggests considering the equilibrium of a wedge of soil above the fixed length (Figure 10.34). If the forces are not in equilibrium, the fixed length must be increased.

10.5 BORED PILE AND DIAPHRAGM WALLS

Bored pile and diaphragm walls are a relatively recent advance in retaining wall construction compared with steel sheet piling. According to Puller (1996), the first diaphragm walls were tested in 1948 and the first full scale slurry wall was built by Icos in Italy in 1950, with bentonite slurry support as a cut-off wall. Icos constructed the first structural slurry wall in the late 1950s for the Milan Metro. Slurry walls were introduced in the US in the mid 1960s by European contractors. The first application in the US was in New York City in 1962, for a 7-m diameter by 24-m deep shaft (Tamaro 1990).

The use of bentonite support in bored piling followed, allowing the construction of progressively larger-diameter piles for foundations, as well as secant and contiguous piled retaining structures. This was followed by a rapid growth in the use of (often more economical) continuous flight auger (CFA) piles as retaining structures. The initial use of bentonite supported, and then CFA, piles was not without problems in the UK (e.g. see Fleming and Sliwinski 1986) largely because of problems of construction control introducing pile defects, but by the 1990s bored pile and diaphragm walls had become widely used in the UK and elsewhere.

CIRIA Report 104 (Padfield and Mair 1984) provided initial design guidance for the design of retaining walls embedded in stiff clay, and was primarily used for relatively stiff bored pile and diaphragm walls. This report has since been superseded by CIRIA Report C580 (Gaba et al. 2003), which has a wider scope, covering the design of temporary and permanent embedded walls 'in stiff clay and other competent soils'. Such walls are often constructed using diaphragm walling and bored cast-*in situ* piling. CIRIA Report C580 is referred to extensively in this book.

This section is intended to draw out issues that are relevant to this class of wall. In particular, it considers the effect of calculation methods on the results obtained for serviceability and ultimate limit states to obtain wall depth, maximum bending moment, and prop load, and the analyses necessary to obtain realistic predictions of wall and ground movements.

10.5.1 Analysis for embedment, bending moments and prop loads

Appendix 10 of CIRIA C580 compares the results of analyses for four simple wall problems: a cantilever wall and a propped wall, each under short-term (undrained) and long-term (drained) conditions. Each was analysed using the following commercially available software:

- Limit equilibrium methods (STAWAL, ReWaRD)
- Subgrade reaction (Winkler spring) and pseudo finite element methods (FREW, WALLAP)
- Numerical modelling using finite element and finite difference methods (SAFE, FLAC)

CIRIA C580 concludes that

- Where there is little or no opportunity for earth pressure redistribution (e.g. for a cantilever walls or flexible singly-anchored [sheet-pile] walls) all methods of analyses are likely to give similar embedment depths and bending moments.
- For propped or anchored walls where earth pressure redistribution can occur (i.e. where the wall and props are stiff, as may be the case with bored pile or diaphragm walls in stiff clays) design using limit equilibrium methods may result in longer walls and greater bending moments than will be required using soil-structure interaction analyses (i.e. discrete spring or continuum modelling).
- Where earth pressure redistribution occurs, prop or anchor loads calculated using limit equilibrium methods may be a significant underestimate and should be treated with caution.

- If calculated prop loads are significantly different from those derived from experience (e.g. using the DPL method in Section 10.4) then the designer should carefully review the assumptions made in the analyses, and carry out a sensitivity analysis.
- For embedded walls where the total horizontal earth pressure is similar on the retained and excavated side near the base of the wall, the results of calculations may be very sensitive to small changes in pressures. As well as reviewing the initial pressure input to the analyses, the designer should check the effects of finite element mesh discretisation, and finite difference grid point spacings.

CIRIA C580 makes a number of other important points about the analysis of this type of wall:

- Even though many walls of this type will take the greatest loads and bending moments during construction, analyses should generally be carried out in terms of effective stress. The rate at which excess pore pressure dissipates is difficult to predict, and short-term analyses give less conservative results compared to long-term analyses.
- Where walls support significant vertical applied loading, the magnitude and direction of any wall friction or adhesion that is assumed in the calculations should be appropriate for each construction stage, and in the long term. Where loads are large and the wall can settle, then wall friction and adhesion should be conservatively assessed over the embedded portion of the wall, and should be assumed to be zero on the retained side above excavation level.
- If total stress analysis is used, flooded tension cracks or a minimum pressure of 5z (kPa) should be assumed on the retained side of the wall.
- Despite a number of studies, it remains difficult to predict the reduction in *in situ* earth pressures caused by wall installation. CIRIA C580 suggests that diaphragm wall installation may cause a 20% reduction and bored piling a 10% reduction in the *in situ* lateral earth pressure distribution, and that in the long term earth pressures will remain largely unchanged.
- It is sensible, if not essential, to carry out simple hand calculations and numerical modelling to make a check on the results of more complex numerical modelling.

10.5.2 Wall deflections and ground movements

Many walls of this type are used to support deep basements or infrastructure such as cut-and-cover tunnels for metros, or excavations for highway underpasses. Prediction of ground movements around such excavations is often critical, if damage to adjacent structures and services is to be avoided.

Case histories of walls embedded in stiff soil, with a traditional lumped factor of safety of at least three against basal heave, indicate that wall deflections and associated ground movements are insensitive to wall thickness and stiffness (see CIRIA C580). It follows that flexible sheet-pile walls can be more economic in stiff soils than equivalent concrete alternatives, without increasing ground movements. In general, flexible walls with many props will give similar displacements to stiff walls with fewer props. The cost of additional propping may or may not outweigh the savings obtained by using a flexible wall section.

10.5.3 Analysis for wall and ground movements

The form of analyses used is important. Limit equilibrium calculations, by their nature, do not model soil-structure interaction, or the movement of the ground. Winkler spring analyses can model wall flexibility, but cannot realistically model the ground. However, they can be used with calibration against similar wall and support geometries, in similar ground conditions. Finite element or finite difference analyses can successfully predict ground movements, but only if a suitable constitutive model can be used with realistic soil parameters.

Clayton (2011) concludes that in order to make reasonable predictions of ground movements the following should be taken into account:

- A realistic stiffness profile with depth is required in order to predict wall movements, and obtain an estimate of the magnitude of ground movements to be expected away from the back of the wall.
- Stiffness degradation with increasing strain level must be modelled, based on small-strain laboratory testing of undisturbed samples, if the pattern of ground movements adjacent to the wall is to be realistically predicted. Even so, minor changes in input parameters may lead to significant changes in predicted ground movements.
- Anisotropy of stiffness needs to be modelled. In stiff clays the horizontal Young's modulus is typically considerably higher than its vertical counterpart. In the London clay, recent research suggests that horizontal stiffness may be twice the vertical stiffness.

10.5.4 Use of berms to control wall displacements

When horizontal displacements around a retaining structure need to be controlled, the choice between the different methods (such as anchoring, strutting, or the use of berms) will depend on cost effectiveness, and the form of the final structure. If, as is common, the permanent structure has not been designed with the restriction of ground movements in mind, then the cost of supporting the sides of the excavation, and preventing damage to adjacent structures, will be included by the contractor as part of the temporary works. In this case, strutting from the basement slab and the use of berms will be attractive. A number of computer studies have been carried out to investigate the effectiveness of berms (Clough and Denby 1977; Potts et al. 1992). Cases investigated include sheet-pile and diaphragm walls, with trapezoidal and triangular berms of variable geometry. Typical results are shown in Figure 10.35, from Potts et al. (1992) who analysed a specific (but representative) case of a diaphragm wall embedded in stiff clay.

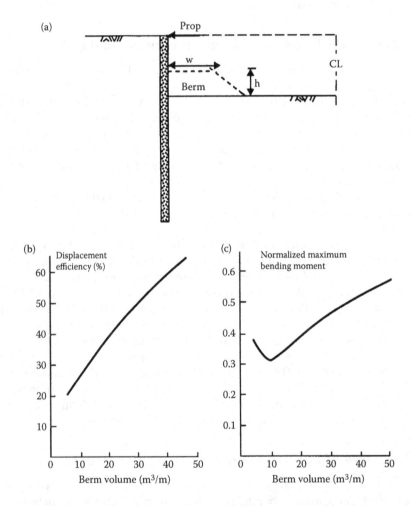

Figure 10.35 Influence of temporary berm on wall displacements and bending moments. (a) Geometry of wall and berm. (b) Displacement efficiency as a function of berm volume. (c) Normalized maximum bending moment as a function of berm volume. (From Potts, D.M. et al., The Use of Soil Berms for Temporary Support of Retaining Walls. Proc. Int. Conf. Retaining Structures, Cambridge, 1992.)

Figure 10.35a defines the geometry of the problem. Figure 10.35b plots displacement efficiency against berm volume. Displacement efficiency is defined such that a wall propped before full excavation (no berm) is 100% efficient, and a fully cantilevered wall (excavated without a berm) is 0% efficient. As one might expect, larger berms result in bigger reductions in movement. However, the largest berm analysed (which represented 30% of the total material to be excavated) only had an efficiency of about 60%. Figure 10.35c plots maximum bending moment (normalised by the maximum moment in the fully propped wall) against berm volume. This shows, perhaps surprisingly, that there is an optimum berm volume giving the smallest maximum moment.

Several simple methods have been suggested for determining the effect of earth berms on the stability of embedded walls, including

- Adding an equivalent surcharge
- Raising the effective formation level
- Enhancing passive earth pressures
- Performing multiple Coulomb wedge analyses

In the *equivalent surcharge method* (Fleming et al. 2008) the berm is replaced by a uniform surcharge of equivalent weight (W), whose extent is defined by the width of a passive wedge rising from the wall's toe, as illustrated in Figure 10.36. The magnitude of the surcharge is given by

$$q_{berm} = \frac{W}{d.\cot\theta} = \frac{\gamma.b.h}{d.\cot\left(45° - \phi'/2\right)} \tag{10.35}$$

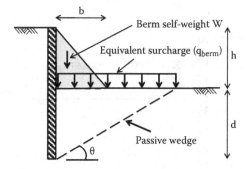

Figure 10.36 Equivalent surcharge method of modelling a berm in limit equilibrium analysis.

where
> b is the breadth of the berm
> h is the height of the berm
> γ is the average bulk (weight) density, and
> ϕ' is the effective angle of friction of the soil in the passive wedge.

In this method, the lateral support provided by the berm over its height is—conservatively—ignored.

The equivalent surcharge method has been shown (Daly and Powrie 2001) to give factors of safety that are between 15% and 25% greater than those obtained using multiple Coulomb wedge analyses (see sub-section below).

In the *raised formation method*, the berm is replaced by raising the level of the formation in front of the wall, as illustrated in Figure 10.37, by a distance Δh given by

$$\Delta h = \frac{b}{6} \tag{10.36}$$

where b is the breadth of the berm. This method effectively ignores that part of the berm that lies above a line rising at 1 in 3 from the berm toe.

The raised formation method has been shown to give factors of safety that are between 5 and 11% greater than those obtained using multiple Coulomb wedge analyses (see sub-section below). This is the method recommended in CIRIA C580 for routine limiting equilibrium analysis of earth berms.

In the *enhanced passive earth pressure method* (Williams and Waite 1993), the simple construction shown in Figure 10.38 is used to determine the increase in passive earth pressure owing to the berm.

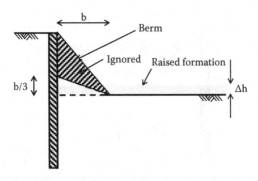

Figure 10.37 Raised formation method of modelling a berm in limit equilibrium analysis.

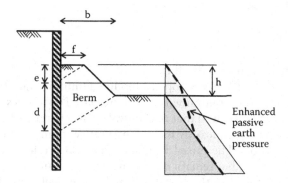

Figure 10.38 Enhanced passive earth pressure method of modelling a berm in limit equilibrium analysis.

When the berm is absent, passive earth pressures acting on the wall $\left(\sigma'_p\right)$ are given by the darker triangle in Figure 10.38, i.e.

$$\sigma'_p = K_p \times \sigma'_v = K_p(\gamma z - u) \tag{10.37}$$

where

K_p is the passive earth pressure coefficient
σ'_v is the vertical effective stress
γ is the bulk unit weight (weight density) of the soil below formation
z is depth below formation and
u is the pore water pressure at that depth.

If the formation level was to be raised by a distance h (i.e. to the top of the berm), then passive earth pressures acting on the wall would be given by the lighter triangle of Figure 10.38, i.e.

$$\sigma'_p = K_p \times \sigma'_v = K_p\left(\gamma[z+h] - u\right) \tag{10.38}$$

Earth pressures are assumed to transition between these two values from a depth, e, below the top of the berm, where

$$e = f.\tan\phi' \tag{10.39}$$

where

f is the breadth of the flat part of the berm and
ϕ' is the effective angle of friction of the soil below formation.

to a depth, d, below formation given by

$$d = b.\tan\left(45° - \phi'/2\right)$$

(10.40)

where b is the overall breadth of the berm.

Daly and Powrie (2001) developed a modified limiting equilibrium calculation of earth berms in undrained conditions, based on a series of Coulomb wedge analyses, and showed that the equivalent surcharge method will lead to designs that are 'significantly more conservative' than the modified limit equilibrium method. The raised formation method produces designs that are less conservative, but remain more conservative than the modified limit equilibrium method. On the basis of centrifuge model tests (Powrie and Daly 2002), it has been concluded that increasing the size of a berm is more advantageous than increasing the depth of embedment of a wall supported by a small berm, and that for a berm of given geometry, increasing wall embedment has little effect on stability.

10.6 KING POST AND SOLDIER PILE WALLS

These types of wall can represent an attractive option when temporary excavation support is required for basement excavations in stiff ground, away from other construction. As noted in Chapter 6, king posts can be installed from ground level in pre-bored holes. As excavation proceeds, timber or concrete planking can be placed in sections, and the king piles can be restrained using bracing, props or anchors.

The braced excavation data originally presented by Peck (1969) was obtained from this type of construction. The earth pressures experienced by a king post wall are significantly affected by workmanship, and these reflect the fact that more or less ground movement may occur. Where timber planking has been used the 'snugness' of fit between the excavated soil profile has a significant influence on the control of ground movements. Where *in situ* reinforced concrete has been used as planking the size of the panel and the speed with which it is constructed are important.

Where soldier piles are used, there will be much less passive resistance available from below excavation level to support the wall, relative to that provided by a diaphragm or bored pile wall. Broms (1965) has shown that the estimation of passive resistance of piles based upon their width and K_p values for continuous walls is too conservative. His recommendations are given in Figure 10.39, where it should be noted that in cohesive soils, resistance should be ignored to a depth of 1.5 pile diameters. In cohesionless soils, once the pile penetration is greater than one pile diameter, K_p values may be trebled. A factor of safety of 1.5 on K_p values is recommended by Broms.

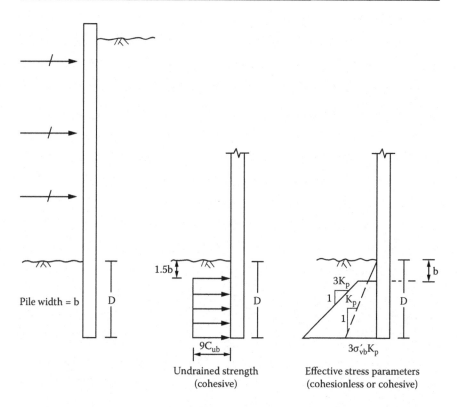

Pile width = b

D

1.5b

9C_ub

Undrained strength
(cohesive)

$3K_p$

K_p

1

1

$3\sigma'_{vb}K_p$

b

D

Effective stress parameters
(cohesionless or cohesive)

Figure 10.39 Passive resistance for soldier piles. (From Broms, B.B., *Lateral Resistance of Piles in Cohesionless Soils*. Proc. ASCE, J. Soil Mech. Found Div., 90, SM3, 123–156, 1965.)

Chapter 11

Composite walls and other support systems

As explained in Chapter 6, Section 6.4 and Figure 6.3, a range of superficially similar techniques exist that use anchors or reinforcement to provide support:

- Soil reinforcement (a.k.a. 'reinforced fill', 'mechanically stabilised earth' (MSE) or 'Reinforced Earth®')
- Anchored earth
- Soil nailing
- Anchored shotcrete or anchored pre-cast facings

These are covered in the following sections.

Eurocode 7 classifies walls that are composed of elements of gravity and embedded walls as 'Composite Walls'. This chapter also includes material on a number of other types of retaining systems, such as bridge abutments and cofferdams.

11.1 REINFORCED SOIL

Reinforced soil creates a gravity wall using light factory-constructed facings and closely-spaced strips or layers of reinforcement within (usually) granular backfill. Known commercially as 'Reinforced Earth' and 'Mechanically Stabilised Earth', it was invented by the French architect and engineer, Henri Vidal, in the 1960s. The first Reinforced Earth wall was built in the United States in 1971, and since then, the advantages of the technique have made it a popular solution for many earth retention applications. According to the Federal Highway Administration (2010), walls up to 30 m high have been constructed in the U.S.A.

11.1.1 Mechanics of reinforced soil

Observations on model and full-scale reinforced-soil walls, and on unit cell tests in the laboratory, have shown that reinforcement acts to alter the pattern of stresses in the soil, thus enabling greater applied loading to be supported. Both compressive and tensile strains occur when soils are subject to shear loading, and reinforcement acts advantageously when placed in directions in which tensile strains occur.

Early idealisations thought of reinforced soil as a new composite material, with the reinforcement providing an 'anisotropic cohesion' in the direction of reinforcement. A more direct idealisation (Jewell and Wroth 1987) considers potential failure surfaces and assesses the resultant reinforcement forces acting across these, as shown in Figure 11.1, for a direct shear box test on reinforced sand. There is a two-fold effect in which

- The horizontal component of the reinforcement force ($P_R \sin \theta$) reduces the shear loading, τ, that the sand must resist.
- The vertical component of the reinforcement force ($P_R \cos \theta$) increases the normal effective stress, σ_n, in the sand, allowing additional frictional resistance to be mobilised on the shear surface.

This leads to the simple expressions shown in Figure 11.1 for the shear strength of reinforced soil. Thus, the shear strength on any potential rupture surface can be assessed once values for the mobilised soil strength and the mobilised reinforcement force have been assumed. This allows a conventional estimate of safety factor to be made.

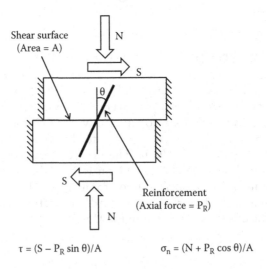

$$\tau = (S - P_R \sin \theta)/A \qquad \sigma_n = (N + P_R \cos \theta)/A$$

Figure 11.1 Modified direct shear box test on reinforced sand.

Equilibrium exists on any potential shear surface when the combination of the mobilised shear strength and mobilised reinforcement force balance the applied shear loading. There are three limiting conditions for the magnitude of reinforcement force:

  i. The tensile strength of the reinforcement must not be exceeded.
 ii. The maximum bond stress between the reinforcement and fill must not be exceeded.
iii. The tensile strains of the reinforcement and the soil must be compatible.

Toward the free end of the reinforcement, the maximum available force is likely to be governed by bond stresses, whereas toward the centre of the reinforcement, the strength or limiting tensile strain is likely to govern. A bond can be developed between reinforcement and soil in both shear and bearing. Shear occurs along those surfaces parallel to the axial direction; bearing occurs on those surfaces normal to the axial direction (such as the cross members in a polymer grid or the ribs on a steel strip) (see Figure 11.2). The generation of bond stresses along the reinforcement is the principal characteristic of reinforced soil that distinguishes it from traditional anchoring.

11.1.2 Detailed design of reinforced fill walls

The objectives of the design are to provide a reinforced-soil mass that is able to provide adequate support to the retained soil, while being internally stable. More specifically, the reinforced-soil mass must have an adequate margin of safety against outward sliding, overturning, bearing failure and deep-seated failure—without any one layer in the reinforced zone becoming overstressed.

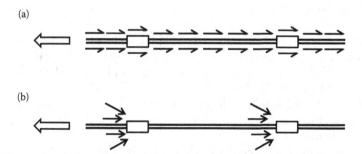

Figure 11.2 Mechanisms of soil-reinforcement bond. (a) Friction on planar surfaces. (b) Bearing on transverse members.

Design methods are covered in detail in BS 8006 Part 1 (2010) and in FHWA-NHI-10-024 (2010). These documents cover the design of reinforced slopes, in addition to the walls and abutments that are the focus of the attention of this section.

At the design stage, the wall height, applied loadings and required facing details will be prescribed. In addition, there may be constraints on the materials that can be used for fill (e.g. see Section 3 of BS 8006—Part 1 and Table 3-1 of FHWA [2010]) and reinforcement respectively. The designer must determine the number, strength, spacing and length of reinforcing elements.

Reinforced-soil walls must be checked with respect to two forms of stability:

    a. External stability
    b. Internal stability

Broadly speaking, external stability will govern reinforcement length, and internal stability will govern reinforcement spacing. External stability checks are very similar in form to those used for gravity walls, whereas internal stability is unique to this type of wall and for soil nailing, and arises from the interactions between the soil and the reinforcement.

Various recommendations exist for the preliminary proportions of a reinforced soil wall—specifically, the width of the reinforced zone ≈ length of reinforcement layers—typically based on wall height. BS 8006—Part 1 takes the 'mechanical height', for which the wall must be designed, as the vertical distance from the toe of the physical wall to the point at which a line drawn at $\tan^{-1}(0.3)$ (= 16.7°) to the vertical outcrops on the ground surface above the wall. A simple example is shown in Figure 11.3, which also defines the preliminary reinforcement length, L.

There will normally be an embedment D_m of the foot of the wall into the foundation soil of between H/20 (walls) and H/10 (abutments). This is not just about removing poorer subsoil, but also so that the wall is founded below the zone of frost penetration, which in the UK requires a depth of about 0.5 m.

The total height of the reinforced soil block will then be 5%–10% greater than H_1, and this is termed the 'mechanical height' of the wall, H. If the ground surface falls away from the horizontal in front of the wall, D_m for walls should be increased to H/7 for a 2:1 slope, and H/5 for a 3:2 slope. In addition, for walls where the crest is not level but the soil surface ramps up at an inclination to the horizontal, the mechanical height is defined as shown in Figure 11.3.

The length of reinforcement L is typically 70%–80% of the mechanical wall height (i.e. L = 0.7–0.8H). For low walls (3–6 m), soil to reinforcement bond is critical and a minimum length of 5 m is used. For reasons of

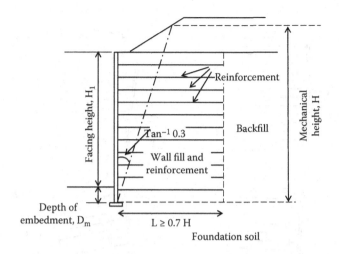

Figure 11.3 Basic definitions and preliminary reinforcement length (L) for a simple reinforced soil wall.

construction expedience, it is normal to use reinforcement layers of constant length, although in some cases, it might be desirable to vary the length of layers to make more efficient use of reinforcement material. Figures 19 and 20 of BS 8006—Part 1 (2010) provide initial sizing for a variety of wall geometries and configurations.

Design must consider both external and internal stability (i.e. those mechanisms that largely avoid intersecting the wall and its components, and those mechanisms that involve failure of components such as reinforcement).

11.1.3 External stability

The external stability of a reinforced-soil wall is assessed in the same way as for a conventional gravity-retaining wall. Considering the reinforced zone as a monolith, the mechanisms of overall failure that must be checked are shown in Figure 11.4. In the general case, the reinforced (wall) fill, the retained (back) fill, and the underlying (foundation) soil might be significantly different, so additional subscripts of w, b and f are used with quantities such as ϕ', γ, K_a, etc. in the following discussions.

11.1.3.1 Outward sliding on base

Outward sliding is initiated by the thrust of the unreinforced backfill, and is most likely to occur on a plane just above or below the lowest level of reinforcement (Figure 11.4a). This is because the coefficient of friction in

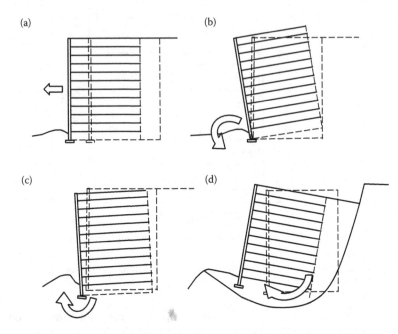

Figure 11.4 Mechanisms of overall failure in reinforced soil walls. (a) Sliding on base. (b) Toppling. (c) Bearing capacity failure. (d) Deep-seated failure.

sliding (μ_s) will generally be lower between soil and reinforcement than soil on soil.

Furthermore, if the bottom reinforcement layer is resting on the foundation soil of lower frictional resistance than the wall backfill, then μ_s must be based on the poorer soil (i.e. $\mu_s = \tan(\phi'_f)$). In traditional design, the factor of safety against outward sliding is based on simple force equilibrium. Neglecting cohesion, which is unlikely to be significant on the base, and assuming that pore pressures are zero, the long-term sliding resistance at the base of the wall is

$$Sliding\ resistance = \mu_s(\gamma_w H + w_s)L \qquad (11.1)$$

where

μ_s $\alpha_s \tan\left(\phi'_f\right)$, the lesser of sliding between soil and soil or between soil and reinforcement

α_s is the interaction factor for sliding between two surfaces (see Section 11.1.5)

γ_w is the bulk unit weight of the wall fill

w_s is the surcharge applied to the top of the wall

L   is the length of the reinforcement

H   is the wall height

$$Driving\ force = \frac{1}{2}K_{ab}\ \gamma_b H^2 + K_a w_s H \tag{11.2}$$

where

K_{ab} is the coefficient of active earth pressure of the retained fill
γ_b is the bulk unit weight of the backfill

On this basis, the factor of safety against sliding, F_s, is

$$F_s = \frac{2\mu_s(\gamma_w H + w_s)}{K_{ab}(\gamma_b H + 2w_s)(H/L)} \geq 2.0 \tag{11.3}$$

Alternatively, the minimum reinforcement length can be expressed as

$$L_{min} \geq \frac{F_s K_{ab}(\gamma_b H + 2w_s)H}{2\mu_s(\gamma_w H + w_s)} \tag{11.4}$$

11.1.3.2 Toppling or limiting eccentricity

Toppling, also known as rotation about the toe, overturning, or 'limiting eccentricity' in the U.S.A., is initiated by the thrust of the unreinforced backfill, causing the reinforced block to topple forward (Figure 11.4b). Traditionally, the factor of safety is calculated from overturning and restoring moments above the toe of the wall:

$$F_s = \frac{3(\gamma_w H + w_s)}{K_{ab}(\gamma_b H + 3w_s)(H/L)^2} \geq 2.0 \tag{11.5}$$

With typical L/H ratios used in practice, overturning is rarely a problem.

11.1.3.3 Bearing failure

Bearing failure occurs if the maximum vertical stress exerted by the reinforced soil block exceeds the bearing capacity of the underlying soil (Figure 11.4c).

Because of the overturning moment applied by the retained fill, and the lower ground level at the wall face, this will occur nearer the facing blocks than the back of the wall.

Conventional practice is to estimate the vertical stress distribution on the base of the wall and compare this with the allowable bearing pressure, q_a.

For a uniform wall with surcharge, and assuming a trapezoidal pressure distribution between the base of the wall and the ground on which it is founded:

$$(\sigma_v)_{max} = (\gamma_w H + w_s) + K_{ab}(\gamma_b H + 3w_s) (H/L)^2 \le q_a \qquad (11.6a)$$

As with a conventional gravity wall, there is also a requirement that the minimum vertical stress remains positive, to ensure the contact pressure distribution is nowhere in tension:

$$(\sigma_v)_{min} = (\gamma_w H + w_s) - K_{ab}(\gamma_b H + 3w_s) (H/L)^2 > 0 \qquad (11.6b)$$

Alternatively, assuming a constant Meyerhof type of pressure distribution, i.e. acting over a reduced length $L' = (L - 2\ e)$:

$$(\sigma_v)_{max} = \frac{\gamma_w H + w_s}{1 - \dfrac{K_{ab}(\gamma_b H + 3\ w_s)(H/L)^2}{3\ (\gamma_w H + w_s)}} \le q_a \qquad (11.7)$$

Allowable bearing pressure may be taken directly from foundation codes (e.g. BS 8004) or estimated using bearing capacity theory (see Chapter 9), with $F_s \ge 2$. If bearing capacity is inadequate, the designer may consider using a reinforced soil 'slab' underneath the wall. This would involve excavation of the subsoil to a particular depth, replaced by horizontal layers of granular fill and reinforcement. Alternatively, a cellular 'mattress' can be constructed with interlocking vertical sheets of geogrid, backfilled with sand and gravel. Smith and Worrall (1991) review the various solutions available.

11.1.3.4 Deep-seated failure

The analysis of deep-seated failure proceeds along identical lines to those for conventional gravity structures (Figure 11.4d). An appropriate stability calculation (using circular or non-circular mechanisms) should be carried out to ensure an overall $F_s \ge 1.5$.

11.1.4 Internal stability

The internal stability of the reinforced zone must be checked with respect to the mechanisms shown in Figure 11.5, namely

- Tensile failure of any of the reinforcements
- Pull-out failure, between the wall fill and any of the reinforcements

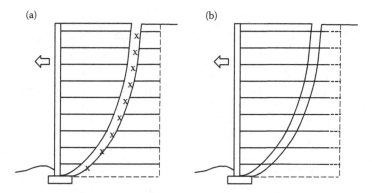

Figure 11.5 Mechanisms of internal failure in reinforced soil walls. (a) Tension failure. (b) Pull-out failure.

Tension failure is checked for each layer taking into account the self weight of the fill, any surcharge or line loads on the crest, and the increase of vertical stress near the wall face caused by horizontal thrust and overturning moment produced by the retained backfill. This latter effect produces a trapezoidal stress distribution along each layer, very much like that assumed under Section 11.1.3 'Bearing failure' (above) for the base of the wall.

Pull-out failure is checked by considering both the pull-out capacity of individual layers, and the equilibrium of planar wedge mechanisms through the reinforced zone that are restrained by several layers. For each individual layer, bond lengths beyond a point of maximum tension are assessed. Bond or anchorage lengths beyond critical wedges must be sufficient to maintain equilibrium and prevent pull-out.

Analytical methods are mostly based on Coulomb and Rankine theories. The two most often used in design are the

a. Tie-back wedge analysis (DTp, 1978)
b. Coherent gravity analysis (MdT, 1979)

Both are described in BS 8006—Part 1 (2010).

Tie-back wedge analysis assumes that in-service wall deformations will cause the face to rotate outward about the toe (Figure 11.6). Active conditions are assumed throughout the reinforced zone, with a constant value of earth pressure coefficient K_{aw}. A Coulomb type wedge inclined at $45° + \phi'/2$ to the horizontal is adopted as the critical failure mode, and is assumed to coincide with the locus of maximum reinforcement tension.

The *coherent gravity analysis* is a semi-empirical method based on the earth pressure theory and measurements taken on many model and full-scale walls. It is the standard method used by the Reinforced Earth

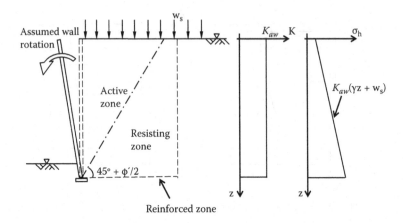

Figure 11.6 Assumptions in the tie-back wedge analysis.

Company for steel strip reinforced walls. In-service wall deformation is assumed to cause an outward rotation of the face about the top (Figure 11.7). A line of maximum reinforcement tension is assumed to divide the reinforced mass into an *active* and a *resistant* zone. A variable earth pressure coefficient is assumed, having a value of K_{ow} at the top, reducing to K_{aw} at 6 m depth and constant thereafter. In addition, empirical rules have been evaluated for the variation of the coefficient of friction μ_p between soil and reinforcement (see Section 11.1.5).

Stability calculations for retaining walls are generally carried out per metre run of wall, which is also the case for pull-out forces in continuous width reinforcements such as geogrids and woven geotextiles. The pull-out

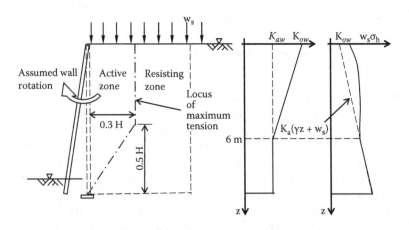

Figure 11.7 Assumptions in the coherent gravity analysis.

force for discrete reinforcing elements (strips, bars, anchors), however, is normally calculated on a per element basis and must therefore be converted to an equivalent force per metre, for each layer. This is readily obtained by dividing the force for a single reinforcement by the horizontal spacing, S_h.

Tension failure will occur if the reinforcement strength is insufficient to carry the horizontal load (T) applied to it. Each layer is assumed to carry the local horizontal stresses, acting over an area equal to half the vertical spacing above and below the layer per metre run of wall (Figure 11.8):

$$T = \sigma'_h S_v$$
$$\ = K_w \sigma'_v S_v$$

(11.8)

where
   S_v is the vertical spacing
   K_w is the earth-pressure coefficient ($K_{aw} \leq K_w \leq K_{ow}$)

The vertical stress acting at a given point in the reinforced soil mass may have a number of components, which will be evaluated separately and then combined.

Self-weight and surcharge. Equations 11.4 and 11.5, derived earlier for maximum vertical stress at the base of the reinforced zone, can be used to determine σ_v at any level in the wall. The depth z below the top of the wall is substituted for the height H:

$$(\sigma_v) = (\gamma_w z + w_s) + K_{ab}(\gamma_b z + 3\ w_s)(z/L)^2$$

(11.9)

The second term represents the enhancement of vertical stress due to the overturning effect of the horizontal thrust acting on the back of the reinforced zone, above the level of reinforcement being considered.

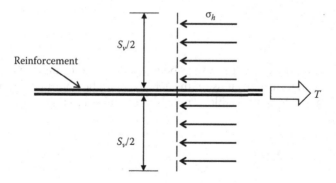

Figure 11.8 Local equilibrium check.

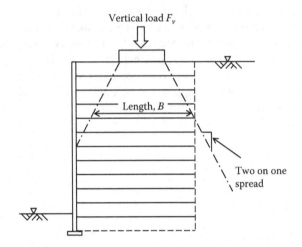

Figure 11.9 Dispersal of vertical strip load.

Vertical line loads. These may be present if the wall is acting as a bridge abutment, supporting a bank seat. Loads are assumed to disperse with depth at a conventional spread of 1:2 (horizontal:vertical), so that at some depth z the load acts over a length B of reinforcement (Figure 11.9). The stress is given by

$$\sigma_v = F_v/B \tag{11.10}$$

BS 8006 provides for eccentric vertical loading on the bank seat.

Horizontal line loads. Any horizontal loads (again, likely to arise in bridge abutment applications) are carried directly by the reinforcements in the upper portion of the wall (Figure 11.10). The horizontal stress distribution that results on the front face of the wall varies linearly from $2F_h/z_h$ at the top of the wall, to zero at a depth of z_h. Accordingly, at some depth z,

$$\sigma_h = \frac{2F_h}{z_h}\left(1 - \frac{z}{z_h}\right) \tag{11.11}$$

Total tension. By summing the above equations and multiplying by the vertical spacing, the total tension in each layer can be obtained. However, if the spacing is a design variable, it may be more useful to rearrange Equation 11.6 to give

$$(S_v)_{max} = P_d/\sum\sigma_h \tag{11.12}$$

where P_d is the design strength of the reinforcement per metre run of wall.

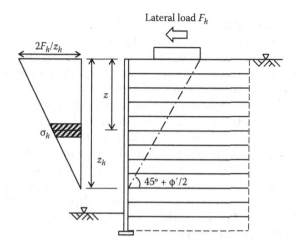

Figure 11.10 Dispersal of horizontal load.

For metallic reinforcements, design strength is taken as the product of permissible stress and cross-sectional area of the element. For non-metallic reinforcements, the convention is to use a characteristic strength reduced by a number of partial factors. Equation 11.12 may be plotted against depth below surface to give a useful indication of the required spacing (Figure 11.11).

Practical reinforcement spacing may be governed by facing connections (as in Reinforced Earth® walls using precast units) or by compaction layer thicknesses (as in geotextile reinforced soil walls with wrap around facing).

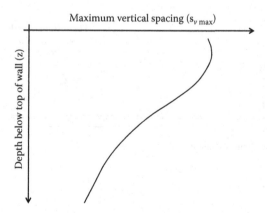

Figure 11.11 Example of required variation of reinforcement spacing to prevent local tension failure.

A constant vertical spacing is most convenient from a construction view-point, but in larger walls it may be desirable (and more efficient) to vary the spacing in two or more zones. In the latter case, spacing can still be based on integer multiples of the maximum compaction layer thickness—for example, with a layer of reinforcement incorporated after every second or third layer of backfill has been placed and compacted.

Pull-out failure. Standard practice is to check the pull-out resistance of the reinforcement both locally and globally. Local checks compare maximum required tension with maximum available bond force in each individual layer. Global checks consider potential failure mechanisms developing in the reinforced zone and compare the overall driving and resisting forces.

Local Bond Analysis. The maximum tension in a given reinforcement layer will occur at some distance from the face, depending on the elevation of the layers (Figure 11.12). At the point of maximum tension, T_{max}, the pull-out resistance per metre width P_p is

$$P_p = P_f + P_b \tag{11.13}$$

where

   P_f is the resistance generated by friction on upper and lower surfaces parallel to the axial direction
   P_b is the resistance generated by bearing on cross members perpendicular to the axial direction (and applicable, therefore, only in the case of meshes or grids, etc.)

In broad terms,

$$P_p = A_f \sigma'_n \mu_p + A_b \sigma'_b \tag{11.14}$$

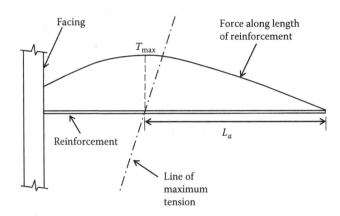

Figure 11.12 Distribution of tensile force in a reinforcing layer.

where

- A_f is the reinforcement area on which friction may develop
- σ'_n is the effective normal stress
- μ_p is the coefficient of friction in pull-out
- A_b is the reinforcement area on which bearing may develop
- σ'_b is the effective bearing stress

The areas A_f and A_b are based on the length of reinforcement that is being pulled out (i.e. the distance from the point at which tension is being applied to the free end). This length is known as the bond or anchorage length, L_a. For the purposes of checking local bond, L_a is the distance from the point of maximum tension to the free end.

For most types of reinforcement, the coefficient of friction in pull-out (μ_p) is expressed as a proportion of the soil friction angle ϕ':

$$\mu_p = \alpha_p \tan \phi' \tag{11.15}$$

where α_p is an interaction factor, $0 < \alpha_p \leq 1$. BE 3/78 (DTp, 1987) requires a minimum factor of safety of 2 on local bond (i.e. $T_{max} \leq P_p/2$). Explicit forms of Equation 11.11 for specific types of reinforcement are given in Section 11.3.

Wedge stability analysis. Trial wedges originating from the face of the wall at several different elevations are considered (Figure 11.13a). If surcharge and line loads are present, the force required to hold a general wedge in equilibrium is (Figure 11.13b)

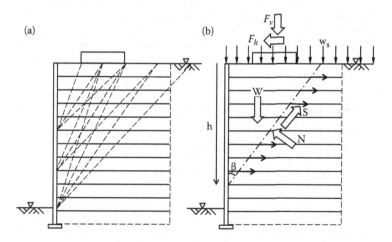

Figure 11.13 Wedge stability analysis (internal failure). (a) Trial wedges. (b) Forces acting on a typical wedge.

$$T_w = \left[\frac{h\tan\beta(\gamma_w h + 2\,w_s) + 2F_v}{2\tan(\phi'_w + \beta)}\right] + F_h \tag{11.16}$$

This is compared with the summation of forces available from each layer of reinforcement intersected by the wedge. These forces will be the lesser of the pull-out resistance (suitably factored) and the design strength for each layer. For a satisfactory design,

$$T_w \le \sum (P_p/F_s \text{ or } P_d) \tag{11.17}$$

The value of L_a used to calculate the pull-out resistance for each layer is simply the length extending beyond the point of intersection with the slip plane. Again, BE 3/78 requires a factor of safety F_s of 2 against wedge pull-out failure.

The above calculations for tension and pull-out failure can be very conveniently programmed for solution by a spreadsheet. Alternatively, computer programs have been developed to facilitate a more interactive approach to designing reinforced-soil structures (e.g. Woods and Jewell 1990).

Example calculation 11.1

An 8-m-high wall is to be built using sand fill and polymer grid reinforcement. The sand has an angle of friction, $\phi' = 33°$, and a unit weight $\gamma = 19$ kN/m^3, whereas the backfill and foundation soil have $\phi' = 27°$ and $\gamma = 18.5$ kN/m^3. A surcharge loading of 15 kN/m^2 is to be allowed for, and the maximum safe bearing pressure for the foundation soil is 280 kN/m^2. Two grids of different long-term design strength are available; grid A at 13 kN/m and grid B at 23 kN/m (both have interaction factors $\alpha_p = 0.9$ and $\alpha_s = 0.85$). The sand fill will be compacted in layers 225 mm thick.

EARTH PRESSURE COEFFICIENTS

$$K_{aw} = (1 - \sin 33°)/(1 + \sin 33°) = 0.295$$
$$K_{ab} = (1 - \sin 27°)/(1 + \sin 27°) = 0.376$$

EXTERNAL STABILITY
Outward sliding (Equations 11.3 and 11.4):

$$\mu = \alpha_s \tan \phi_f = 0.9 \times \tan 27° \approx 0.433$$

For a factor of safety of 2.0 against sliding, the minimum length of layers is

$$L_{min} \geq \frac{2.0 \times 0.376 \times (18.5 \times 8 + 2 \times 15) \times 8}{2 \times 0.433 \times (19 \times 8 + 15)} \geq 7.39 \text{ m}$$

Therefore, adopt a length of 7.4 m and check other external stability criteria.

Overturning (Equation 11.5):

$$F_s = \frac{3 \times (19 \times 8 + 15)}{0.376 \times (18.5 \times 8 + 15)(8/7.4)^2} = 5.92 \, (\geq 2.00)$$

Bearing failure (Equation 11.6):

$$(\sigma_v)_{max} = (19 \times 8 + 15) + 0.376 \times (18.5 \times 8 + 3 \times 15)(8/6)^2$$
$$= 167 + 84.7 = 251.7 (< 280 \text{ kPa})$$

Check that contact stresses at the base of the reinforced zone are compressive everywhere (i.e. no tension):

$$(\sigma_v)_{min} = 167 - 84.7 = 82.3 \text{ kPa} \, (> 0)$$

∴ a length of 7.4 m is satisfactory.

INTERNAL STABILITY

Combining Equations 11.8 and 11.9 and substituting values, the maximum tension in a layer at depth z below the top of the wall is given by

$$T_i = K_{aw}[(\gamma_w z + w_s) + K_{ab}(\gamma_b z + 3 w_s)(z/L)^2] S_v$$
$$= 0.295 [(19z + 15) + 0.376 (18.5z + 45)(z/7.4)^2] S_v$$

or

$$(S_v)_{max} = \frac{P_d}{0.295 [(19z + 15) + 0.376(18.5z + 45)(z/7.4)^2]}$$

For the two different grids, this equation plots as shown in Figure 11.14. In this figure, the 'spacing' axis has been divided into multiples of compaction layer thickness V to indicate which spacing is appropriate to given zones of the wall. Grid type B is able to carry local stresses at $S_v = V$ (225 mm) near the base of the wall, changing to $S_v = 2V$ (450 mm) further up, then $3V$ (675 mm), $4V$ (900 mm), etc. Grid type A would be

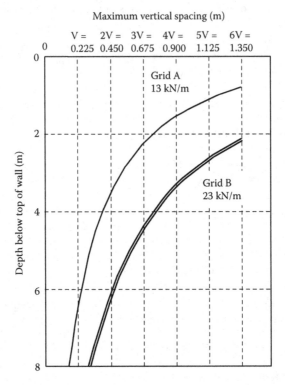

Figure 11.14 Maximum vertical spacing of reinforcement as a function of depth below top of wall.

overstressed at S_v = V near the base, but would be satisfactory further up the wall, followed by increases to 2V, 3V, and so on. At spacings in excess of 1 m, the beneficial effects of reinforcement are less well distributed throughout the soil mass, so S_v = 4V (900 mm) would be considered a practical upper limit in this example (for either grid). An efficient arrangement is often achieved by using stronger grids in the lower regions of the wall, switching to the lighter grids toward the top. One such layout is shown in Figure 11.15.

WEDGE STABILITY CHECK

Trial wedges are selected, in this case originating from the wall face at depths of 2, 4, 6 and 8 m below the top of the wall. The total required force T_w is calculated. For each reinforcement intersected by the trial wedge, the available force is less than the pull-out resistance P_p and the design tensile strength P_d.

From Equation 11.16, the critical wedge angle $\beta = 45° - \phi'_w/2 = 28.5°$ (Figure 11.16). For a wedge of height, h, is

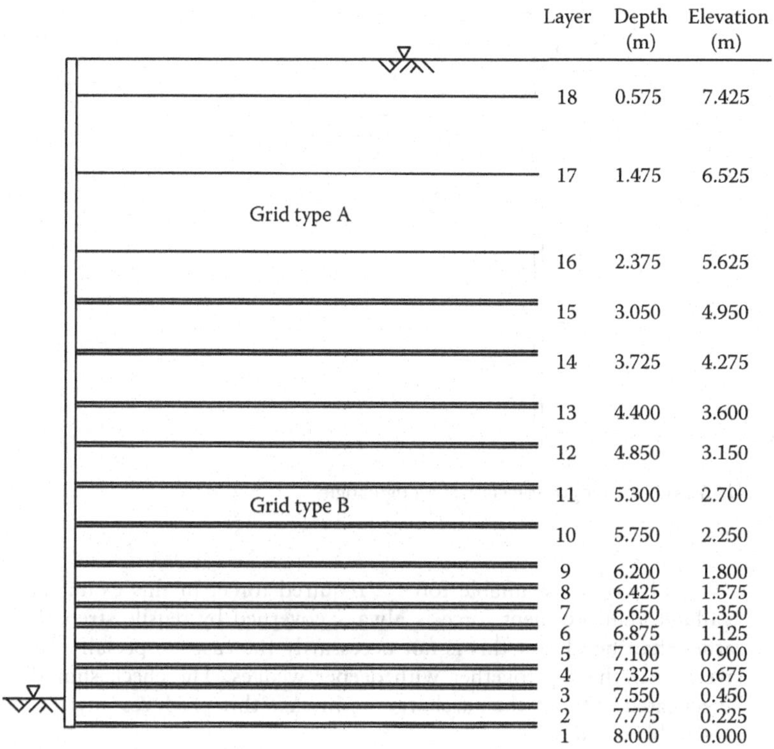

	Layer	Depth (m)	Elevation (m)
Grid type A	18	0.575	7.425
	17	1.475	6.525
	16	2.375	5.625
	15	3.050	4.950
	14	3.725	4.275
	13	4.400	3.600
	12	4.850	3.150
Grid type B	11	5.300	2.700
	10	5.750	2.250
	9	6.200	1.800
	8	6.425	1.575
	7	6.650	1.350
	6	6.875	1.125
	5	7.100	0.900
	4	7.325	0.675
	3	7.550	0.450
	2	7.775	0.225
	1	8.000	0.000

Figure 11.15 Layout of reinforcement.

$$T_w = \frac{h \tan 28.5°(19h + 2 \times 15)}{2 \tan(33° + 28.5°)} = 2.8h^2 + 4.42h$$

For a reinforcing layer at depth z below the top of the wall, the pull-out resistance is given by (Equation 11.20)

$$P_p = 2[7.4 - (h - z) \tan 28.5°] \times (19z + 15) \times 0.9 \times \tan 28.5°/2$$

The results are summarised as follows:

Wedge depth (m)	Required force T_w (kN/m)	Available force $\sum min(P_p/F_s, P_p)$ (kN/m)	F	Grids involved
2.0	20	26	1.31	0A+2B
4.0	62	85	1.37	2A+3B
6.0	127	177	1.39	6A+3B
8.0	214	361	1.69	14A+3B

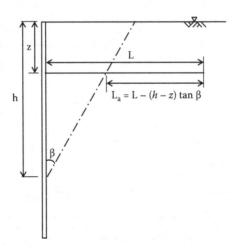

Figure 11.16 Anchor length for critical wedge angle.

For all wedges, available force ≥ required force. In this example, available reinforcement force is always governed by tensile strength rather than bond, but this is not necessarily the case—especially for layers near the top together with deeper wedges. The check should be repeated without the surcharge w_s in case this produces any less favourable results.

Example calculation 11.2

Gabions may be used in the place of conventional reinforced earth facing units. The free-draining nature of the rock-filled wire mesh baskets means that there will be no build up of pore water pressure on the wall. BS 8002 (1994) recommends that the density of the stone fill is taken as 60% of the solid material.

The wall should be proportioned so that the resultant force lies within the middle third of every horizontal section. Because of the external roughness of the gabions, it can be assumed that $\delta = \phi'$ on the back of the wall, and that sliding at the base is resisted by the full angle of friction of the underlying soil.

The possibility of horizontal shear failure between adjacent gabions should be investigated at several elevations in the wall, ignoring any beneficial effects of the wire mesh.

Figure 11.17 gives the dimensions for a proposed wall. The various properties are as follows:

Soil

$\gamma = 18 \text{ kN/m}^3$
$w_s = 18 \text{ kPa}$

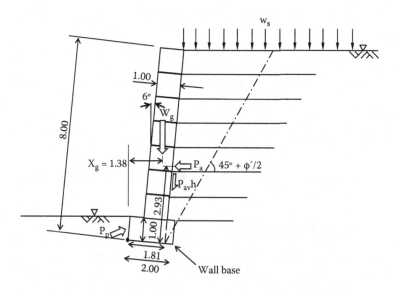

Figure 11.17 Reinforced earth with gabion facing (dimensions in metres).

$\phi' = 30°$
$\delta = 30°$

Gabions

Maccaferri (8 × 10)
Wire diameter = 3 mm
P_{ult} = 53 kN/m
γ_g = 1650 kg/m³
c_g = 21 kPa
$\phi^* = 25 \cdot \gamma_g - 10° = 25 \cdot 1650/1000 - 10 = 31.25°$ (see Equation 9.24)

Wall

Retained height = 7 m
Width of gabions in wall (B_w) = 1.00 m
Height of gabions and vertical reinforcement spacing = 1.00 m
Back tilt $\alpha = 6°$ ($\beta = 90° + \alpha = 96°$)
Width of wall base (B_b) = 2.00 m
Depth of wall base = 1.00 m

Pull-out tests demonstrate that only 2–3 m anchorage length is required to provide a pull-out resistance ≥ breaking strength of the mesh under as little as 1 m of overburden.

Active thrust with $\beta = 96°$ and $\phi = \delta = 30°$, $K_a = 0.2535$

$$P_a = \left(\frac{1}{2}\gamma H^2 + w_s H\right)K_a = 182.5 \text{ kN/m}$$

$$P_{ah} = P_a \cos(\delta - \alpha) = 166.7 \text{ kN/m}$$

$$P_{av} = P_a \sin(\delta - \alpha) = 74.2 \text{ kN/m}$$

Moment of active thrust about toe:

$$M_a = P_{ah} \frac{H}{3} \frac{H + 3 w_s/\gamma}{H + 2 w_s/\gamma} - P_{av} x_v = 2.93 P_{ah} - 1.81 P_{av} = 354.6 \text{ kN m/m}$$

Passive pressure with $\beta = 90°$ and $\phi = \delta = 30°$, $K_p = 10.10$, base 1.0 m deep

$$P_p = \frac{1}{2}\gamma H_p^2 K_p = 90.9 \text{ kN/m}$$

$$P_{ph} = P_p \cos\delta = 78.7 \text{ kN/m}$$

Weight of gabion wall

$$W_g = A_g \gamma_g = 148.5 \text{ kN/m}$$

Moment about toe (restoring):

$$M_g = W_g x_g = 1.38 W_g = 204.9 \text{ kN m/m}$$

TENSION IN REINFORCING LAYERS

Calculate tension in each layer assuming uniform distribution of load. If the tension in each layer is T, the moment due to all the layers about the toe is $(1+2+...+7) \times T = 28T$. Assuming that the resultant load acts through the centre of the base,

$$e = \frac{B_b}{2} - (M_g - M_a + 28T)/N = 0$$

where N is the normal force at the base of the gabion wall = $W_g + P_{av}$ = 221.9 kN/m.

Therefore, for equilibrium,

$$T = 13.3 \text{ kN/m}$$

LOCAL EQUILIBRIUM CHECK

Examine layer 2 (second from bottom) as being the most highly stressed. S_v is the vertical spacing of the reinforcement.

$$T_{(2)} = (\gamma z + w_s)K_a S_v = (18.6 + 18)(0.2535)1 = 31.9 \text{ kN/m}$$

$(< 53 \text{ kN/m})$

GABION INTERNAL STABILITY

Calculate shear stress between gabions at the most critical section (7 m below top of wall):

$$S = \left(\frac{1}{2}\gamma z^2 + w_s z\right) K_a \cos\delta - 6T = 44.7 \text{ kN/m}$$

$$N = \left(\frac{1}{2}\gamma z^2 + w_s z\right) K_a \sin\delta - 7\gamma_g = 187.3 \text{ kN/m}$$

$$\tau_{ag} = \frac{N}{B_w}\tan\phi_g + c_g = \frac{187.3}{1}\tan(31.25°) + 21 = 134.7 \text{ kPa}$$

$$\tau = S/B_w = 44.7 \text{ kPa}(< \tau_{ag})$$

Factors of safety

Against outward sliding

$$F_s = \left(P_{ph} + N\tan\phi' + \sum T\right)/P_h = 1.80$$

Against overturning

$$F_s = \left(M_g + \sum T\,\bar{y}\right)/M_a = 1.63$$

Against bearing failure

$$\sigma_v = N/B_b = 111 \text{ kPa}$$

$$F_b = q_{ult}/\sigma_v$$

Against tension failure

$$F_s = P_{ult}/T_{ave} = 53/13.3 = 3.98$$

Notes

1. The assumption of uniform load in layers is a simplification, but is compensated for by a relatively high safety factor (3.98) on tension failure.
2. Reinforcement layers should extend through gabion wall for at least 1 m and be securely tied to the gabion (in this case they can be used as gabion lids).

11.1.5 Pull-out capacity of different forms of earth reinforcement

A wide range of materials may be used as the reinforcement in reinforced earth systems.

11.1.5.1 Steel strip reinforcement

Strips derive their pull-out resistance entirely from friction on upper and lower surfaces. The pull-out force of a single strip of width b is given by

$$P_p = 2b \int_0^{L_a} \sigma'_n \mu_p \, dL \tag{11.18}$$

where L_a is the bond length.

To obtain the pull-out force per metre run of wall, the equivalent width of reinforcement per metre run, W_R, is defined in terms of the width and horizontal spacing S_h of the strips:

$$W_R = b/S_h \tag{11.19}$$

Hence, the pull-out resistance (per metre) for a layer of strips at depth z below a level ground surface with uniform surcharge is

$$P_p = 2 \, W_R \, L_a(\gamma \, z + w_s)\mu_p \tag{11.20}$$

For *steel strips in granular soil*, the Reinforced Earth Company use two different types of steel strip, smooth and ribbed (Figure 11.18a). For the ribbed (also called high adherence) strip in compact granular soil, σ'_n is enhanced by dilatancy (the increase of volume due to shearing) in the surrounding soil. Tests can thus lead to implied values of α_p $(= \mu_p/\tan \phi')$ in excess of unity if σ'_n is taken to be the overburden stress. Coefficients of friction used for Reinforced Earth walls are as follows (Figure 11.18b):
Smooth

$$\mu_p = \tan \delta = 0.4 \tag{11.21}$$

Ribbed

$$\mu_{p0} = \mu_0 = 1.2 + \log C_u \; (z = 0) \tag{11.22}$$

$$\mu_p = \tan \phi' \; (z \geq 6 \text{ m}) \tag{11.23}$$

(a)

(b)

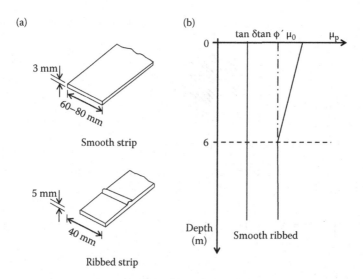

3 mm

60–80 mm

Smooth strip

5 mm

40 mm

Ribbed strip

Depth (m)

Smooth ribbed

$\tan \delta \tan \phi'$ μ_0 μ_p

0

6

Figure 11.18 Steel strip reinforcement. (a) Surface forms. (b) Coefficient of friction.

where δ is the angle of skin friction between steel and soil, and C_u is the coefficient of uniformity of the fill. A sacrificial thickness will generally be required to allow for corrosion over the design life.

11.1.5.2 Synthetic strip reinforcement

Synthetic strips are generally made of glass-reinforced plastic or high-tenacity polyester or nylon fibres encased in a polypropylene sheath.

The interaction coefficient for pull-out α_p for synthetic strips depends on the surface finish and the grading of the soil. Jones (1985) quotes values of α_p between 0.75 and 0.85 for GRP strips in granular materials and 0.53–0.64 in clay fill. Specialist manufacturers' literature should be consulted, or laboratory tests carried out, for other values.

11.1.5.3 Steel bar reinforcement

Bar reinforcement (Figure 11.19) is similar to strip reinforcement in many ways, but because it has a circular cross-section, normal stresses act around the circumference of a bar rather than just on upper and lower surfaces. One approach (John 1987) is to assume that the vertical effective stress, σ'_v, acts over the plan width (top and bottom) of 2d, and that σ'_h acts over the

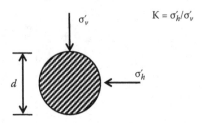

$K = \sigma'_h / \sigma'_v$

Figure 11.19 Stresses acting on bar reinforcement.

remainder of the circumference, $d(\pi - 2)$. Hence the pull-out resistance of a single bar is:

$$P_p = \int_0^{L_a} \left[\sigma'_v 2\, d + \sigma'_h\, d(\pi - 2) \right] \mu_0\, dL \tag{11.24}$$

11.1.5.4 Woven and non-woven geotextiles

Both *woven* and *non-woven geotextiles* have been used successfully to construct reinforced-soil walls. Pull-out resistance is very straightforward, as a geotextile acts as a simple full-width rough sheet:

$$P_p = 2 \int_0^{L_a} \sigma'_n \alpha_p \tan \phi'\, dL \tag{11.25}$$

Hence, the pull-out resistance per metre width for a layer of geotextile at depth z below a level ground surface with uniform surcharge is

$$P_p = 2\, L_a\, (\gamma\, z + w_s) \alpha_p \tan \phi' \tag{11.26}$$

In practice, the interaction coefficient α_p is determined from pull-out tests on reinforced soil; values in the range 0.7–1.0 have been measured. Geotextile manufacturers' literature should be consulted for product-specific values.

11.1.5.5 Wire mesh reinforcement

Wire meshes tend to have a relatively large aperture size and it is normal to distinguish the shear and bearing components of bond stress. The pull-out resistance per metre run is

$$P_p = m \int_0^{L_a} \pi\, d\sigma'_{ave} \alpha_p \tan \phi'\, dL + \sum_1^n d N_q \sigma'_v \tag{11.27}$$

where

  m is the number of longitudinal wires per metre width
  d is the diameter of steel wire
  σ'_{ave} is the average normal stress
  N_q is the bearing capacity factor
  n is the number of transverse wires in the length L_a

If σ'_{ave} is expressed as $\dfrac{1}{2}\left(\sigma'_v + \sigma'_h\right) = \dfrac{1}{2}\sigma'_v(1+K)$, the pull-out resistance (per metre) for a layer of wire mesh at depth z below a level ground surface with uniform surcharge is

$$P_p = m\,\pi\,d\,L_a\,\frac{1}{2}(1+K)(\gamma\,z + w_s)\alpha_p\,\tan\phi' + nd\,N_q(\gamma\,z + w_s) \tag{11.28}$$

Some workers have proposed using Terzaghi's values for N_q; others suggest that end-bearing factors for a pile are more appropriate. Alternatively, factors proposed by Jewell et al. (1984) could be used.

11.1.5.6 Polymer grid reinforcement

Polymer grids ('geogrids') generally have a smaller aperture size than wire meshes, and interact with soil in a highly efficient (though rather more complex) manner. Jewell et al. (1984) have demonstrated that friction and bearing components may be satisfactorily combined into a single bond coefficient:

$$\alpha_p = \alpha_a\,\frac{\tan\delta}{\tan\phi'} + \alpha_b\,\frac{d}{2s}\,\frac{N_q}{\tan\phi'} \tag{11.29}$$

where

  α_a is the fraction of grid surface area that is solid
  α_b is the fraction of grid width available for bearing
  δ is the angle of skin friction
  d is the diameter/thickness of transverse members
  s is the spacing of transverse members
  N_q is the bearing capacity factor

Theoretical upper and lower bounds for N_q have been derived that conveniently bound published data on grid pull-out tests (note: all angles in radians):

$$N_q \geq \exp[(\pi/2 + \phi')\tan\phi']\tan\phi']\tan(\pi/4 + \phi'/2) \tag{11.30a}$$

$$N_q \leq \exp[(\pi\tan\phi')]\tan^2(\pi/4 + \phi'/2) \tag{11.30b}$$

For s/d ratios in the range 10–20, geogrids behave as perfectly rough sheets with $\alpha_p = 1.0$. In any case, the pull-out resistance (per metre) for a geogrid at depth z below a level ground surface with uniform surcharge is defined in the same way as for a geotextile:

$$P_p = 2\,L_a\,(\gamma\,z + w_s)\,\alpha_p\,\tan\phi' \qquad\qquad (11.26\ bis)$$

11.1.6　Factors for avoiding limit states in reinforced fill

Previous sections have followed the traditional approach using factors of safety against failure that differ depending on the mechanism being investigated. More recent codes have adopted a partial factor approach.

Since the partial factors and load factors given in BS EN 1997-1:2004 (Eurocode 7) have not been calibrated against performance and experience, for reinforced fill, BS 8006: Part 1 does not recommend the use of the factors it contains. The approach of BS 8006: Part 1 is to classify walls according to the economic ramifications of failure (Table 9 and Section 5.3.2). For structures with high consequences (e.g. those acting as abutments or supporting roads and railways), a partial factor, f_n, of 1.1 should be applied. Partial factors are also applied to

- The ultimate tensile strength of reinforcement (Section 5.3.3)
- Soil parameters (Section 5.3.4)
- Soil/reinforcement interaction (Section 5.4.5)
- Soil self-weight (Section 5.3.6)
- External dead and live loads (Section 5.3.6)

Detailed values can be found in Section 6: Walls and abutments of BS 8006-1:2010, in Tables 11, 12 and 13.

FHWA-NHI-0-024 (FHWA 2010) also adopts partial factors, or 'Load and Resistance Factor Design' (LRFD), and is an update to FHWA-NHI-00-043 (2001) (which used global safety factors in an Allowable Stress Design (ASD) approach). According to FHWA (2010), 'Load and resistance factors for MSE walls are currently calibrated by fitting to ASD results'. A full list of load combination and load table factors are contained in Appendix A of FHWA (2010). For most walls, the following should be considered:

Permanent loads	Transient loads
Horizontal earth loads	Vehicular collision force
Earth surcharge load	Earthquake load
Vertical pressure from dead load of earth fill	Vehicular live load
	Live load surcharge

Typical MSE wall load combinations and load factors are given in Tables 4.1 and 4.2 of FHWA (2010). Resistance factors for external stability are given in Table 4.5. Resistance factors for internal stability are given in Table 4.7.

11.2 MULTI-ANCHORED EARTH RETAINING STRUCTURES

Reinforced soil (a.k.a. MSE) (see Section 11.1) in principle works by improving the properties of soil by including closely-spaced reinforcement in selected backfill, thus creating a gravity retaining structure that supports the ground behind it. The reinforcement is placed in backfill, as it is constructed, and is not pre-stressed. The stresses on the facing system are very small. In contrast, multi-anchored walls use more widely spaced elements which function by connecting the (structural) wall facing to more stable, generally *in situ* ground, and by improving the behaviour of more weathered and weaker near-surface ground.

Multi-anchored systems typically comprise a facing of some sort which is connected by the anchor tendon to a fixed length, grouted in natural soil, or an anchor block or deadman, placed in backfill. The anchors may be active (pre-stressed) or passive (picking up load as the wall moves). For a passively anchored wall, there may not be sufficient movement to reduce earth pressures to their active values. And where anchors are pre-stressed, to protect adjacent structures from ground movements, the loads to be taken by the anchors and facing units could exceed those derived from at-rest (K_0) conditions.

Earth pressure distributions, modified from those recommended by Terzaghi and Peck to allow for long-term conditions and multiple anchor levels are recommended in report FHWA-IF-99-015 (1999), and are given in Figures 11.20 to 11.22. For sands, water pressures and horizontal stresses due to surface loadings on the surface of the retained soil should be added to the pressures shown in Figure 11.20 to calculate the total stresses to be supported by the anchors.

For stiff to hard clays, the modified earth pressure distributions for temporary walls are shown in Figure 11.21. For a temporary wall, the total load per metre run of wall must be assumed to exceed $3H^2$. It can be expected, as a result of the dissipation of short-term negative pore pressures caused by unloading the retained soil during excavation, that the loads to be carried in the long-term may be greater than those shown in Figure 11.21. FHWA (1999) states that 'The apparent earth pressure diagram for stiff to hard clays under temporary conditions should only be used when the temporary condition is of a controlled short duration and there is no available free water. If these conditions are not met, an apparent earth pressure diagram

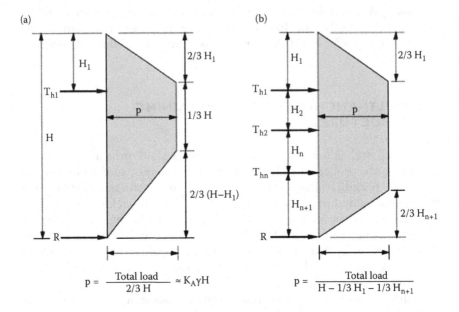

(a)   (b)

$$p = \frac{\text{Total load}}{2/3\,H} \approx K_A \gamma H$$

$$p = \frac{\text{Total load}}{H - 1/3\,H_1 - 1/3\,H_{n+1}}$$

H_1 = Distance from ground surface to uppermost ground anchor

H_{n+1} = Distance from base of excavation to lowermost ground anchor

T_{hi} = Horizontal load in ground anchor i

R = Reaction force to be resisted by subgrade (i.e. below base of excavation)

p = Maximum ordinate of diagram

Total load = $0.65\,K_A \gamma H^2$

Figure 11.20 Recommended apparent earth pressure distributions for the design of walls retaining sand and supported by ground anchors (US FHWA 1999). (a) Walls with one level of ground anchors. (b) Walls with multiple levels of ground anchors.

for long-term (i.e. permanent) conditions using drained strength parameters should be evaluated'.

For long-term or permanent anchored walls, FHWA (1999) recommends that the total load per metre run of wall calculated using the pressure distribution for temporary walls should be compared with a total load per metre run of $0.65\,K_a\,\gamma H^2$, where K_a is based on the effective angle of friction of the clay soil, and the larger of the two values should be used in the design. The report notes that an effective angle of friction of 22° (a reasonable value for high plasticity clays) results in an equivalent pressure, p, in the Terzaghi and Peck's (1967) envelope of $0.4\gamma H$. The effective angle of

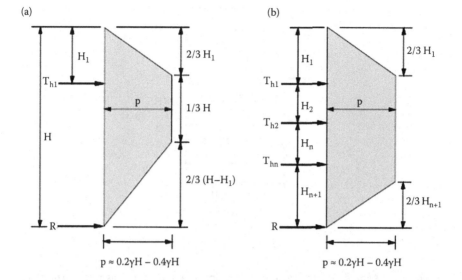

(a)

(b)

$p \approx 0.2\gamma H - 0.4\gamma H$

$p \approx 0.2\gamma H - 0.4\gamma H$

H_1 = Distance from ground surface to uppermost ground anchor

H_{n+1} = Distance from base of excavation to lowermost ground anchor

T_{h1} = Horizontal load in ground anchor i

R = Reaction force to be resisted by subgrade (i.e. below base of excavation)

p = Maximum ordinate of diagram

Total load (kN/m run of wall) = $3H^2$ to $6H^2$ (H in metres)

Figure 11.21 Recommended apparent earth pressure distributions for the design of temporary walls retaining stiff to hard clays and supported by ground anchors (US FHWA 1999). (a) Walls with one level of ground anchors. (b) Walls with multiple levels of ground anchors.

friction equivalent to a pressure of $0.2\gamma H$, 39°, is unlikely to be reasonable for clays, whatever their plasticity is.

The use of anchored walls for the temporary support of soft clays may be possible provided that a competent ground is to be found within a small depth below the base of the excavation. As with the Terzaghi and Peck (1967) recommendations for braced excavations, the apparent earth pressure is expected to be a function of the stability number, N_s, where

$$N_s = \gamma H/c_u \qquad (11.31)$$

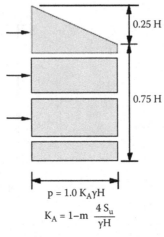

$$p = 1.0\, K_A \gamma H$$

$$K_A = 1 - m\,\frac{4\,S_u}{\gamma H}$$

m = 1.0 Except as noted

Figure 11.22 Apparent earth pressure distribution for temporary walls in soft to medium clays (FHWA [1999] after Terzaghi and Peck [1967]).

where

γ is the total unit weight of the clay

c_u is the average undrained shear strength of the clay below the base of the excavation

H   is the excavation depth

For soft to medium clays, FHWA (1999) recommends the use of the Terzaghi and Peck (1967) diagram (Figure 11.22). Where N_s < 6, m can be taken as 1.0, as recommended by Peck (1969). For N_s > 6, where large areas of soil can be expected to yield significantly as excavation proceeds, the use of m = 0.4 (as per Terzaghi and Peck 1967) is considered unconservative. For $4 < N_s < 5.14$, FHWA (1999) recommends the use of K_a = 0.22. Where N_s > 5.14, the use of pressures based on Henkel's (1971) solution is recommended.

The loads developing on the wall facing are supported by anchor rods and ties that carry them back to fixed lengths in stable ground. There are two principal types of ground anchor:

- Grouted ground anchors
- Mechanical ground anchors

11.2.1 Grouted ground anchors

The components of a typical grouted ground anchor are shown in Figure 11.23.

Grouted ground anchor walls came into widespread use in the late 1970s because of their ability to provide unobstructed working space, particularly

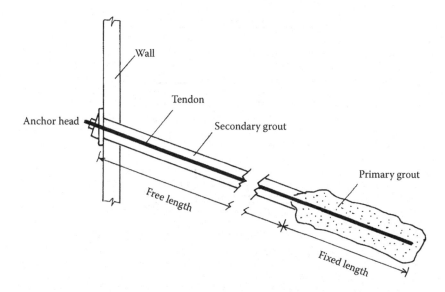

Figure 11.23 Grouted ground anchor components. (Corrosion protection omitted for clarity.)

in top-down construction, reduced construction time and reduced cost. Each type of anchor (Figure 11.24) (Littlejohn [1990]) is formed in and develops pull-out capacity in a slightly different way. The differences are mainly in the geometry of the fixed anchor length. The principal types are listed below, with details of pull-out resistance in the following section. Further advice on design and construction of grouted anchors can be found in (for example) FHWA (1999) and BS 8081 1989.

Type A anchors are formed using straight shaft boreholes (Figure 11.24a). The borehole may need to be lined during drilling and tendon placement if the ground is heavily fissured, and borehole collapse is likely. These anchors are used in rock and also in very stiff to hard cohesive deposits. Resistance to pull-out is governed by shear at the ground/grout interface. Fixed anchor lengths are typically around 8 m with a borehole diameter of approximately 110 mm.

Type B anchors are most commonly used in weak fissured rock and coarse granular alluvium. They use low-pressure grouting in a lined borehole which is steadily withdrawn so that the grout permeates through the soil pores around the lower part of the casing to form a grouted enlargement with minimum disturbance to the surrounding ground (Figure 11.24b). In this case, pull-out resistance is developed principally from side shear and partly from end bearing.

Type C anchors are similar to type B, except that low-pressure grouting is applied first and allowed to stiffen followed by high-pressure grouting.

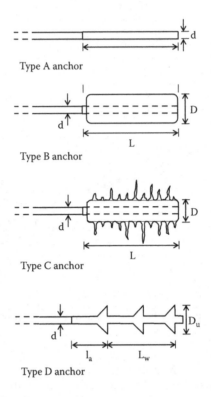

Figure 11.24 Classification of ground anchors.

This causes 'hydro-fracturing' of the ground mass which gives a grout root system through the ground and beyond the core diameter of the borehole (Figure 11.24c). This technique is used in fine granular soils and sometimes in stiff cohesive deposits.

Type D anchors use an under-reamed borehole to enhance end bearing resistance to the anchor (Figure 11.24d). Five under-reams are usually drilled up to a diameter of 600 mm. These anchors are most commonly used in firm to stiff cohesive soils, and pull-out resistance is developed through a combination of side shear and end bearing.

More recent developments include *jet-grouted anchors* and single-bore multiple anchors (SBMA). The former are used to apply high loads through a short fixed length in clayey, sandy, silty soils. A lined borehole is drilled and then a carefully controlled rotating water jet is used to form a water-filled enlargement along the fixed length. High-pressure grouting is used to flush out the water and fill the enlargement with grout, following which the anchor tendon is stressed and the grout enlargement is pressurised.

SBMA give greater anchor capacity by making the bond stress distribution along the fixed length more uniform and therefore more efficient. This is achieved by having a number of individual doubly-protected strands, each with its own fixed length, terminated in a staggered manner down the length of the borehole. Each strand is then stressed and locked off individually to create more even bond stress distribution. Barley (1997) describes SBMA in greater detail.

11.2.2 Mechanical ground anchors

A number of mechanical systems are in use for anchoring walls, for example, those using anchor beams and blocks for anchored sheet-pile retaining walls. The use of mechanical anchors is more restricted in anchored earth. The two examples that follow show that they have their place, however.

The *Duckbill* is a form of low-capacity anchor that can easily be installed in looser and softer ground conditions using light-weight plant, to support pre-cast concrete panels (Figure 11.25). Each anchor consists of a longitudinal sharp-edged aluminium alloy tube, with a polypropylene-coated metallic wire strand attached to its centre. The tube and strand are driven into the ground to the required depth of penetration with a rod and hammer. Tension is then applied to the strand which causes the tube to rotate and lock in place. This type of anchor may be useful for smaller walls, in difficult locations.

The T-bar anchor reinforcement system developed for the UK Road Research Laboratory Anchored Earth system may superficially resemble reinforced fill, but in fact the system derives most of its stability from the pull-out resistance from the 'deadman' action of the triangular hoop at the remote end of the bar. Murray (1983) assumed a particular mechanism of soil slip around the triangular hoop and derived:

$$P_p = K_p \, \sigma'_v \, w \, d \, sec \, \alpha \exp[2(\pi - \alpha)\tan\phi'] \qquad (11.32)$$

where w, d and α are defined in Figure 11.26.

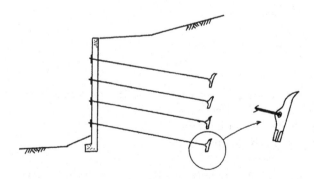

Figure 11.25 Anchored earth wall using Duckbill anchors and pre-cast facing panels.

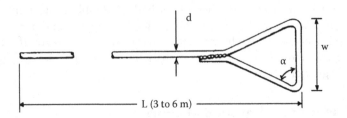

Figure 11.26 Geometry of TRL Anchored Earth deadman.

Alternatively, pull-out may occur either as a bearing failure on the full frontal area of the hoop (perpendicular to the axis of the anchor), or by shearing of the wedge of soil enclosed within the hoop. This gives

$$P_p = 2 w d \left(4 K_p \sigma_v' \right) \tag{11.33}$$

or

$$P_p = w d \left(4 K_p \sigma_v' \right) + 2 A_t \sigma_v' \tan \phi' \tag{11.34}$$

where A_t is the area of the triangle $= \dfrac{1}{4} w^2 \tan \alpha$. Pull-out force should be evaluated with both Equations 11.29 and 11.30, and the lower value taken for design.

The Austrian loop-anchored wall, possibly the most successful type under this category, is designed along similar lines to a 'deadman' anchored wall. Each unit comprises a precast facing panel, a synthetic strip loop, and a semicircular anchor unit. Horizontal stress acting over the area of the facing panel is transmitted by the loop back to the anchor unit, which has a pull-out resistance of

$$P_p = N_q \sigma_v' w_A h_A \tag{11.35}$$

where
   N_q is the bearing capacity factor (use Equation 11.30a)
   σ_v' is the vertical effective stress
   w_A is the front width of anchor unit
   h_A is the height of anchor unit

Frictional resistance generated along the synthetic strip is usually ignored.

11.3 SOIL NAILING

Most soil nailing uses steel bars that are grouted into pre-drilled holes in order to strengthen the ground, thus permitting the construction of steep excavated slopes. Satisfactory performance relies on mobilising tensile resistance in the nails, which in turn requires significant bond between soil and nail—which grouting ensures.

The US Federal Highway Administration has published extensive guidance on the design, construction and monitoring of soil nailing, in reports FHWA0-IF-03-017(2003) and FHWA-SA-96-69R (1998). Other important publications include Recommendations Clouterre (1991), and CIRIA report C637 (Phear et al. 2005). Recently, the British Standards Institution has provided a code of practice for the soil nail design BS 8006-2:2011, and for components and construction BS EN 14490:2010.

For new slopes, following excavation of each level, ground, nails are installed without pre-stress, and at a downward inclination (Figures 6.38, 6.39 and 11.27). Key components of a soil nail are

- The tendon—often a steel bar typically 10–30 mm in diameter
- Head plate, tapered washer and locking nut
- Duct, spacers, sheath and coating to prevent corrosion
- A grout annulus (Figure 11.27)

Nails are typically installed in rows that are between 1.5 m and 2.0 m apart, vertically.

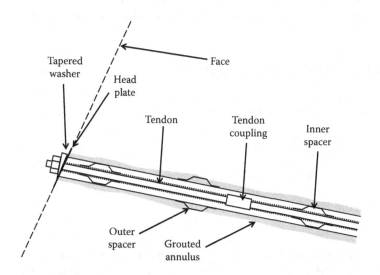

Figure 11.27 Components of a grouted soil nail.

Most of the length of the nails is protected by the grout, but the excavated face and the nail head are protected by a temporary and subsequently a permanent facing. The temporary facing, usually composed of shotcrete and light mesh, is placed shortly after excavation of each level, and before the nails for that level have been installed. These types of wall are convenient in highway construction because they are constructed from the top down, as general excavation takes place. Despite its name, soil nailing is commonly used in materials other than soil, such as weak and/or weathered rock.

Soil nailing can also be used to improve the stability of existing retaining structures, natural slopes and embankments. When using them in these applications, it should be borne in mind that this is a passive system, and therefore further displacement will be necessary to mobilise any support.

A major consideration when deciding whether or not to use this technique is the suitability of the ground for soil nailing:

- It must be possible to form benches 1–3 m high, from which to install the nails.
- Soft clays and silts are generally unsuitable for soil nailing. Clays with undrained shear strengths >50 kPa will be suitable, but water flush drilling should be avoided, for nail installation.
- Coarse materials, such as cobbles and boulders, may prevent nail installation.
- Granular soils will not maintain stability during nail installation, unless they are naturally cemented.
- Preexisting instability may require long nails.
- A high groundwater table may threaten or prevent installation, and groundwater may contain unfavourable chemicals, such as sulphates.

11.3.1 FHWA recommendations

FHWA (2003) suggests that the design of a soil nail wall must ensure a margin of safety against the ultimate limit states associated with (Figure 11.28)

- 'Global' failure
- Sliding stability
- Bearing failure (basal heave)
- Failure of the soil/nail system
- Facing failure
- Corrosion

whilst also ensuring that serviceability limit states, such as excessive movement of the wall toward the excavation, or excessive differential settlement, are avoided.

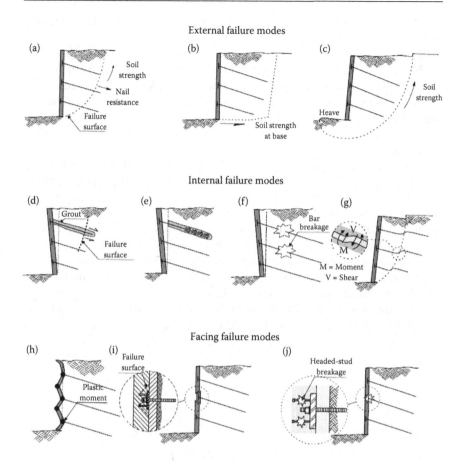

Figure 11.28 Principal failure modes for soil nailing. (a) Global stability failure. (b) Sliding stability failure. (c) Bearing failure (basal heave). (d) Nail-soil pull-out failure. (e) Bar-grout pull-out failure. (f) Nail tensile failure. (g) Nail bending and/or shear failure. (h) Facing flexure failure. (i) Facing punching shear failure. (j) Headed-stud failure. (From FHWA, Soil Nail Walls, Report No. FHWA0-IF-03-017, FHWA, Washington, DC, USA, 2003. Available at http://www.fhwa.dot.gov/engineering/geotech/library_sub.cfm?keyword=020.)

11.3.1.1 'Global' failure

Byrne et al. (1998) consider three types of block failure:

- 'External failure', where the soil nail wall mass is treated as a block, which trial failure surfaces do not intersect
- 'Internal failure', where all surfaces intersect nails
- 'Mixed failure', where failure surfaces intersect some nails

All three, plus sliding and bearing failure, are termed as 'global failure' in the following discussion.

FHWA (2003) suggests that for global failure modes, trial surfaces are analysed, treating the soil nail wall mass as a block, and evaluating its stability. Where a trial surface intersects one or more soil nails, the nails add to the stability of the block by providing external forces that are added to the soil resisting forces along the failure surface. A search is made to find the surface with the minimum factor of safety, as in slope stability analysis. This is termed the 'critical failure surface'.

Figure 11.29 gives an example of a trial surface that might be analysed to search for global failure. The driving forces that may lead to failure are the self-weight of the block abcd (W), and the resultant of any surcharge loads (Q). These are resisted by the shear forces, S_1 and S_2, on the trial surface (in this case chosen as a two-part wedge). At failure, S_1 and S_2 are related to N_1 and N_2 by the tangent of the effective angle of friction, so the directions of the resultants of these forces are known. The maximum available nail tensile forces at the trial surface (T_1 and T_2) can be calculated for the lengths of nail outside the block (see below). In this case, the upper nails do not contribute to stability because the trial failure surface does not pass through them. In a simple hand check, calculation stability can be assessed

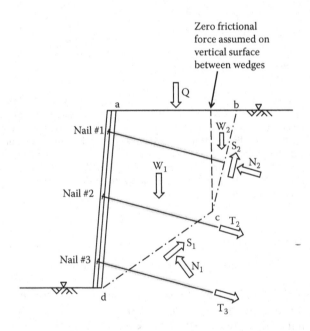

Figure 11.29 Example of forces on a trial global failure surface.

on the basis of either force equilibrium (perhaps using a force diagram) or moment equilibrium.

Global failure analyses of soil nail systems are normally carried out using specialist software, for example the SNAIL, TALREN, GOLDNAIL or FHWA SNAP (Soil Nail Analysis Program) 2010,* which uses two-dimensional limit state analyses similar to traditional slope stability analyses. SNAIL uses two-part planar wedges, GOLDNAIL uses circular failure surfaces, and TALREN offers circular, non-circular and logarithmic spiral failure surfaces. Different shapes of failure surface, assumptions and numerical solutions have been adopted over the years in a number of different methods, but comparisons suggest that (at least in granular soils) these do not produce significantly different computed factors of safety, and therefore produce similar design nail lengths. The individual analyses for global failure, sliding and bearing failures may be carried out using traditional hand calculation methods.

11.3.1.2 Internal failure—nail-soil pull-out

Strain measurements in instrumented walls have shown that in the upper part of the wall the maximum tensile force occurs approximately 0.3–0.4 H behind the facing, whilst at the bottom of the wall, this reduces to 0.15–0.2 H.

From a review of maximum nail tensile forces measured in in-service walls (Byrne et al. 1998), it was concluded that the average maximum tensile force in the upper two-thirds of a soil nail wall is

$$T_{max} = 0.75 \, K_a \, \gamma H \, S_v \, S_h \qquad (11.36)$$

where
  K_a is the coefficient of active earth pressure
  γ is the unit weight of the retained soil
  H is the wall height
  S_v and S_h are the vertical and horizontal nail spacings

The tensile force in the lower part of the wall reduces to approximately 50% of this. Briaud and Lim (1997) state that in the top row of nails

$$T_{max} = 0.65 \, K_a \, \gamma H \, S_v \, S_h \qquad (11.37)$$

For lower nails, they suggest a force of 50% of this value.

The available resistance provided by soil nails can be determined by pull-out testing, or from experience. The results of pull-out tests should be

---

* Downloadable from http://www.cflhd.gov./programs/techDevelopment/geotech/SNAP/.

interpreted with caution, since they create a different bond stress distribution than will be experienced in service. Elias and Juran (1991) provide the values given in Table 11.1.

For hand check calculations, these can be used to calculate the required additional nail length required to support the nail forces given above.

Table 11.1 Estimated bond strength of nails in soil and rock

Material	Construction method	Soil/Rock type	Ultimate bond strength. Qu (kPa)
Rock	Rotary Drilled	Marl/limestone	300–400
		Phyllite	100–300
		Chalk	500–600
		Soft dolomite	400–600
		Fissured dolomite	600–1000
		Weathered sandstone	200–300
		Weathered shale	100–150
		Weathered schist	100–175
		Basalt	500–600
		Slate/Hard shale	300–400
Cohesionless Soils	Rotary Drilled	Sand/gravel	100–180
		Silty sand	100–150
		Silt	60–75
		Piedmont residual	40–120
		Fine colluvium	75–150
	Driven Casing	Sand/gravel	
		low overburden	190–240
		high overburden	280–430
		Dense Moriane	380–480
		Colluvium	100–180
	Augered	Silty sand fill	20–40
		Silty fine sand	55–90
		Silty clayey sand	60–140
	Jet Grouted	Sand	380
		Sand/gravel	700
Fine-Grained Soils	Rotary Drilled	Silty clay	35–50
	Driven Casing	Clayey silt	90–140
	Augered	Loess	25–75
		Soft clay	20–30
		Stiff clay	40–60
		Stiff clayey silt	40–100
		Calcareous sandy Clay	90–140

Source: Elias, V. and Juran, I., Soil nailing for stabilization of highway slopes and excavations. *Report FHWA-RD-89-198*, FHWA, Washington, D.C., 1991.

11.3.1.3 Factors of safety

The FHWA minimum recommended factors of safety are shown in Table 11.2, and calculations for internal failure modes are given in FHWA (2003).

11.3.2 BS 8006-2 Recommendations

In contrast to the FHWA (2003) recommendations, and with the exception of near vertical walls with hard facings, BS 8006 Part 2 (2011) does not require the sort of checks for

- – Bearing capacity
- – Forward sliding on the base
- – Overturning

Table 11.2 FHWA recommendations for factors of safety for soil nailing

			Minimum recommended factors of safety		
			Static loads [1]		Seismic loads[2]
Failure mode	Resisting component	Symbol	Temporary structure	Permanent structure	(Temporary and permanent structures)
External Stability	Global Stability (long-term)	FS_G	1.35	1.5[1]	1.1
	Global Stability (excavation)	FS_G	1.2–1.3[2]		NA
	Sliding	FS_{SL}	1.3	1.5	1.1
	Bearing Capacity	FS_H	2.5[3]	3.0[3]	2.3[3]
Internal Stability	Pull-out Resistance	FS_P	2.0		1.5
	Nail Bar Tensile Strength	FS_T	1.8		1.35
Facing Strength	Facing Flexure	FS_{FF}	1.35	1.5	1.1
	Facing Punching Shear	FS_{FP}	1.35	1.5	1.1
	H.-Stud Tensile (A307 Bolt)	FS_{HT}	1.8	2.0	1.5
	H.-Stud Tensile (A325 Bolt)	FS_{HT}	1.5	1.7	1.3

Source: FHWA, Soil Nail Walls. Report No. FHWA0-IF-03-017, FHWA, Washington DC, USA. http://www.fhwa.dot.gov/engineering/geotech/library_sub.cfm?keyword=020, 2003.

Note: (1) For non-critical, permanent structures, some agencies may accept a design for static loads and long-term conditions with $FS_G = 1.35$ when less uncertainty exists due to sufficient geotechnical information and successful local experience on soil nailing; (2) The second set of safety factors for global stability corresponds to the case of temporary excavation lifts that are unsupported for up to 48 hours before nails are installed. The larger value may be applied to more critical structures or when more uncertainty exists regarding soil conditions; (3) The safety factors for bearing capacity are applicable when using standard bearing-capacity equations. When using stability analysis programs to evaluate these failure modes, the factors of safety for global stability apply.

that are carried out for gravity walls. It is suggested that because nailed soil does not act as a rigid block, its analysis is more comparable to that for slope stability.

BS 8006 Part 2 2011 therefore recommends that the designer should select initial dimensions based on slope angle (see CIRIA C637 and Table 11.3), and then should check against

- Rotational failure through soil and nails
- Sliding failure through soil and nails
- Deep-seated failure, around the nails
- Failure at the soil/grout interface (short nails)
- Failure of the nails themselves, due to inadequate cross-section

Rotational, sliding and deep-seated failure can be checked by hand using two-dimensional limit equilibrium analyses to assess stability, and empiricism to assess serviceability. Two forms of stability analysis are carried out:

1. Internal stability, where failure surfaces are either fully contained within the soil nail zone, or pass through some part of it
2. External stability, where failure surfaces do not pass through nail positions

It is suggested that both forms of analyses can be carried out using either Bishop's slope stability method (see Chapter 5), or a wedge method. A two wedge analysis (see Figure 11.29) is suggested as satisfactory for testing internal stability, but a three wedge analysis may be necessary to check for sliding on weak layers beneath the base of the wall.

BS 8006 Part 2 2011 uses a partial factor approach to BS EN 1997 Part 1 (2004)—as elsewhere, the characteristic values of material properties are 'a cautious estimate of the value affecting the occurrence of the limit state'. The partial factors, given in Table 5 of BS 8006 Part 2, are identical to those in the UK National Annex (Table 7.3) for actions and material properties but, in addition, partial factors are given for soil nail resistances (see Table 11.4), and there is mention of a model factor for other stability

Table 11.3 Initial dimensions for design

Slope angle	<45°	45° to 60°	>60°
Nail length	0.5–2.0 H	0.5–1.5 H	0.5–1.2 H
Nail spacing			
– vertical	1.5–3.0 m	1.0–2.0 m	0.75–1.5 m
– horizontal	1.5–3.0 m	1.0–2.0 m	0.5–2.0 m
Facing type	Soft non-structural for erosion control	Flexible to maintain face stability	Hard—high forces at the facing connection

Table 11.4 Partial factors specifically for soil nail resistances

Method of determining ultimate bond stress, τ_{bu}	Factors (γ_k) for determining characteristic bond stress (τ_{bk}) from ultimate values (τ_{bu})	Factors ($\gamma_{\tau b}$) for determining design bond stress (τ_{bd}) from characteristic values (τ_{bk}) $\tau_{bd} = \tau_{bk}/\gamma_{\tau b}$	
		set 1	set 2
Empirical pull-out test data	1.35–2.0 Based on degree of similarity of structure, soils, construction methods, etc.	1.11	1.5
Effective stress (using $\phi'_k, c'_k = 0$)	1.0–1.35 Based on potential for dilation and slope deformation in active zone	1.11	1.5
Total stress (using $c_{u,k}$)	1.35–2.0 Taking into account plasticity, potential for strain softening and shrink/swell	1.11	1.5
Pull-out tests	see BS EN 14490:2010 Cautious estimate of test data, based on number and location of tests and consistency	1.1–1.3 for coarse-grained soils 1.5–1.7 medium and high plasticity soils	1.5–1.7 coarse-grained soils 1.0–2.25 medium and high plasticity soils

calculation methods, that must be determined through calibration. These factors do not appear, explicitly, to take into account whether the structure is permanent or temporary.

BS 8006 Part 2 suggests that the designer establish a nail strength 'envelope' (Figure 11.30) so that during stability calculations the available nail resistance can easily be established for any given nail and assumed failure surface.

Figure 11.31 shows a trial global failure surface. It is evident that for the simplistic example shown:

- Nail #1 will not contribute to the stability along this failure surface.
- The stabilising force T_2 that can be contributed by Nail #2 will be limited by pull-out from the soil to the right of the trial failure surface.
- The stabilising force T_3 that can be contributed by Nail #3 will be limited by its pull-out from the soil and the facing to the left of the trial failure surface.

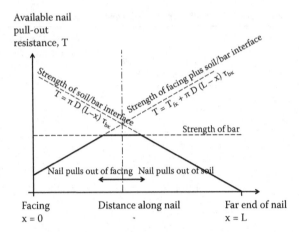

Figure 11.30 Example nail strength envelope.

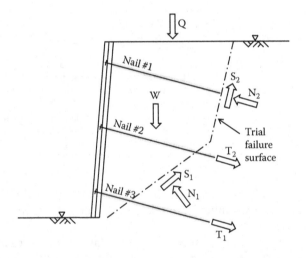

Figure 11.31 Example of forces on a trial global failure surface.

Despite considerable work in the past, according to BS 8006 Part 2, design of nails is based upon 'at best over simplified and conservative models' (Clause 4.3.2). Factors affecting pull-out resistance include groundwater conditions, installation effects, nail length and soil/nail geometry, and relative stiffness effects. Estimates of ultimate bond stress (τ_{bu}) can be based on

- Empiricism (Table 11.1)
- Total stress analysis (see below)

- Effective stress analysis (see below)
- The results of pull-out tests (see Tables 11.2 and 11.4)

Total stress analysis is only relevant for nails in clay. The method is similar to that for bored piles in clay, where the available shear strength on the pile shaft is calculated by multiplying the measured undrained shear strength by a factor (α, which is of the order of 0.5 for piles).

$$\tau_{bu} = \alpha \, c_u \tag{11.38}$$

BS 8006 Part 2 suggests that the value of α 'is likely to be in the range 0.5–0.9 for bond lengths ranging from 7–3 m'.

Effective stress analysis requires knowledge of the effective stress levels around the grout/soil interface. Whilst the vertical total stress may be relatively easy to estimate, equilibrium pore water pressures and the horizontal stresses are not. The ultimate bond stress can be calculated from

$$\tau_{bu} = \lambda_f \sigma_r' \tan\phi' + \lambda_c c' \tag{11.39}$$

where

$$\sigma_r' = \frac{\left(\sigma_v' + \sigma_h'\right)}{2} = \frac{\sigma_v'(1+K)}{2} \tag{11.40}$$

BS 8006 Part 1 suggests $K = 1$, whilst HA68/94 suggests $K = K_a$.

c' is conservatively taken as zero

λ_f and λ_c are interface factors for friction and cohesion, respectively. BS 8002: Part 2 suggests that the frictional interface factor, λ_f, should be between 0.7 (for smooth interfaces) and 1.0 (for rough interfaces).

The available nail pull-out resistance based upon characteristic bond strengths (Figure 11.30) should be multiplied by the 'set 1' and 'set 2' partial factors in Table 11.4, during design.

11.4 DESIGN OF BRIDGE ABUTMENTS FOR · EARTH PRESSURE

Traditional bridge abutments have decks that are supported on bearings, which are intended to protect the deck from longitudinal forces due to backfill, and the abutment wall from the effects of thermal deck expansion, which will occur due to heat of hydration after deck pouring, and as a result of daily and seasonal air temperature changes.

Two types of traditional abutment that have earth pressures exerted upon them are shown in Figures 11.32a and 11.32b. Both have decks that are supported on bearings, which allow the deck to slide, relative to the abutments. A third type, the 'integral bridge abutment' (see Chapter 3 and later in page 456), is increasingly used. In this arrangement, each end of the deck is fixed to the top of the abutment.

The retaining wall abutment shown in Figure 11.32a supports the full height of fill, and pressures are normally calculated on the basis that the top

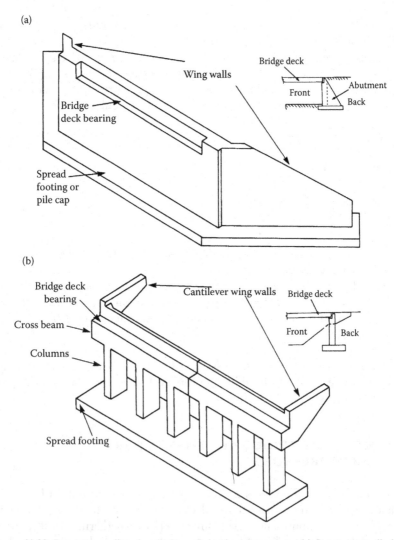

(a)

(b)

Figure 11.32 Retaining wall and spill-through bridge abutments. (a) Retaining wall abutment. (b) Spill-through or skeleton abutment.

of the abutment may move sufficiently toward the bridge deck to achieve active conditions. Provided that the abutment retains a free-draining granular fill, it is a common practice in the United Kingdom (Hambly 1979) to adopt an 'equivalent fluid pressure' approach to estimate the lateral earth pressure so that any depth, z, below the top of the embankment

Lateral pressure = $5z$ (kN/m^2), where z is in metres.

For this assumption to be valid, the wall should be able to move sufficiently to generate active conditions, and the wall must have drainage to prevent the build-up of positive pore pressures. This condition is roughly equivalent to an active earth pressure coefficient of 0.25, the Rankine value for $\phi' = 37°$. For abutments on piles, where lateral movements may be restricted, a coefficient of lateral earth pressure intermediate between active and at-rest is sometimes assumed, depending on the estimated degree of lateral restraint. Compaction pressures, which are greater than active pressures near ground surface, should be used to calculate the maximum bending moment at the base of the wall (see Chapter 3, Part I).

For other types of fill, the preliminary design charts of Terzaghi and Peck (1967) are sometimes used. If only poor-quality fill is available, the problem of high lateral pressures may be avoided by using bank seats on piles near the top of the fill to support the bridge deck, rather than a retaining wall abutment.

A further form of abutment in common use is the so-called 'spill-through' or 'skeleton' abutment (Figure 11.32b). In this form of structure, the abutment columns are embedded in the embankment, which slopes down away from the cross-beam beneath the bridge deck. For this type of structure,

   i. The influence of the soil in front of the abutment should be considered.
  ii. Soil arching and the effects of friction between the soil and the sides of the columns may be important.

In the Building Research Establishment survey of UK design practice carried out by Hambly (1979), it was reported that out of 20 statements of current design methods given by experienced soil engineers for the design of spill-through abutments, there were 12 different methods in use. The four most common design criteria in order of increasing magnitude of resulting earth pressure, according to Hambly, were

   a. For stable embankments with side and end slopes of 1 or 2 or less it is argued that the spill-through columns will not reduce the stability of the embankment, nor significantly influence movements of the embankment. It is assumed that there will be no net lateral earth pressures on the columns.
   b. Chettoe and Adams (1938) recommend that active pressures should be assumed to act on the rear face of the columns, but that the area of

the column face should be arbitrarily doubled. Shear on the column sides is ignored. This approach is based upon the assumption that soil behind the abutment flows forward between the columns, leading to increased column loads due to arching and column side shear. No pressure is taken on the front face of each column, on the basis that it is difficult to compact the soil in this zone. Stability is improved by the action of the self weight of the fill acting downward on the base slab.

c. Huntington (1967) recommended that full active pressure should be assumed to act only on the rear face area of the columns if they are widely spaced, but should be assumed to act over the gross area if the space between the columns is less than twice the column width. In effect, this approach means that the soil is assumed either not to arch or to arch perfectly between the columns. The fill in front of the columns is assumed to sustain up to active pressures, taking account of the batter slope, acting on the area of the column face alone.

d. Full active pressure is taken over the gross width of the rear of the abutment, with no allowance for the soil resistance on the front.

Little information is available from the monitoring of full-scale abutment structures. Work by Lindsell et al. (1985) has demonstrated, however, that most of the conventional assumptions are not justified. In reality, lateral pressures on spill-through abutments are balanced until the fill reaches the level of the top of the columns. If the bridge deck bearings are not free to slide during the deck pour, or the bridge deck is carried on some other part of the cross-beam during construction, large lateral pressures may result from thermal deck expansions immediately after the deck is poured. In any case, out-of-balance forces will result from the compaction of fill behind the cross-beam. Experiments by Moore (1985), coupled with observations of the deflections of the Wisley spill-through abutment (Lindsell et al. 1985), which indicate lateral movements at the top of the abutment that are never greater than about 5 mm, suggests that

a. Neither the active state nor the passive state will be approached.
b. Friction on the side of the columns is more likely to be mobilised than any significant proportion of passive resistance in front of the columns. This friction is likely to be dependent on the lateral pressure on the sides of the columns produced by compaction.

There is now an increasing trend for bridges to have integral abutments, where the top of the abutment wall is structurally connected to the bridge deck. Observations of the changes in prop loads and the pressures on integral abutments as a result of temperature effects have been discussed in

Section 3.3.9. It was concluded that two types of behaviour have been observed in the backfill to integral abutments:

1. In some materials (e.g. *in situ* clays supported by an embedded abutment) the pressures induced by cycling are controlled solely by the stiffness of the soil and the imposed strain, with no build-up of horizontal total stress with time.
2. In other materials (e.g. backfilled sands and gravels retained by frame abutments), the horizontal stresses progressively increase, until the active state is reached each time that the wall moves away from the soil, and the passive state is reached each time the wall moves toward the soil.

In the first case, a conservative rough estimate of the lateral earth pressure induced by thermal cycling can be made by calculating the thermal movement at the top of the wall (as described above), calculating the horizontal strain in the soil (the thermal movement divided by the distance from the top of the wall to the estimated point of rigid wall rotation), and multiplying this by the estimated horizontal Young's modulus of the soil. Young's modulus will vary with position down the wall, typically increasing with effective stress (and therefore depth).

A more precise prediction of earth pressures developed by soil supported by embedded abutments requires numerical modelling of soil-structure interaction (Bloodworth, Xu, Banks and Clayton 2011). The stiffness of the soil used in such a model should take into account anisotropy, strain-level dependence and non-linearity. Two approaches are possible:

a. Use a non-linear constitutive model that faithfully reflects the degradation of soil stiffness with strain.
b. Incorporate soil stiffness values that are appropriate for the strain levels expected, given the geometry of the structure and its foundations, and the predicted temperature-induced movement range.

In the second case, for frame integral abutments, the change in earth pressure can be estimated from the vertical effective stress down the back of the wall, combined with active and passive earth pressures. It is clearly possible that a bridge deck and abutment backfilling might be completed in either the summer or the winter months. However, Springman, Norrish and Ng (1996) have demonstrated that the initial direction of loading has no influence on the behaviour of granular soil during subsequent cyclic loading. The use of limiting active and passive pressures on frame integral abutments may seem conservative, but there is growing evidence from field monitoring that these are indeed achieved (Barker and Carder 2006).

11.5 COFFERDAMS

Cofferdams and caissons may be built through water and/or soft ground, in order to allow construction in relatively dry conditions in sound material (rock or soil) below. The objectives of their construction may, to give some examples, be to

- Provide a safe working area within which to construct the foundations for the pier of a bridge crossing a river
- Gain access to a tunnel, or provide a working space within which to launch tunnel boring machines
- Allow excavation for the recovery or installation of plant
- Install an impervious core for an earth dam

According to Puller (2003), the main difference between a cofferdam and a caisson in this application is that a caisson forms part of the permanent works, whilst the cofferdam is temporary. Caissons have been constructed in a range of materials (cast iron, steel, reinforced concrete) but cofferdams are normally constructed using steel sheet piling, which can be extracted and reused after each job is completed.

Cofferdams tend to be preferred to caissons when

- The area of excavation is large, as compared with the depth of water/soil to be supported.
- Sheet piling can easily be driven to the required depth.
- Dewatering is practical.
- Open excavation can be carried out.

In cases where there are obstructions, such as boulders, that may compromise the driving of steel sheets, or highly permeable soil, that may require excavation under pressure, in a so-called 'pneumatic caisson', cofferdams are unlikely to provide the optimal solution. However, Puller (2003) states that 'Improvements to construction methods and mechanical equipment have tended to increase the use of cofferdam construction to greater depths in recent years. Typically in the 1960s and 1970s, the economical depth of cofferdams would have been in the range 15–20 m of water, depending on subsoil type. In later years, however, the economical limit of cofferdam construction frequently extends 30–40 m or more, and caisson work continues to decline in popularity'.

Figure 11.33 shows an example of cofferdam construction, based on the St. Germans pumping station near King's Lynn, in Norfolk. Replacement of an existing pumping station required the construction of a new one within the drainage system. The river was diverted, and a double-skinned parallel walled steel sheet-pile structure, filled with sand, was constructed.

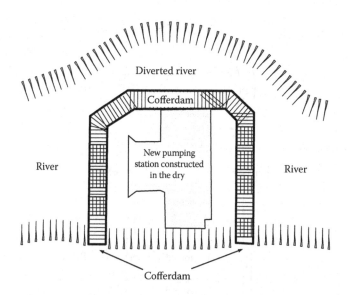

Figure 11.33 Example of cofferdam construction. Based on St. Germans pumping station, Norfolk. (Redrawn from http://www.mlcpumping.info/How.html.)

This allowed the ground to be excavated, and the new pumping station to be constructed in the dry.

Cellular and double sheet-pile walls are also used as quay walls and jetties. In these applications, BS6349 refers to 'cellular' and 'double-wall' sheet-pile structures. Such walls are backfilled and used essentially as gravity type structures. Other types are excavated into existing ground (Figure 11.34). General reviews of cofferdam design can be found in Packshaw (1962) and Williams and Waite (1993). Puller (2003) categorises cofferdams as

1. 'Sheeted types' that have shoring, anchors or internal bracing to provide stability. These can be divided into
   a. Circular cofferdams, where external earth and water pressures are supported by a combination of hoop compression and the bending stiffness of circular wallings (Figure 11.34)
   b. Braced cofferdams, where external earth and water pressures are largely supported by a system of wallings, struts or anchors, king piles and puncheons
2. 'Double skin' types use steel sheet piling to form cells that either are circular or have parallel walls, each containing fill material (as in Figure 11.35). Puller notes that 'the strength of these structures depends on the composite action of fill, the interlock strength of the sheeting and the underlying soil support'.

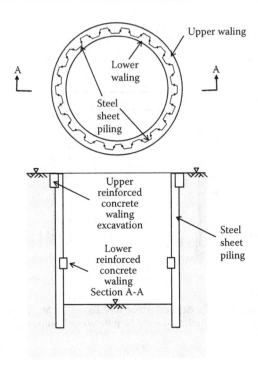

Figure 11.34 Sheeted circular cofferdam. (After Puller, M.B. Deep excavations: A practical manual. Thomas Telford, London, 2003.)

3. 'Gravity and crib' types, where—as the name implies—the lateral forces due to water are resisted by the shear force developed at the base of a gravity structure, such as an embankment dam, concrete block, gabion or crib wall. These simple types were used to build some of the earliest form of cofferdam, and rely amongst other things on a plentiful supply of suitable fill and slope protection materials.

This section is concerned primarily with the geotechnical design of cellular and double-wall sheet-pile cofferdams. Limit states and earth pressures applied to the other two classes of wall have been dealt with in earlier chapters of this book. Iqbal (2009) provides a useful summary of the design and performance of cofferdams. This has been used as the basis of the sections below.

Some currently available design standards and recommendations include

• US Army Corps of Engineers (USACE) design guidelines for sheet-pile cellular structures, cofferdams and retaining structures (1989)

(a)

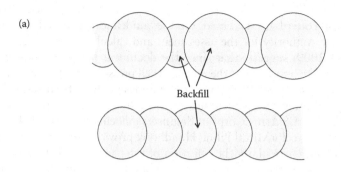

Backfill

(b)

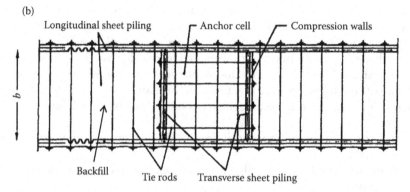

Longitudinal sheet piling   Anchor cell   Compression walls

b

Backfill   Tie rods   Transverse sheet piling

Figure 11.35 Plan views of double-skin cofferdams. (a) Circular. (b) Parallel-walled. (From Puller, M.B. Deep excavations: A practical manual. Thomas Telford, London, 2003; see also Figure 6.31.)

- NAVFAC DM7 guidelines for soil mechanics and foundation design (1971)
- United States Steel (USS) Steel Sheet Piling Design Manual (1984)
- CIRIA Special publication SP95 (1993) on 'The design and construction of sheet-pile cofferdams'
- BS 8002: 1994, the UK Code of Practice for earth retaining structures
- ArcelorMittal (British Steel) Piling Handbook (2008)
- BS6349-2:2010 'Maritime works—Part 2: Code of practice for the design of quay walls, jetties and dolphins', Sections 7.7 and 7.8

These documents have different origins, and therefore different strengths and weaknesses Iqbal (2009). The USACE design guidelines are the result of experience and laboratory studies of cofferdam construction in the USA, particularly in the Mississippi in the 1960s and 1970s. They use early design methods proposed by Terzaghi (1944), Brinch Hansen (1953), Cummings (1957) and Schroeder and Maitland (1979). There is also some limited discussion of the use of numerical modelling in cofferdam design.

The NAVFAC DM-7 document is widely used by geotechnical designers. The section on cofferdams is based on the guidelines proposed by the Tennessee Valley Authority for the assessment and calculation of cofferdam stability. Iqbal (2009) suggests that since the document has apparently not been updated to take into account the modified cell pressure profiles suggested by Maitland and Schroeder (1979), nor does it appear to have been validated against field monitoring results or numerical modelling, it should be used in conjunction with the US Army Corps of Engineers guidelines described above.

Section 7 of the ArcelorMittal Piling Handbook provides guidance on cofferdam design, and particularly the selection of type and size of sheet-piles, and detailing, from a structural capacity viewpoint. It also provides tables for the selection of sheet-pile embedment depth on the basis of pile section modulus, soil strength and type of construction and installation technique. Since it provides little in the way of guidance on overall stability, Iqbal suggests that it is used for pile selection and to identify the installation technique.

The ArcelorMittal Piling Handbook provides a useful list of the causes of failure of cofferdams. Noting that there can be many causes of failure, it states that failure can generally be attributed to one or more of the following:

- Lack of attention to detail in the design and installation of the structure
- Failure to take the possible range of water levels and conditions into account
- Failure to check design calculations with information discovered during excavation
- Over excavation at any stage during the construction process
- Inadequate framing (both quantity and strength) provided to support the loads
- Actual loading on frame members (e.g. from materials, pumps, etc.) not taken into account in the design
- Accidental damage to structural elements not being repaired
- Insufficient sheet-pile penetration to prevent piping or heave
- Failure to allow for the effects of groundwater (excess pore pressure, piping, heave) on soil pressures
- Lack of communication between temporary and permanent works designers, and between designers, site managers and operatives

Iqbal (2009) lists the following mechanisms of failure of cofferdam cells:

- Insufficient interlock strength of sheet-piles
- Internal shear failure within the cell
  - Vertical shear failure
  - Horizontal shear failure
  - Sheet-pile penetration capacity exceeded

- External instability
  - Sliding failure
  - Bearing failure
  - Overturning failure
  - Seepage failure

BS 6349-2:2010 summarises the modes of failure for cellular sheet-piled structures as shown in Figure 11.36.

Iqbal notes the following shortcomings in current cofferdam design:

1. In current design guidelines, the various mechanisms of failure are considered independently without the possibility of unfavourable interactions.

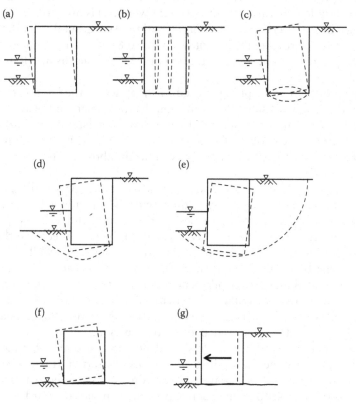

Figure 11.36 Modes of failure of cellular sheet-piled structures. (a) Shear failure within fill. (b) Sheet interlock failure. (c) Tilting on base rupture surface. (d) Bearing capacity failure. (e) Global instability. (f) Rotation about toe. (g) Sliding on base. (Based on BS 6349-2:2010, Maritime works—Part 2: Code of practice for the design of quay walls, jetties and dolphins. British Standards Institution, London, 2010.)

2. Much analysis and numerical modelling has been developed only for circular cell cofferdams, and may not be suitable for parallel-walled cellular cofferdams.
3. Only the USACE document recommendations include a flooded tension crack (Bolton and Powrie 1987).
4. External stability checks are required at the main stages of construction.
5. There is no guidance on the sizing of berms, which are sometimes used to improve lateral stability. It may be possible to optimise design using the methods described by Daly and Powrie (2001) and Smethurst and Powrie (2008).

According to BS6349, *cellular sheet-pile structures* may be founded on weak rock, granular material and very stiff clay. Soft material should be dredged prior to construction. Cellular structures use less steel than double wall sheet-pile structures because straight web steel is used to resist backfill pressure in hoop tension rather than bending. As a general rule, the width of these structures should be not less than 0.8 times the retained height. Installation of straight web sheet piling may be an issue, as it will not resist high driving stresses.

Where cellular sheet piling is used for quay walls, the lateral stress acting on the landward side of the cell should approximate to the active earth pressure as a result of structural deformation. The lateral pressure within the cells should be taken as the 'at rest' (K_0) value. The maximum pressure may be assumed to act at 1/4 of the wall height above the toe of the sheets (Figure 11.37b).

Double-wall sheet-pile structures may also act as gravity walls, provided the walls are prevented from spreading apart. The combined action of the two walls is achieved by anchor ties near the top of the wall and penetration of sheeting at the toe. If spreading cannot be prevented then they can be designed as two independent anchored steel sheet-pile walls. They are less efficient than cellular steel sheet-pile structures because earth pressure must be resisted using sheet-pile bending. They are most suited to founding in medium or dense granular soil, firm to stiff clay, or chalk. In rock, it may be difficult to obtain sufficient toe penetration. As with cellular structures, soft clays should be removed before construction, perhaps by dredging. The distance between the walls should not be less than 0.8 times total wall height above the sheet piling toe, or a hard stratum. If lateral stability against sliding relies on passive pressure on the toe of the seaward wall, then it is reasonable to assume active pressures on the landward side of the structure. If there is higher lateral resistance, perhaps due to penetration into rock, the wall may have to support earth pressures between the active and at-rest values.

A summary of failure conditions is given in Figure 11.36. The following sections provide guidance on design to prevent these.

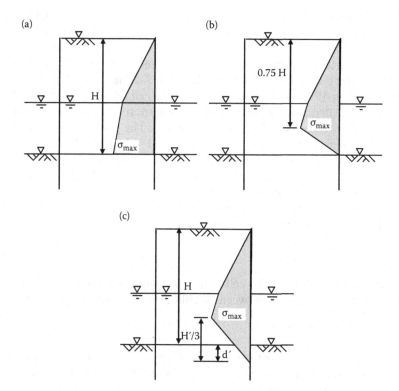

Figure 11.37 Lateral earth pressure distributions assumed in the calculation of sheet-pile interlock forces. (a) Terzaghi (1944). (b) TVA (1957). (c) Maitland and Schroeder (1979).

11.5.1 Interlock strength of sheet-piles

Sufficient sheet-pile interlock strength (Figure 11.36b) is essential in all cofferdams, and particularly in circular double-skin cofferdams. The maximum interlock force is determined from the sum of the earth pressure acting inside the cofferdam plus any out-of-balance water pressure (Δu). The effective earth pressure at any depth is found by multiplying the effective vertical stress at any level by a suitable earth pressure coefficient (see Figure 11.37).

$$\sigma_h = K\sigma_v' + \Delta u \tag{11.41}$$

Terzaghi (1944) used an earth pressure coefficient, K, of 0.4. This was subsequently increased to 0.5 by the US Army Corps of Engineers. The pressure distribution recommended by Terzaghi (1945) and by the US Army Corps of Engineers assumed that the maximum pressure acted at the base of the cofferdam cell (Depth = H, see Figure 11.37a). This distribution was modified by the Tennessee Valley Authority (TVA) guidelines, which placed the maximum at 0.75 H below the top of the cell, the pressure reducing to zero at the base of the cell, for cells founded on rock (Figure 11.37b). The pressure distribution suggested by TVA was modified in the light of field and laboratory data from circular cells embedded in sand/clay (Maitland and Schroeder 1979). It was proposed that the cell fill pressure should drop to zero at the depth where plastic hinge formation was observed in model tests, d' below the dredge level (Figure 11.37c). The point of maximum lateral pressure was found to be at $H'/3$, where $H' = H + d'$. The lateral earth pressure coefficient was calculated as 1.2–1.6 times the active earth pressure coefficient, K_a. This distribution is considered suitable for small cells, whether on rock, clay or sand foundations.

The maximum interlock force per unit length in a circular cell can be calculated from horizontal equilibrium as

$$t_{max} = \sigma_{max}\, r \tag{11.42}$$

where r is the radius of the cell. The maximum interlock force per unit length at connecting arcs between neighbouring cells can be calculated as

$$t_{max} = \sigma_{max}\, L \sec \theta \tag{11.43}$$

where L is the distance between the joint and the centreline of the cofferdam, and θ is the angle between the joint and centreline of the cofferdam, subtended at the centre of the cell.

The factor of safety against interlock failure can then be calculated as

$$F = \frac{T}{t_{max}} \tag{11.44}$$

where T is the interlock strength of the sheet-piles.

11.5.2 Internal shear failure within the cell

Terzaghi (1944) noted that excessive deflections or rotations of cofferdam cells may be caused by internal shear failure of the soil and that this can be avoided by ensuring an adequate factor of safety on assumed shear planes within the fill.

11.5.2.1 Failure on a vertical plane at the centre of the cell (Terzaghi 1944)

Terzaghi suggested that, because of its flexibility, a cofferdam is likely to fail because of shear in the fill it contains, before it fails by overturning or sliding. He proposed a simple method of calculation that compares the applied shear stress with the available strength on the vertical plane passing through the centre of the cofferdam. The total shear force acting on the vertical plane passing through the centreline can be calculated from the assumed triangular pressure distribution at the base of the cell. For an applied bending moment, M, acting on a weightless cell of width B, the contact stresses at the base of the wall vary from $+6M/B^2$ to $-6M/B^2$ (see Figure 11.38). The shear stress, Q, applied on the cell centreline is therefore

$$Q = \frac{3M}{2B} \tag{11.45}$$

The shear resistance is the sum of the resistance offered by the cell fill and the sheet-pile interlock sliding resistance. The shear resistance of the cell fill is equal to the effective lateral pressure at the cell centreline multiplied by the effective angle of friction of the cell fill material. Using the centreline earth pressure distribution proposed by Terzaghi (Figure 11.39b), the resultant horizontal force, P_s, can be found to be

$$P_s = \frac{1}{2}\gamma K(H - H_1)^2 + \gamma K H_1(H - H_1) + \frac{1}{2}\gamma' K H_1^2 \tag{11.46}$$

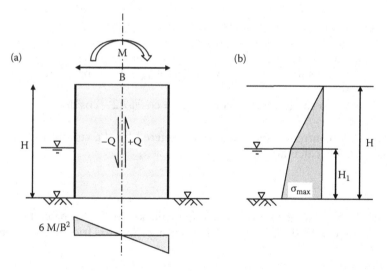

Figure 11.38 Vertical shear resistance in a cofferdam. (a) Applied bending moment and contact stresses. (b) Centreline earth pressure profile. (Redrawn from Terzaghi, K., *Proc. ASCE*, 70, 1015–1050, 1944; *Proc. ASCE*, 71, 980–995, 1945; *ASCE Transactions*, 110, 1083–1202, 1945; Harvard Soil Mechanics Series, 26.)

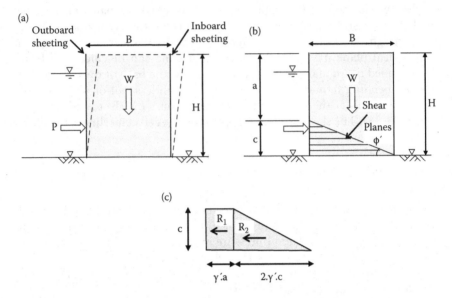

Figure 11.39 Horizontal shear resistance in a cofferdam. (a) Applied forces and cell movement. (b) Resisting wedge. (c) Pressure diagram. (Redrawn from Cummings, E.M., *Trans. ASCE*, 125, 13–34, 1957.)

where

H is the height of the cell

H_1 is the height of water above the base

γ is the unit weight of cell backfill above the water

γ' is the submerged unit weight $(\gamma - \gamma_w)$ of cell backfill, below the water

K is the earth pressure coefficient $= \cos^2\phi'/(2 - \cos^2\phi')$

The total centreline shear resistance offered by the cell fill (S_s) can be conservatively calculated as

$$S_s = P_s \tan \phi' \tag{11.47}$$

The shear resistance provided by the interlocks can be estimated by multiplying interlock tension (P_T), calculated above, by the steel friction coefficient, f

$$S_f = f\,P_T \tag{11.48}$$

Therefore the total shearing resistance, S_T is

$$S_T = P_s \tan \phi' + f\,P_T \tag{11.49}$$

and the factor of safety on shear failure on the vertical plane at the centreline, FoS_{vs} can be calculated by dividing the total shear resistance (S_T) by the ultimate load (Q) on the centreline. It has been suggested that this approach is satisfactory for cells founded on rock, sand and stiff clay, where sufficient foundation stiffness is available. On soft clay, it may be necessary to rely on the resistance of the interlocks alone. The factor of safety for a cell founded on soft clay will be

$$FoS_{vs} = \frac{PRf\left(\dfrac{B}{L}\right)\left(\dfrac{L+0.25B}{L+0.50B}\right)}{M} \tag{11.50}$$

where
- P   is the pressure difference of the inboard sheeting
- R   is the cell radius of the cell
- f   is the coefficient of friction of steel (0.3)
- B   is the effective width of the cell
- L   is the distance between centrelines of adjacent cells
- M   is the net overturning moment

The Terzaghi method was found to be in reasonable agreement with experimental results when the earth pressure coefficient was taken as 0.4–0.5, but experience shows that for conventional sized cofferdams, the values of K are generally above that limit. Hence Terzaghi's method cannot theoretically be applied until the cell width is large enough to allow mobilisation of active and passive zones (Brinch Hansen 1953). TVA (1957) suggested a modified pressure distribution, based on Schroeder and Maitland's (1979) work (Figure 11.37) and the use of an earth pressure coefficient (K) of 1. The total shear resistance (backfill plus steel sheet interlock) is

$$S_T = \frac{1}{2}\gamma K H'^2 (\tan\phi' + f) \tag{11.51}$$

where H' is the height of the cell over which the vertical shear resistance is calculated.

11.5.2.2 Failure on horizontal planes at the base of the cofferdam (Cummings 1957)

The horizontal shear method was developed by Cummings (1957) because of inconsistencies that had become apparent in the methods based on Terzaghi (1944). The method assumes that horizontal shear planes can develop within the fill. According to Puller (2003), 'This [method] implies that fill on the unloaded side of the cell could be reduced without affecting stability', which is not practically sound and it 'should not be used in design to reduce fill levels within cells'.

The concept of the Cummings' method is that the fill offers resistance in the form of a soil wedge rising at an angle ϕ' to the horizontal from the inboard side of the cell (Figure 11.39), where ϕ' is the effective angle of friction of the backfill. The fill above this wedge acts as an overburden, and the soil fails on horizontal shear planes. The total resistance to sliding, R, is (Figure 11.39c)

$$R = R_1 + R_2 = \gamma' \, B \, H \tan \phi' \tag{11.52}$$

Substituting

$$H = a + c \tag{11.53}$$

and

$$\tan\phi' = \frac{c}{B} \tag{11.54}$$

Cummings determined that

$$R = R_1 + R_2 = a \, c \, \gamma' + c^2 \, \gamma' \tag{11.55}$$

He treated Figure 11.39c as a lateral pressure diagram, from which the resisting moment about the base can be computed. The total moment of resistance per unit length of wall is therefore

$$M = R_1 \frac{c}{2} + R_2 \frac{c}{3} = \frac{ac^2\gamma'}{2} + \frac{c^3\gamma'}{3} \tag{11.56}$$

Interlock friction provides additional resistance and can be calculated as

$$M_f = P_T f B \tag{11.57}$$

The total resisting moment provided by horizontal shear failure is $M + M_f$, and the factor of safety against failure will be

$$F = \frac{M + M_f}{M_o} \tag{11.58}$$

where M_o is the net overturning moment. If a berm is used on the inboard side of the cofferdam then

$$F = \frac{M + M_f + P_B\left(\dfrac{H_B}{3}\right)}{M_o} \tag{11.59}$$

where

P_B is the passive resistance provided by the berm, and
H_B is the height of the berm

The pressure diagram assumed by Cummings and shown in Figure 11.39c is hard to explain on the basis of earth pressure theory. It is justified on the basis of case records and considerable experience of its use, particularly in the USA.

11.5.3 Pull-out of outboard piles/Failure of inboard piles

The outboard piles must be checked for pull-out, and the inboard for bearing capacity failure, resulting from the overturning moment on the cell, using the USACE Design Guidelines (1989).

Although the outboard piles are commonly lengthened to increase the seepage path on the outboard side, to reduce uplift pressures, they should also be checked for pull-out. For a cofferdam founded on granular soil, the Guidelines calculate the ultimate pull-out capacity, Q_u, as

$$Q_u = \frac{1}{2} K_a \gamma' D^2 \tan\delta\, p \tag{11.60}$$

where

K_a is the active earth pressure coefficient
γ' is the buoyant unit weight of the soil in contact with the outboard pile
D is the embedment depth
$\tan\delta$ is the coefficient of friction on the soil/pile interfaces (see Table 4.3 of the Guidelines)
p is the sheet-pile perimeter

For sheets embedded in clay, the Guidelines suggest a total stress (short-term) approach, using undrained shear strength. The results of this should be checked using a long-term effective stress calculation.

The overturning moment applied to the cofferdam will result in a pull-out force (Q_p). The factor of safety against the pull-out of the outboard piles is then

$$F = \frac{Q_u}{Q_p} \tag{11.61}$$

For cellular cofferdams on sand, the inboard sheet-piles should be driven to a sufficient depth to counteract the vertical downward friction force F_1 caused by the interaction of the cell fill and the inner face. This friction force is given by

$$F_1 = P_T \tan \delta \tag{11.62}$$

where
   P_T is the interlock force
   δ is the coefficient of friction between steel sheet piling and cell fill

Generally, a factor of safety of 1.5 applied to F_1 is adequate, according to the USS Steel Sheet Piling Design Manual.

11.5.4 External failure

As noted above, external shear failure can occur in a number of ways:

- Sliding failure
- Bearing failure
- Overturning failure
- Seepage failure

11.5.4.1 Sliding failure

Force equilibrium should be checked for kinematically admissible presumed failure surfaces at the base of the cell. These surfaces can be curved, or straight, or a combination, and should take into account any known weaker layers in the intact ground beneath the cell. A program such as 'Limit State: GEO' (Smith and Gilbert 2010) may help the search for critical failure surfaces, and the determination of factor of safety. The USS Steel Sheet Piling Manual states that 'in general, horizontal sliding of the cofferdam at its base will not be a problem on soil foundations' but caution should be exercised in unfavourable ground conditions, such as in interlayered sands and clays, and laminated/varved clays.

11.5.4.2 Bearing capacity failure

Cofferdams are often founded on sand or rock, which will provide adequate capacity against bearing failure. However, when cells are underlain by clay, care should be taken. Bearing capacity can be calculated using any method that takes into account inclined eccentric loading (see, for example, Chapter 9) suitable for gravity retaining structures.

11.5.4.3 Overturning failure

The overturning stability of rigid gravity structures has traditionally been assessed by considering moments about the toe of the wall, with the moment applied by external forces being resisted by the moment provided by the self weight of the structure. Pennoyer (1934) proposed a similar check for cofferdam cells, and it was this that led Terzaghi to propose his check on internal stability. However, Brinch Hansen (1953) raised objections to Terzaghi's method in two respects: first, that his mechanism of internal failure was kinematically inadmissible, and second, that the earth pressure coefficient relevant to failure on a vertical surface ($\cos^2\phi'$) should be much higher than the values used in practice, suggesting that Terzaghi's vertical shear surface is not the critical one.

Brinch Hansen suggested a more critical, kinematically and statically admissible mode of external failure, involving overall rotation about the centre of the cell, with a circular rupture surface passing through the toes of the inboard and outboard sheeting (Figure 11.40). The failure circle is considered to be concave downward for a cell founded on rock, but may be either concave upward or concave downward for a cofferdam founded on clay/sand strata. It is suggested that where the sheet-pile driving depth is shallow the failure surface will be concave downward, as in Figure 11.40a,

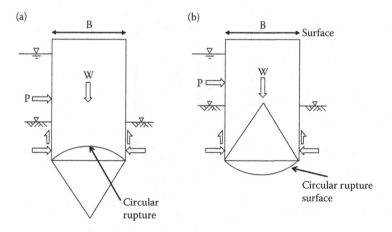

Figure 11.40 Rupture surfaces for Brinch Hansen's (1953) 'Equilibrium Method'. (a) Shallow sheet-pile driving depth. (b) Deeper sheet-pile depth.

and where the driving depth is greater, it may be concave upward (Figure 11.40b), or concave downward, or a combination of both. Where the driving depth is considerable, plastic hinges may develop in the sheets.

In this method, a number of failure surfaces are considered and the factor of safety is calculated for each, in a search for the minimum factor of safety. Although this method is kinematically admissible, it involves complex calculations of the internal forces on the failure plane.

For this reason, Krebs Oveson (1962) suggests using a log spiral surface to replace the circular surface, in what is termed as the 'extreme method' (Brinch Hansen 1953). The log spiral obeys the polar equation

$$r = r_0 \, e^{\theta \tan \phi} \tag{11.63}$$

where

r_0 is the radius at the start of the log spiral (Figure 11.41)
r is the radius after rotation θ about the centre of the log spiral
ϕ' is the angle of friction of the soil

The radius vector at any point on the log spiral makes an angle ϕ' with the normal to the spiral. In frictional soil, with an angle of friction ϕ', the resultant of the internal forces on the log spiral failure surface therefore passes through the centre of the log spiral, and need not be taken into account in any calculation of moment equilibrium about the centre of the log spiral. In addition, because the resultant of the forces on the caisson (Q and the self-weight W in Figure 11.41) must be opposed by the resultant on the log spiral, R, the centre of the log spiral must lie on this line. The search for the critical log spiral is therefore restricted to this line.

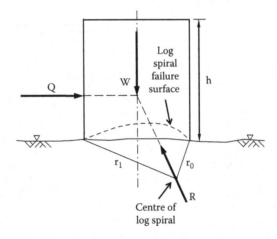

Figure 11.41   Log spiral rupture surface for Brinch Hansen's (1953) 'Extreme Method'.

The moment stability calculation therefore is carried out as follows:

1. Draw the cofferdam to scale, as shown in Figure 11.41.
2. Compute the external (Q) and gravity (W) forces acting on the body above the rupture line.
3. Plot an arbitrary log spiral through the toes of the inboard and outboard sheeting. Locate the centre (pole) of the log spiral on the resultant of Q and W.
4. Take moments about the pole of the log spiral:

$$F = \frac{M_{stabilising}}{M_{driving}} \tag{11.64}$$

5. Draw a number of other log spirals, centred on the line of the resultant, for different values of r_0.
6. Recalculate the factor of safety, F, for each.
7. Plot the factor of safety as a function of the initial radius, r_0, of the log spirals. If the minimum factor of safety, F_{min}, >1, the cofferdam is stable.

11.5.4.4 Seepage failure

For most caissons, the groundwater levels will be high, and the seepage failure needs to be investigated. A flow net is required to calculate the uplift pressures around the structure. If the hydraulic gradient due to upward flow at the inboard edge of the caisson exceeds unity then there is significant risk of piping failure. In addition, high seepage forces will significantly reduce sliding resistance on the base of the cofferdam, and reduce the bearing capacity of the ground on which it is founded.

11.5.5 Numerical modelling

The complex interactions between existing soil, cofferdam and anchors, and backfill, described in brief above, suggest that numerical modelling, using the finite element or finite difference methods, may be the way forward for design. Early numerical modelling is reported by Clough and Hansen (1977) and Shannon and Wilson (1982).

Iqbal (2009) provides an excellent case record of the design, construction, and monitoring of a large cofferdam complex at St. Germans in Norfolk, UK. Displacements and bending moments derived from two- and three-dimensional finite difference analyses are compared with values obtained from data obtained from monitoring the structures. Considerable detail is given of the assumptions made during analyses, which will be extremely helpful for engineers carrying out future design for similar projects.

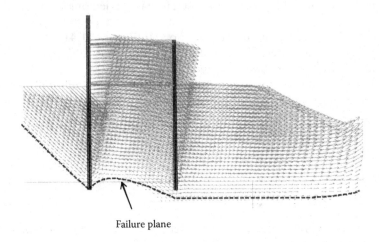

Failure plane

Figure 11.42   Failure surface for St. Germans cofferdam, deduced from displacements predicted using finite difference analysis (factor of safety = 1.25) (Iqbal 2009).

As an example of the results from numerical modelling, Figure 11.42 shows the displacement vectors obtained as the strengths of the existing soil, the fill and the wall friction were gradually reduced during analyses. The failure plane under the cofferdam is reminiscent of the log spiral form suggested by Brinch Hansen's 'Extreme Method' (Figure 11.41). It was found that two-dimensional finite difference predictions in general provided a better fit with monitored bending moments and wall displacements.

Appendix A: Classical earth pressure theory

A.I DEFINITION OF COEFFICIENTS OF EARTH PRESSURE

The earth pressure coefficient, K, is the ratio of effective stress on the back of a wall to the vertical effective stress resulting from self-weight of backfill and/or external surcharge. When a retaining wall yields away from the soil, it is termed the coefficient of active earth pressure, K_a, while the maximum value (when the wall is pushed toward the soil) is termed the coefficient of passive earth pressure, K_p.

A.2 THE DEVELOPMENT OF EARTH PRESSURE THEORY

The development of earth pressure theory is concisely reviewed in Heyman's excellent book, *Coulomb's Memoir on Statics* (Heyman 1972). Developments of fortification and defensive systems at the turn of the 18th century produced structures with deep excavations in soil with near-vertical faces retained by walls (Vauban 1704) (Figure A.1). The earth pressure problem dates from the beginning of the 18th century, since in 1717 Gautier lists five areas requiring research, one of which was the dimensions of gravity-retaining walls needed to hold back soil. A number of workers (Bullet 1691; Couplet 1726, 1727, 1728; de Belidor 1729; Rondelet 1812) appear to have worked on the problem and published their findings. It was Coulomb, in a paper read to the Académie Royals des Sciences in Paris on 10th March and 2nd April 1773, who was to make the lasting impression in this field.

A.2.I Coulomb equations

Coulomb's *Essaí sur une application des règles de maximis & minimis à quelques problèmes de statique, relatifs à l'architecture* was published in

Figure A.1 Military revetments in Malta.

1776. It followed, and was no doubt partly based upon, the experience of a nine-year period which he had spent in Martinique constructing Fort Bourbon (Kerisel and Persoz 1978). It is clearly the work not only of a first-rate applied mathematician, but also of an experienced practising engineer. It contains two ideas of vital importance in soil mechanics. First, he divided the strength of materials into two components, namely cohesion (strength independent of applied forces, and a function only of the area of rupture) and friction (proportional to the compressive force on the rupture plane). This concept, although developed by Coulomb in terms of total stress and subsequently modified for effective stress by Terzaghi, remains the basis of soil-strength theory to this day.

Coulomb's other contribution to soil mechanics relates directly to earth retaining structures. His paper contains two Articles IX, the second of which, 'On earth pressures, and retaining walls', considers a rigid mass of soil sliding upon a shear surface (Figure A.2).

The forces considered were

   i. The weight of the soil, W
  ii. The cohesive and frictional forces acting on the shear surface, bc
 iii. The restraining force, Q, acting normal to the back of the wall

By resolving parallel and perpendicular to the shear surface, bc and considering that forces parallel to the shear surface must sum to zero, Coulomb obtained an expression for the force on the wall, Q. He then recognised that the inclination of the shear surface was not known, and that the shear surface giving the largest wall force, Q, would be required for design. The value of x (Figure A.2) giving the maximum value of Q was obtained by

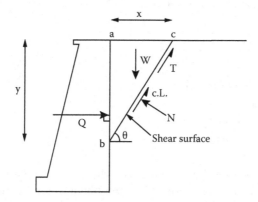

Figure A.2 Geometry for Coulomb's analysis.

differentiation, and this was then back-substituted to give the maximum value of thrust on the wall, Q_{max}.

Thus, for a granular soil ($c' = 0$),

$$R(\searrow)N = W \cos \theta + Q \sin \theta \tag{A.1}$$

$$R(\nearrow)T = W \sin \theta - Q \cos \theta \tag{A.2}$$

From (A.1) and (A.2)

$$T/N = \frac{W \sin \theta - Q \cos \theta}{W \cos \theta + Q \sin \theta}$$

$$= \tan \phi \text{ at failure.} \tag{A.3}$$

As

$$W = \frac{1}{2} \gamma H^2 \cot \theta \tag{A.4}$$

$$Q = \frac{1}{2} \gamma H^2 \cot \theta \tan(\theta - \phi). \tag{A.5}$$

Differentiating gives

$$\frac{\delta Q}{\delta \theta} = \frac{1}{2} \gamma H^2 [\cot \theta \sec^2(\theta - \phi) - \cosec^2 \theta \tan(\theta - \phi)]$$

$$= 0 \tag{A.6}$$

to give max Q. which is the active force. Hence, for Q_a,

$$\theta = 45 + \frac{\phi}{2} \tag{A.7}$$

Substituting (A.7) in (A.5) gives

$$Q_a = \frac{1}{2}\gamma H^2 \frac{(1-\sin\phi)}{(1+\sin\phi)}. \tag{A.8}$$

Coulomb in fact derives an expression directly for the force Q_a for a soil with both cohesion and friction. The component of force causing sliding on be (Figure A.2) is

$$W \sin\theta - Q \cos\theta - (W\cos\theta + Q\sin\theta)\tan\phi - cL \tag{A.9}$$

and, for equilibrium, Coulomb points out that this expression must equal zero, and hence,

$$Q = \frac{W[\tan\theta - \tan\phi] - cL/\cos\theta}{[1+\tan\theta\tan\phi]}. \tag{A.10}$$

By differentiating, in order to find the maximum value of Q, Coulomb determined that the critical shear surface occurred when

$$x = -y.\tan\phi + y(1 + \tan^2\phi)^{1/2} \tag{A.11}$$

Using trigonometrical identities, from Equation A.11 it can be shown that

$$\frac{y}{x} = \frac{1}{(\sec\phi - \tan\phi)} = \frac{\cos\phi}{(1-\sin\phi)} = \frac{(1+\tan(\phi/2))}{(1-\tan(\phi/2))}$$
$$= \tan(45° + \phi/2) \tag{A.12}$$

and since $\tan\theta = y/x$, the critical angle to give a maximum value of Q is

$$\theta_{crit} = 45° + \phi/2. \tag{A.13}$$

Thus in 1773, Coulomb noted that the orientation of the critical shear surface was a function only of the angle of friction of the soil, and was unaffected by its cohesion. This result was not to be discovered in Britain until nearly 150 years later (Bell 1915; Fitzgerald 1915). Coulomb also noted that if the angle of friction is zero (as in the short term, $\phi = 0$, analysis), the critical shear surface would be at 45°.

By back-substituting the orientation of the critical shear surface into the equation for Q (Equation A.10), Coulomb obtained an expression for the active force

$$Q_a = a.y^2 - byc \tag{A.14}$$

where a and b are constants, in terms of $\tan \phi$ only.

From Equation A.10 it can be seen that if $W = \frac{1}{2}\gamma H^2 \cot\theta$, and L = y/sin θ, then

$$Q_a = \frac{1}{2}\gamma y^2 \cot\theta . \tan(\theta - \phi) - cy/(\sin\theta \cos\theta(1 + \tan\theta \tan\phi)). \tag{A.15}$$

Substituting $\theta = 45° + \phi/2$ yields

$$Q_a = \frac{1}{2}\gamma y^2 . \frac{\tan(45° - \phi/2)}{\tan(45° + \phi/2)} - \frac{2cy.\tan(45° - \phi/2)}{\cos\phi(\tan(45° + \phi/2) - \tan\phi)} \tag{A.16}$$

and since $\tan(90° - \alpha) = \cot\alpha$, and $\cos\phi(\tan(45° + \phi/2) - \tan\phi) = 1$,

$$Q_a = \frac{1}{2}\gamma y^2 \tan^2(45° - \phi/2) - 2cy\ \tan(45° - \phi/2) \tag{A.17}$$

or

$$Q_a = \frac{1}{2}\gamma y^2 \frac{(1 - \sin\phi)}{(1 + \sin\phi)} - 2cy\sqrt{\frac{(1 - \sin\phi)}{(1 + \sin\phi)}} \tag{A.18}$$

(compare with Equation A.14).

Differentiating Q_n with respect to y gives the rate of force increase per unit length of wall with depth, which is the active earth pressure at depth y. Thus

$$\frac{dQ}{dy} = \gamma y\left(\frac{1 - \sin\phi}{1 + \sin\phi}\right) - 2c\sqrt{\frac{1 - \sin\phi}{1 + \sin\phi}}. \tag{A.19}$$

This equation was later derived by Français (1820) and Bell (1915).

Coulomb was well aware that the critical shear surface might not be planar, but he rightly noted that 'experience shows that when retaining walls are overturned by earth pressures, the surface which breaks away is very close to triangular', and therefore that a planar shear surface was a sufficiently good approximation.

Coulomb also noted that the contact between the back of the wall and the soil (ab) could be subject to friction and that the coefficient of friction on this surface would be less than on the shear surface (bc). He developed an expression for the active thrust, taking into account wall friction, and found that the orientation of the critical shear surface was changed.

Coulomb's paper clearly shows his practical experience of retaining-wall performance and design. He notes the following circumstances which may cause problems to a retaining wall:

a. Water collecting between the soil and the wall will increase the thrust on the wall. Coulomb states that even though walls are provided with drains, 'these drains get blocked, either by soil carried along by the water, or by ice, and sometimes become useless'.
b. Water penetrating into the soil will change its properties. Coulomb noted that the strength of Fuller's earth was significantly decreased by increasing moisture content. He also noted that when dry adhesive soil is placed behind retaining walls and subsequently wets up, it will swell, leading to 'a thrust on retaining walls that can only be determined by experiment'.
c. Not only may frost block the drains, but expansion due to frost heaves in the soil will lead to pressure increases on the back of the wall.

Coulomb's analysis can be extended to predict passive pressures and forces, when the wall is forced against the soil. Figure A.3 shows the case for a cohesionless frictional soil, supported by a smooth wall without wall adhesion.

Resolving parallel to the shear surface, bc,

$$Q \cos \theta - W \sin \theta - T = 0 \text{ for equilibrium.} \tag{A.20}$$

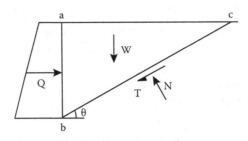

Figure A.3 Passive Coulomb wedge for a simple frictional soil supported by a smooth wall.

Resolving normal to bc

$$N = W \cos \theta + Q \sin \theta \qquad (A.21)$$

and, at failure, on bc

$$T = N.\tan \phi. \qquad (A.22)$$

Substituting Equation A.21 into Equation A.22, and the result into Equation A.20, yields

$$Q \cos \theta - W \sin \theta - (W \cos \theta + Q \sin \theta) \tan \phi = 0 \qquad (A.23)$$

and thus

$$Q = \frac{1}{2} \gamma H^2 \cot \theta \tan(\theta + \phi). \qquad (A.24)$$

Differentiation with respect to θ gives a minimum value of Q (the passive force, Q_p), when

$$\theta = 45° - \phi/2 \qquad (A.25)$$

and back-substituting into Equation A.24 yields

$$Q_p = \frac{1}{2} \gamma H^2 \left(\frac{1 + \sin \phi}{1 - \sin \phi} \right). \qquad (A.26)$$

In 1808 Mayniel extended the work of Coulomb (1776), Woltmann (1794) and Prony (1802) to give a general solution for a fractional, non-cohesive soil, with wall friction. Based on the equilibrium of a wedge of soil with planar boundaries (Figure A.4) as before,

$$Q = \frac{1}{2} \gamma H^2. \cot \theta. \sin(\theta - \phi) \sec(\delta + \phi - \theta). \qquad (A.27)$$

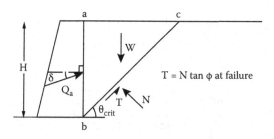

Figure A.4 Force diagram for Mayniel solution for a frictional cohesionless soil.

By differentiating with respect to θ, to maximize Q, it is found that

$$\theta_{crit} = \tan^{-1}\left[\tan\phi + \sec\phi\sqrt{\frac{\tan\phi}{\tan(\phi+\delta)}}\right] \tag{A.28}$$

Back-substitution into Equation A.27 yields

$$Q_a = \frac{1}{2}\gamma H^2\left[\frac{\cos\phi}{\sqrt{\cos\delta} + \sqrt{\sin(\delta+\phi)\sin\phi}}\right]^2 \tag{A.29}$$

$$= \frac{1}{2}\gamma H^2 \frac{\cos^2\phi}{\cos\delta\left[1 + \sqrt{\dfrac{\sin(\delta+\phi)\sin\phi}{\cos\delta}}\right]^2} \tag{A.30}$$

Ingold (1978a) has shown that the Coulomb solution for θ, θ_{crit} = 45 + φ/2, can be substituted into Equation A.27 without significant inaccuracy, and gives the much simpler equation

$$Q_a = \frac{1}{2}\gamma H^2\left[\frac{1}{\cos\delta + \sin(\delta+\phi)}\frac{\sin\phi}{}\right] \tag{A.31}$$

It is therefore evident that although the introduction of wall friction modifies the critical failure plane giving Q_a, the change is small. The major effect of the introduction of wall friction is to reduce Q_a through a reduction in the angle between Q_a and the failure plane, bc.

The Mayniel solution was further extended in 1906 by Müller-Breslau, to give a general solution for a frictional cohesionless soil which allows for sloping backfill, sloping back of wall and a frictional wall (Figure A.5). The solution, obtained on the same basis as the Coulomb solution, was found to be

$$Q_a = \frac{1}{2}\gamma H^2 \frac{f_1}{\sin\alpha.\cos\delta} \tag{A.32}$$

where

$$f_1 = \frac{\sin^2(\alpha+\phi).\cos\delta}{\sin\alpha.\sin(\alpha-\delta)\left[1 + \sqrt{\dfrac{\sin(\phi+\delta)\sin(\phi-\beta)}{\sin(\alpha-\delta)\sin(\alpha+\beta)}}\right]^2}. \tag{A.33}$$

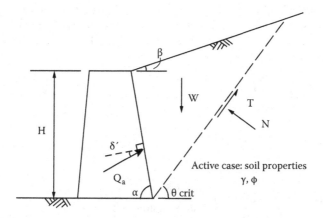

Figure A.5 Force diagram for Müller-Breslau solution for a frictional cohesionless soil.

The active force is inclined to the horizontal at $(90 - \alpha + \delta)$, and of course is inclined to the normal to the back of the wall at δ. Thus the component of active force normal to the back of the wall is

$$Q_{an} = \frac{1}{2}\gamma H^2 . \frac{f_1 \cos\delta}{\sin\alpha.\cos\delta} = \frac{1}{2}\gamma H^2 \frac{f_1}{\sin\alpha} \qquad (A.34)$$

and the horizontal component of Q_a is $\frac{1}{2}\gamma H^2 f_1$.

It can be seen that in special cases this equation can be reduced. For example, if

$$\alpha = \beta = 0$$

then

$$Q_a = \frac{1}{2}\gamma H^2 \left[\frac{\cos\phi}{\sqrt{\cos\delta} + \sqrt{\sin(\delta + \phi)\sin\phi}} \right]^2 \qquad \text{(see Equation A.29)}$$

which is the Mayniel solution, while if, in addition,

$$\delta = 0$$

then

$$Q_a = \frac{1}{2}\gamma H^2 \frac{\cos^2\phi}{(1 + \sin\phi)^2} = \frac{1}{2}\gamma H^2 \frac{1 - \sin\phi}{1 + \sin\phi} = \frac{1}{2}\gamma H^2 \tan(45° - \phi/2)$$

(see Equation A.8)

which is the Coulomb solution.

The general Müller-Breslau solution can also be obtained for a wall in the passive state. Here, the direction of wall friction is reversed, so that Q_p is inclined downward relative to the normal to the back of the wall. It is found that

$$Q_p = \frac{1}{2}\gamma H^2 . \frac{f_2}{\sin\alpha\cos\delta} \tag{A.35}$$

where

$$f_2 = \frac{\sin^2(\alpha - \phi)\cos\delta}{\sin\alpha\sin(\alpha + \delta)\left[1 - \sqrt{\dfrac{\sin(\phi + \delta).\sin(\phi + \beta)}{\sin(\alpha + \delta).\sin(\alpha + \beta)}}\right]^2} . \tag{A.36}$$

The Coulomb, Mayniel and Müller-Breslau solutions were all developed

  i. In terms of total stress, with no allowance for the inclusion of pore water pressures
 ii. For rigid (i.e. incompressible) soil
iii. For failure on a critical discrete planar, shear surface

A.2.2 Rankine's approach

In 1857, Rankine published his paper 'On the stability of loose earth' in the *Philosophical Transactions* of the Royal Society. He extended earth pressure theory by deriving a solution for a complete soil mass in a state of failure, as compared with Coulomb's solution which had considered a soil mass bounded by a single failure surface. His solution for the failure of cohesionless mass of soil with a horizontal ground surface can be obtained as follows, in terms of the stresses p and q.

By considering the force equilibrium of an element of soil of size dx.dy (Figure A.6)

$$R(\searrow)\sigma\,ds = p.dx.\cos\theta + q.dy.\sin\theta$$
$$dx = ds.\cos\theta; \quad dy = ds.\sin\theta \tag{A.37}$$

Therefore

$$\sigma = p.\cos^2\theta + q.\sin^2\theta. \tag{A.38}$$

$$R(\nearrow)\,\tau.ds = p.dx.\sin\theta - q.dy.\cos\theta \text{ for the active case.} \tag{A.39}$$

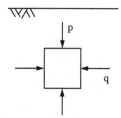

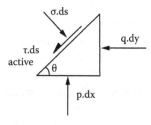

Figure A.6 Pressures and dimensions for Rankine's analysis.

Therefore

$$\tau = p.\sin\theta.\cos\theta - q.\sin\theta.\cos\theta \qquad\qquad (A.40)$$

At failure

$$\tau/\sigma = \tan\phi$$

$$\tau/\sigma = \frac{(p-q)}{(p.\cot\theta + q.\tan\theta)} \qquad\qquad (A.41)$$

$$\frac{\partial(\tau/\sigma)}{\partial\theta} = \frac{(p-q)(-p.\cosec^2\theta + q.\sec^2\theta)}{f(p,q,\theta)} \qquad\qquad (A.42)$$

Therefore for τ/σ_{max},

$$p/q = \tan^2\theta. \qquad\qquad (A.43)$$

Substituting (A.43) into (A.41)

$$\tau/\sigma = \frac{(p-q_a)}{\left[p\sqrt{q_a/p} + q_a\sqrt{p/q_a}\right]} = \frac{(p-q_a)}{\left(2\sqrt{p.q_a}\right)} = \tan\phi. \qquad\qquad (A.44)$$

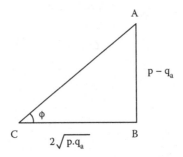

Figure A.7 Pressures and dimensions for Rankine's analysis.

Equation A.44 can be shown to be true from triangle ABC, Figure A.7

$$AC = [p - 2p.q_a - q_a + 4p.q_a]^{1/2}$$
$$= p + q_a$$

(A.45)

Therefore

$$\sin\phi = \frac{p - q_a}{p + q_a}$$

(A.46)

Defining the coefficient of active earth pressure, as q_a/p, therefore $K_a = q_a/p = (1 - \sin\phi)/(1 + \sin\phi)$ (see Equation A.8).

It can be seen that, for conditions identical to those in Coulomb's analysis, Rankine obtains the same solution as Coulomb, since for any depth below ground surface, z,

$$Q_a = \int_{z=0}^{z=H} q_a.dz$$

(A.47)

and since $p = \gamma.z$ and $q_a = K_a p$

$$Q_a = \frac{1}{2}\gamma H^2 K_a = \frac{1}{2}\gamma H^2 \frac{1 - \sin\phi}{1 + \sin\phi}$$

(see Equation A.8)

In 1882, Mohr's paper on the representation of stresses and strains showed how the stresses on and within an element in a solid in plastic equilibrium could be represented by a circle.

Consider, once again, the element of soil shown in Figure A.6. As we found in Equations A.38 and A.40, resolving perpendicular and normal to the plane on which the normal and shear stresses σ and τ act, gives

$$\sigma = p.\cos^2 \theta + q.\sin^2 \theta \tag{A.48}$$

and

$$\tau = (p - q) \sin \theta \cos \theta \tag{A.49}$$

These values of σ and τ, if plotted as abscissa and ordinate of a point (σ, t), yield a circle with centre at (p + q)/2 and radius (p − q)/2 as θ goes from 0 to 180°. Thus the full set of states of stress within a two-dimensional element can be represented by a circle (Figure A.8). If the element of soil is not failing, then α will be less than φ.

It can be seen in Figure A.8 that there are two points which lie on the normal stress axis, i.e. they represent the stresses on planes where the shear stress is equal to zero, and the normal stress is either at a maximum or a minimum. These stresses are known as the major and minor 'principal stress', and the planes upon which they act are termed as the 'principal planes'. (It can be seen that p and q in Rankine's analysis act upon the principal planes, since there is no shear stress on the horizontal, or the vertical plane.)

In soil mechanics it is normal to denote compressive stresses, and volumetric and linear strain decreases, as positive. The normal convention is to term the major principal stress as σ_1, and the minor principal stress as σ_3.

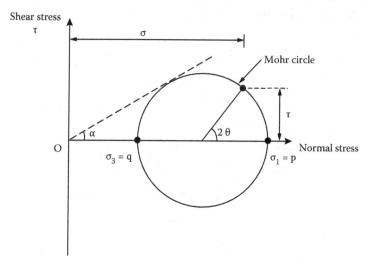

Figure A.8 Mohr circle of stress.

At failure, Coulomb proposed that shear force is related by a constant to the normal force, or

$$\tau_f = \sigma_n \tan \phi$$

if the forces are brought to stresses by dividing by the area of contact, for a frictional, non-cohesive material (Figure A.9). When this occurs, the Mohr circle of stress touches the failure envelope, and the shear and normal stress on the failure plane are given by points f. The direction of the plane upon which a given combination of normal and shear stress acts can be readily determined from a Mohr circle using the 'pole method'.

Figure A.10 shows an example of the use of the pole method, in this case to find the orientation of the failure planes. To find the pole:

a. The magnitude of the shear stress and normal stress on two planes of known orientation must be given, in order to plot the Mohr circle.
b. A line is drawn through one stress point, parallel to the plane upon which the particular shear and normal stresses act.
c. The procedure is repeated for the other stress point—the two lines will intersect at a point on the Mohr circle, and this point is known as the pole.

To find the inclination of the plane upon which any possible combination of shear and normal stress acts, the procedure is reversed.

a. Determine the stress point on the Mohr circle
b. Join the stress point to the pole—the stresses act on a plane in the actual material parallel to this line

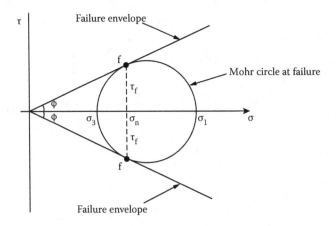

Figure A.9 Mohr circle for Coulomb failure conditions.

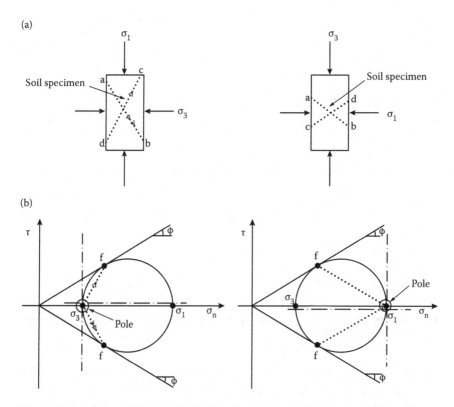

Figure A.10 Examples of the pole method for determining the orientation of failure planes. (a) Triaxial compression (left) and extension (right) specimens. (b) Mohr circle representations of imposed stresses.

An example is given in Figure A.10. In (a), two soil specimens are subjected to compressive stresses. The left-hand specimen has the major principal stress acting vertically, upon the horizontal plane. The right-hand soil specimen has the major principal stress acting horizontally, on the vertical plane. In (b), the Mohr circles are drawn. The magnitude of σ_1 and σ_3 must be known (and, of course, since they are principal stresses, $\tau = 0$). For the left-hand specimen σ_1 acts on the vertical plane, and a vertical line is therefore drawn through σ_3. The two lines in fact intersect at σ_3 which therefore is also the pole. For the right-hand specimen, the same construction produces a pole at σ_1.

To find the orientation of the failure planes in the specimen it is only necessary to join the pole to the stress points representing failure, i.e. points f where the Mohr circle touches the Coulomb failure envelope. Since there are two points f, failure surfaces can be expected in two directions in each case. These are shown dotted in Figure A.10(b) and are superimposed on the specimen in Figure A.10(a).

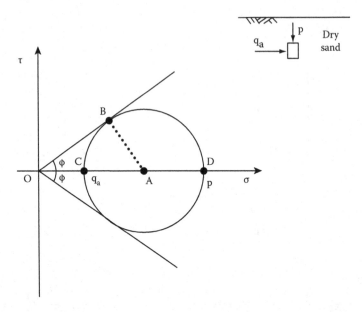

Figure A.11 Mohr circle for active failure state. Horizontal ground surface. Rankine condition ($\delta = 0$).

Mohr circles may be used with great effect to derive solutions for the Rankine analysis. Figure A.11 shows the Mohr circle for the active failure condition for a granular soil. For the active condition, the vertical stress (p) due to the weight of soil above the element of soil being considered remains constant, and the horizontal stress (q) is reduced until failure occurs. Since, in the simple Rankine solution, there is no shear stress on either the vertical or the horizontal planes, p and q are principal stresses. The magnitude of p is known, and the problem is to determine q in terms of p.

$$q_a = OA - AC \qquad\qquad\qquad\qquad (A.50)$$

$$p = OA + AD \qquad\qquad\qquad\qquad (A.51)$$

but

$$AC = AD = AB$$

and

$$\frac{AB}{OA} = \sin\phi \qquad\qquad\qquad\qquad (A.52)$$

$$K_a = \frac{q_a}{p} = \frac{OA - AC}{OA + AD} = \frac{OA - AB}{OA + AB} = \frac{1 - AB/OA}{1 + (AB/OA)}$$

$$= \frac{1 - \sin\phi}{1 + \sin\phi} \quad \text{(see Equation A.8)}. \tag{A.53}$$

A similar derivation of q_a in terms of p can be carried out for the much more complex ease of a mass of soil with its ground surface inclined to the horizontal.

For the element shown at the top of Figure A.12, resolve perpendicular to q_a and τ.

$$R(\searrow)p\cos\beta. \quad 1 = \sigma(1/\cos\beta)$$
$$\sigma = p\cos^2\beta. \tag{A.54}$$

Since p is the resultant of σ and τ

$$\tau = \sigma \tan\beta = p \sin\beta \cos\beta. \tag{A.55}$$

At failure, the Mohr circle must touch the failure envelope. It is assumed that the force on the vertical, q_a, acts parallel to the ground surface (i.e. $\delta = \beta$). Therefore, on the Mohr circle, point X represents the shear and normal stresses on a plane parallel to the ground surface. Since these forces act on a plane inclined at β to the horizontal, the pole is at P. The stresses acting on a vertical plane are therefore represented by point C.

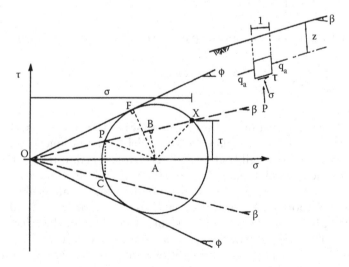

Figure A.12 Active failure, Rankine analysis—inclined ground surface ($\delta = \beta$).

Since $K_a = q_a/p$, and $q_a = OC = OP$, and $p = \sigma/\cos^2 \beta = OX/\cos \beta$

$$\frac{q_a}{p} \cdot \frac{1}{\cos\beta} = \frac{OB - BP}{OB + BX} \tag{A.56}$$

and $BP = BX$.

The earth pressure coefficient, K_a, is found by obtaining OB and BP in terms of OA, from the various triangles:

$$OB = OA \cos \beta \tag{A.57}$$

$$AF = OA \sin \phi \tag{A.58}$$

$$AB = OA \sin \beta \tag{A.59}$$

and therefore, from equations and (A.58) and (A.59)

$$BP = OA \, (\sin^2\phi - \sin^2\beta)^{1/2} \tag{A.60}$$

and substituting Equations A.57 and A.60 into Equation A.56 yields

$$
\begin{aligned}
K_a = \frac{q_a}{p} &= \cos\beta \, \frac{\cos\beta - (\sin^2\phi - \sin^2\beta)^{1/2}}{\cos\beta + (\sin^2\phi - \sin^2\beta)^{1/2}} \\
&= \cos\beta \, \frac{\cos\beta - (\cos^2\beta - \cos^2\phi)^{1/2}}{\cos\beta + (\cos^2\beta - \cos^2\phi)^{1/2}}
\end{aligned}
\tag{A.61}
$$

If the analysis is repeated for the passive state, the result obtained is

$$K_p = \frac{q_p}{p} = \cos\beta \, \frac{\cos\beta + (\cos^2\beta - \cos^2\phi)^{1/2}}{\cos\beta - (\cos^2\beta - \cos^2\phi)^{1/2}} \tag{A.62}$$

It should be noted that in all Rankine analyses, it is assumed that the resultant force on the vertical plane acts parallel to the ground surface. The value of the angle of wall friction (δ) is therefore equal to that of the inclination of the ground surface (β) and cannot be varied. Consequently, the Rankine condition for a horizontal ground surface is applicable only to walls with smooth backs, or walls unable to take shear, since $\delta = 0$.

In 1915, Bell extended Rankine's solution to allow for the effects of soil cohesion. The original work was carried out analytically, and will not be repeated here. It is much more straightforward to derive the equations from Mohr circles.

Bell was engaged on the design of some monoliths to provide a sea-wall at H.M. Dockyard, Rosyth, Scotland, when he realized that the wide range of angles of repose for soils quoted in books of reference at the time (between 1° and 45°) required further investigation. He carried out a number of 'undrained' direct shear tests, and deduced that for clay the law of shearing resistance was

$$q = k + p_n \tan \alpha \tag{A.63}$$

or, as some would now say

$$\tau_f = c_u + \sigma_n \tan \phi_u. \tag{A.64}$$

For a soil with cohesion and friction, with a horizontal ground surface, and a vertical smooth supporting wall with no adhesion to the soil, the stresses at failure may be represented by the Mohr circle in Figure A.13.

Let $K_a = (1 - \sin \phi)/(1 + \sin \phi)$ as for the cohesionless Rankine state.

$$p = OE = OA + AE = OA + AB = CA - CO + AB \tag{A.65}$$

$$q_a = OD = OA - AD = OA - AB = CA - CO - AB \tag{A.66}$$

but

$$CO = c \cot \phi \tag{A.67}$$

and

$$CA = AB \ \text{cosec} \ \phi. \tag{A.68}$$

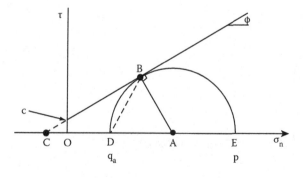

Figure A.13 Active failure, Rankine-Bell analysis.

Therefore

$$p = AB \left(\mathrm{cosec}\ \phi + 1\right) - c \cot \phi \tag{A.69}$$

and

$$AB = \frac{\left(p + c \cot \phi\right)}{\left(\mathrm{cosec}\ \phi + 1\right)}. \tag{A.70}$$

Hence, from Equations A.66–A.69 and A.70

$$
\begin{aligned}
q_a &= \frac{\left(p + c \cot \phi\right)}{\left(\mathrm{cosec}\,\phi + 1\right)}\left(\mathrm{cosec}\,\phi - 1\right) - c \cot \phi \\
&= \left(p + c \cot \phi\right)\frac{\left(1 - \sin\phi\right)}{\left(1 + \sin\phi\right)} - 2c.\frac{\cos\phi}{\left(1 + \sin\phi\right)} \\
&= K_a p - 2c\sqrt{K_a} \ \text{(compare with Equation A.19).}
\end{aligned} \tag{A.71}
$$

Bell was able to demonstrate, and Fitzgerald to prove mathematically as Coulomb had done in 1773, that the introduction of a soil cohesion intercept has no effect on the orientation of the failure planes in the soil. This can be seen from the Mohr circle in Figure A.13. Since q_a acts on the vertical plane, and p acts on the horizontal plane, the pole is at D, and OB gives the orientation of the failure plane.

Now

$$A\hat{C}B = \phi$$
$$A\hat{B}C = 90°$$
$$B\hat{A}C = 90° - \phi$$

Since

$$A\hat{B}D = A\hat{D}B$$

then

$$A\hat{D}B = \frac{1}{2}\left(180° - \left(90° - \phi\right)\right) = 45° + \phi/2, \tag{A.72}$$

as found in Equation A.7.

A.3 CLASSICAL SOLUTIONS AND EFFECTIVE STRESS

The results of Bell's shear tests were obtained in terms of total stress, and all the analysis discussed in this chapter was carried out in terms of total stress. As noted in Chapter 2, however, following the work of Terzaghi in the early 1920s and the introduction of the concept of effective stress as the controlling influence on strength and compression, coefficients of earth pressure are defined in terms of effective stress, and are a function of the effective shear strength parameters c' and ϕ'.

In effect, therefore, since none of the preceding analyses has taken account of pore water pressure, it has implicitly been assumed to be zero. Despite this, these classical solutions can readily be applied to problems involving groundwater.

By differentiating with respect to depth, the solutions obtained by Coulomb, Mayniel and Müller-Breslau for active and passive forces can be converted to horizontal pressure. For example, since the active force $Q_a = \frac{1}{2}\gamma H^2 K_a$ for the entire wall height H, then the pressure $q_a = \gamma.z\, K_a$ at any depth, z, below ground surface. Rankine and Bell's solutions are already in this form.

If the coefficient of earth pressure is now to be calculated from the effective strength parameters c' and ϕ', then the effective horizontal stress must be calculated from the effective vertical stress at every level, or

$$q_a' = (\gamma.z\, u)K_a \tag{A.73}$$

or

$$\sigma_h' = \sigma_v'.K_a \tag{A.74}$$

and the total horizontal stress will be

$$q_a = q_a' + u = (\gamma z - u)K_a + u \tag{A.75}$$

or

$$\sigma_h = \sigma_h' + u = (\gamma z - u)K_a + u. \tag{A.76}$$

As Coulomb noted in 1773, groundwater increases the active thrust on a wall. This is demonstrated in the simple cases illustrated in Figure A.14 where a rise in groundwater to 3 m below ground level raises the total thrust on the wall by almost 70%. Thus, an accurate assessment of groundwater

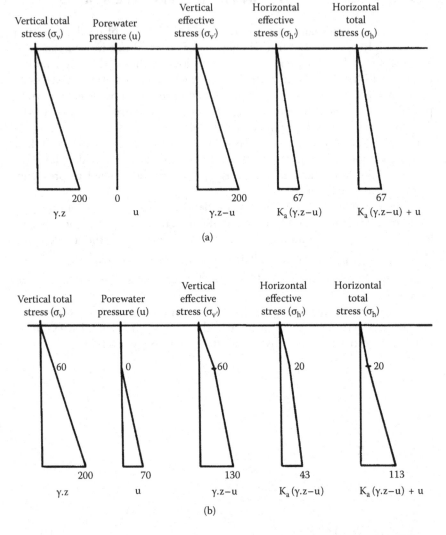

Figure A.14 Example showing the effect of groundwater on active earth pressures. (a) Pressure distributions for a 10m-high smooth wall supporting dry granular soil (γ = 20 kN/m³, c' = 0, φ' = 30°, Kₐ = 0.33). (b) Pressure distributions for the same wall, with groundwater (γ_w = 10 kN/m³) 3m below the top of the wall.

conditions is vital for realistically assessing the forces on walls. Whereas in the active case, the force is increased by the presence of groundwater, in the passive case the available soil resistance is reduced. The effects of groundwater are summarised in Figure A.15 where the dashed lines show the horizontal earth pressure in the absence of groundwater.

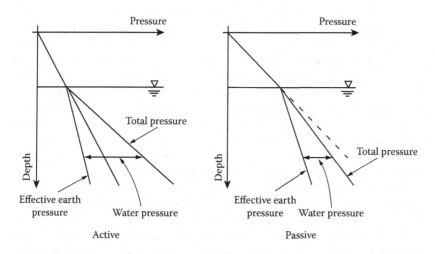

Figure A.15 Summary of the effect of groundwater on active and passive earth pressures.

A.4 GRAPHICAL TECHNIQUES

Section A.3 considered the classical solutions for earth pressure, which were based on analytical techniques. While complete analytical solutions are attractive, they are limited, principally because

a. Only simple soil or wall geometries may be considered—it is sometimes necessary to find solutions to problems which are too complex to allow a simple analytical solution.
b. Analytical solutions are not available for non-planar failure surfaces or non-uniform stress fields—some passive solutions require the use of a curved failure surface, when wall friction is present.

In the first case, active pressures can only be calculated for planar soil boundaries, and a planar failure surface. Even when these conditions exist, the most complex analysis produced by Müller-Breslau will not allow for wall adhesion or an effective cohesion intercept as a soil property. While soil can have both effective cohesion and friction in Coulomb's or Bell's analyses, in this case no wall friction or wall adhesion may be included.

If the groundwater table is not horizontal, or the ground surface is uneven, or soil conditions are complex, then analytical solutions are unlikely to be available. A graphical technique must then be used.

In Section A.3, the classical solutions for earth pressure problems were derived, based on either planar failure surfaces, or uniform stress fields. In reality, the introduction of wall friction modifies the stress field at the boundary, but this modification does not extend throughout the soil mass. As a result, the principal stress directions are rotated with increasing distance from the wall, the stress field is no longer uniform, and any plane of failure will be curved.

Figure A.16(a) shows two elements of soil adjacent to retaining walls. For a smooth wall (i.e. no wall friction, $\delta' = 0$) the stresses on the horizontal and vertical planes are principal stresses. The right-hand figure shows that when wall friction exists, complementary shear stresses also occur on the horizontal planes, and the principal stress directions are no longer vertical and horizontal.

Figure A.16(b) shows the Mohr circle construction for the active case, with wall friction. The Mohr circle is tangential to the failure line. The stresses on the vertical plane are represented by point A, since $\tau/q_{an} = \tan\delta'$. The pole is therefore at P, and the failure planes are in the direction of PF_1 and PF_2. If $\delta' = 0$, then point C would represent both the stresses on the vertical plane, and the pole, and the failure planes would be parallel to CF_1 and CF_2. It can therefore be seen that wall friction causes a clockwise rotation in the active failure surface close to the wall. As shown in Figure A.16(c), however, this rotation does not occur far from the wall where the effects of

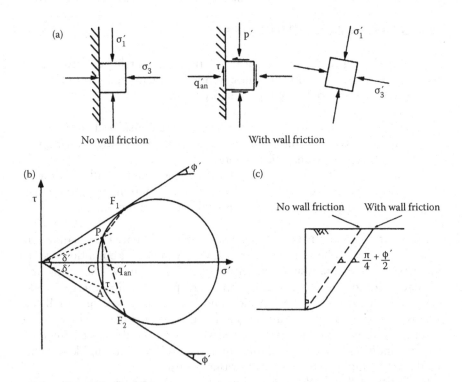

Figure A.16 Effects of wall friction on principal stress directions and failure surfaces. (a) Stresses applied to an element of soil adjacent to the wall. (b) Mohr circle for active case with wall friction. (c) Effect of wall friction on active failure surface (exaggerated).

wall friction are negligible. The failure surface will therefore curve from the bottom of the wall, reaching the ground surface at the same angle as would have been found without wall friction ($45° + \phi'$). The other limiting condition, when $\delta' = \phi'/2$, can be seen from Figure A.16(c) to give a failure plane orientated at ϕ'. If, for example, a granular soil had an effective angle of friction of 30°, a smooth wall would give a failure surface inclined at 60°, and a very rough wall would, close to the wall, have a failure surface inclined at 30°. In reality it is unlikely that the angle of wall friction would exceed one half of the effective angle of friction of the soil. For the case of $\phi' = 30°$, and $\delta' = 15°$, the critical failure surface for the Mayniel solution lies at about 55° from the horizontal, only 5° from the Coulomb solution for a smooth wall and $\phi' = 30°$. For the active case, therefore, solutions involving planar failure surfaces are sufficiently accurate for engineering purposes.

The introduction of wall friction to the classical solutions for a passive planar shear surface causes unrealistically large increases in the predicted forces on the wall. This is because the mass of soil being failed increases dramatically as the inclination of the critical shear surface is reduced. Figure A.17(a) shows

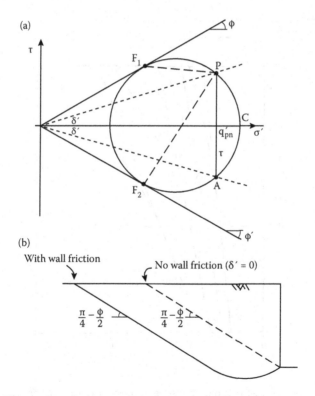

Figure A.17 Effects of wall friction on passive failure. (a) Mohr circle for passive case with wall friction. (b) Effect of wall friction on passive failure surface.

the Mohr circle construction for the passive case with wall friction. In the absence of wall friction, the shear surface would leave the base of the wall in Figure A.17(a) parallel to F_1C, which has been shown to be inclined at $45° - (\phi'/2)$ above the horizontal. As the angle of wall friction (δ') increases, the failure surface reduces its inclination, until eventually when $\delta' = \phi'$ the failure surface leaves the base of the wall ϕ' below the horizontal. Therefore it becomes necessary to analyse the problem using a curved failure surface, such as that shown in Figure A.17(b).

When analysis is carried out using both planar and curved shear surfaces, it is found that when the angle of wall friction exceeds about 10° the classical solutions give higher passive earth pressure coefficients than those based on a curved surface (Figure A.18). Unlike Coulomb's active case, it is of course the failure surface amongst all possible failure surfaces which yields the minimum passive force available to support the structure which is required for design, Thus, for problems involving passive pressure and significant values of wall friction, it is necessary to use an analytical or graphical technique which models the curved failure surface.

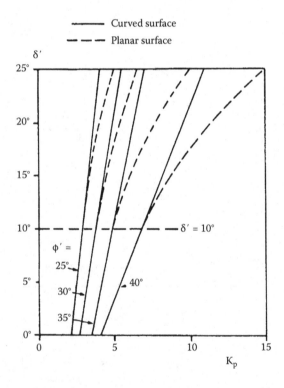

Figure A.18 Effect of wall friction on passive earth pressure coefficient ($c' = 0$, $c'_w = 0$, vertical back of wall and horizontal ground surface).

A.4.1 Graphical techniques for the active case

In the active case, two conditions can arise. For temporary works, the soil may require support only in the short-term, and it may be tempting to analyse using the undrained shear strength parameter, c_u, for the soil strength. A number of graphical solutions have been presented for the short-term active state, for example, to predict the height to which unsupported vertical faces of soil will stand. They are not presented here, because short-term analysis may present a danger to life, since, as discussed in Chapter 1, it is not possible to predict the rate at which pore pressures dissipate, and the soil moves from the end-of-construction state to the long-term state.

In this section, only the second, long-term, condition is considered, and effective stress solutions are given for the active case. In the active case, as we saw for the classical analytical solutions, a planar failure surface such as that used by Coulomb gives a solution that is sufficiently accurate for engineering purposes, even when wall friction is introduced.

Figure A.19 shows the graphical procedure for a simple case which would normally, in fact, be solved analytically. The steps in finding a solution are as follows:

a. Determine the geometry of the problem (e.g. wall height, ground surface profile, wall profile, groundwater position).
b. Determine the soil properties (in this simple case: there is only one granular soil, but the soil might well contain layers, with each layer having a different set of properties, c', ϕ' and bulk density, γ).
c. Draw space diagrams, and select a trial failure surface, such as bc.
d. Consider all the forces acting on the trial wedge, and determine those which are known in both direction and magnitude. (In Figure A.19 only W is known in magnitude and direction.) Start to draw the force diagram with the weight of the wedge, W, followed by any other forces known in both magnitude and direction.
e. At failure, the relationship between the normal and shear forces on either the failure plane or the back of the wall is known. (For example, in Figure A.19 failure on bc, $T = N \tan \phi'$.) On these planes, the direction of the resultant is therefore known. (For example, in Figure A.19, the resultant of T and N acts at ϕ' to the normal to bc, as shown on the space diagram, but since there is no wall friction in this particular problem, Q acts normal to the back of the wall.)
f. These resultant forces, such as R and Q in the simple case in Figure A.19, are then known in direction but not in magnitude. They can be drawn out from d and e on the force diagram to find the intersection at f. The forces can then be scaled from the force diagram.
g. The procedure is repeated for different failure surfaces, until the failure surface giving the largest value of Q (i.e. the active force Q_a) is found.

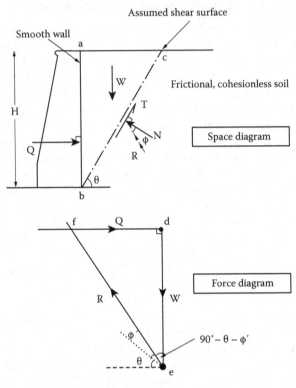

NB. θ is varied to find Q_n ($\simeq Q_{max}$).

Figure A.19 Graphical procedure to determine the active force on a wall (simple case).

Since, for a simple case, it is known that the critical shear surface is inclined at 45° +(ϕ'/2) to the horizontal, it seems reasonable to use approximately such an angle for θ in the first trial; θ can then be varied on either side to determine the critical shear surface for the particular case being investigated.

Figure A.20 shows a complex active case involving most complexities that could be envisaged. These are

   i. An irregular ground surface
   ii. Tension cracks, in this case full of water
   iii. Wall friction (δ') and wall adhesion c'_w
   iv. A sloping groundwater profile

In addition, one might sometimes expect

   v. A sloping back of wall.

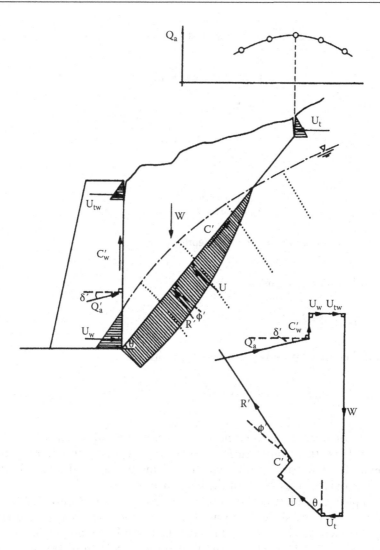

Figure A.20 Wedge analysis for the long-term active case.

The problems in determining the geometry of the problem are principally involved with the selection of a suitable tension crack depth, and the determination of groundwater pressures on the back of the wall and the chosen failure surface.

Tension cracks will occur in cohesive soils. They can be seen at ground surface during a dry summer, formed as a result of ground shrinkage. If, for example, Bell's equation is used to predict the active pressure in a soil with both effective friction and effective cohesion, such as a stiff, overconsolidated clay, the pressure diagram shown in Figure A.21 results.

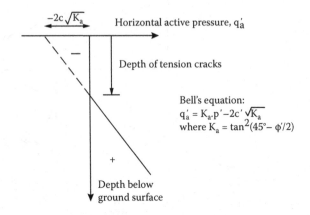

Figure A.21 Prediction of tension-crack depth, z_0.

For $q_a' = 0$ (i.e. no tension)

$$K_a(\gamma.z - u) = 2c'\sqrt{K_a} \qquad (A.77)$$

and, if there is no pore water pressure (u) present

$$z_0 = \frac{2c'}{\gamma}.\frac{1}{\sqrt{K_a}} = \frac{2c'}{\gamma}.\tan\left(45° + \frac{\phi'}{2}\right) \qquad (A.78)$$

Soil cannot sustain tension, and it is therefore logical to expect cracks to develop from the ground surface down to a depth z_0. In reality, little is known of the actual depth to which cracking occurs, and Equation A.78 simply provides an estimate of the likely depth.

Complex groundwater conditions, such as shown in Figure A.20, must be determined by seepage analysis. The easiest method of carrying this out is by flow net sketching, the details of which can be found in Chapter 4 of this book, and in Harr (1962) and Cedergren (1989). In reality, it is rarely worth assuming such a complex groundwater geometry, because groundwater conditions are normally rather poorly known during design. It will generally be sufficient to assume a horizontal groundwater surface, with hydrostatic pore water pressures below. When determining the groundwater regime, the object of the work is to find the pore water pressure distributions on the back of the wall, and on the shear surface.

As before, the angle θ is varied to find the maximum value of Q_a, the total active thrust on the wall. In the case shown in Figure A.20, W, C', C_w', U, U_w, U_t and U_{tw} are all known in both magnitude and direction, W can be calculated from the soil density and its area, U_t and U_{tw}, equal

$\frac{1}{2}\gamma_w\,z_0^2$, where z_0 is the depth of the tension crack. It is assumed that during a severe storm tension cracks would fill with rainwater U, and U_w are found by integrating the pore water pressure on the shear surface and the back of the wall C' and C'_w are equal to c'.l and c'_w.h, respectively, where c' is the effective cohesion intercept, l is the length of the shear surface, c'_w is the effective adhesion between the wall and the soil, and h is the length of the wall/soil contact.

The unknown forces are the effective normal force and the shear force on the back of the wall (which combine to produce Q' acting at the angle of the wall friction (δ') to the normal to the back of the wall), and the effective normal force and the shear force on the shear surface, which combine to produce R' acting at ϕ' to the normal to the shear surface. The direction of these two forces is known, and so the force polygon can be closed. The total force on the wall (Q), which must be maximized, is the resultant of Q'_a, C_w. U_w and U_w, can be scaled. The maximum value of Q is the active force Q_a, and is found by varying θ.

A.4.2 Determination of line of thrust for complex geometries

For a frictional soil, when the wall is planar and there is a linear increase of horizontal stress with depth, the resultant force will act at one-third of the height. When effective cohesion is introduced, or the back of the wall is irregular, then the position of the resultant force will be unknown. The pressure distribution on the back of the wall must be determined before the position of the resultant active force can be found, and with graphical methods only the resultant force is known.

One method of determining the pressure distribution on the back of the wall is to compare the total force for different heights of supported soil and obtain a crude pressure distribution by subtraction. Figure A.22 shows such a system, and it can be seen that the amount of computation involved would normally be prohibitive, since the critical shear surface must be found for several levels below the top of the wall.

Terzaghi (1943) suggests that for soil in the active condition, with a complex wall and soil surface geometry, the approximate point of application of the active force can be found as follows:

i. Determine the critical shear surface, to give the active force.
ii. Find the centroid for the area of soil sliding on the critical shear surface.
iii. Draw a line parallel to the critical shear surface, through the centroid, to the back of the wall—the active force is applied at this point.

This method is illustrated in Figure A.23.

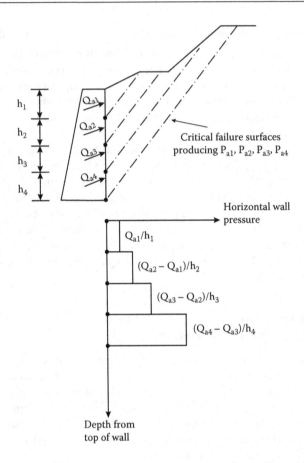

Figure A.22 Method of determining approximate pressure distribution with depth.

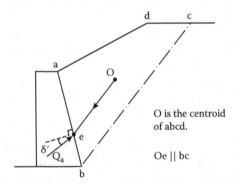

Figure A.23 Terzaghi's approximate method for determining the point of application of the active force.

A.4.3 Graphical techniques for the passive case

Two graphical methods using curved failure surfaces were described by Terzaghi in his *Theoretical Soil Mechanics* (1943). The more complex of the two, due to Ohde (1938), is the 'logarithmic spiral' method; this is rarely used due to its complexity, but in fact gives a more accurate solution than the 'friction circle' method (Krey 1936), which is more commonly used and is described below.

The 'friction circle' method or 'ϕ-circle' method need only be used when wall friction is sufficiently high to require a curved failure surface to be used. The method is very time-consuming, and so should be avoided if possible.

An assumed failure surface is made up from two parts:

a. Furthest from the wall, a planar surface inclined at $45° - (\phi'/2)$ to the horizontal, in the Rankine zone where stresses remain unaffected by wall friction (cd_1 in Figure A.24)

b. Nearest to the wall, a circular arc tangential to, and passing through the lowest point of the Rankine zone, and also passing through the bottom of the wall (d_1b in Figure A.24)

The failure surface is constructed as follows:

a. The position of f_1d_1 is arbitrarily selected, and d_1c is then drawn from d_1 at an angle of $45° - (\phi'/2)$ to the horizontal.

b. For the circular arc (bd_1) to be tangential to d_1c at d_1, the centre of the circle must lie on the normal to d_1c at d_1, i.e. on d_1e (O is, as yet, unknown).

c. For the circular arc to pass through b, bO must equal d_1O. From this, the precise location of O and d_1e can be found.

For a frictional soil, the resultant of the normal and shear forces on each small element of the arc bd_1 must act at ϕ' to the normal to the surface, i.e. at ϕ' to the normal to the radius, gO—see Figure A.25. From triangle Ogh, it can be seen that Oh equals $r \sin \phi'$, for all elements along the arc bd_1. The assumption made in the friction circle method is that the resultant of all these elemental forces will also be tangential to a circle of radius $r \sin \phi'$. This assumption leads to the largest error in the method, and gives an estimate of passive force which is too low (i.e. on the safe side). Taylor (1937) has provided a correction chart, which is shown in Figure A.26. The length $r \sin \phi'$ does not need to be calculated, since the friction circle must be tangential to ad_1 (Figure A.24).

The method involves the stability of prism abd_1f_1. The following forces are calculated:

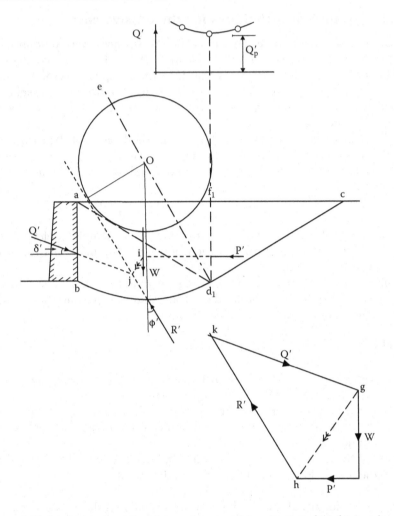

Figure A.24 Friction circle method for passive-state determination of the frictional component of passive force.

i. The weight per unit length of wall (W) of abd_1f_1, from its area and bulk unit weight, γ
ii. The effective force acting on d_1f_1, which will be the Rankine force since d_1f_1 lies outside the influence of wall friction

Thus

$$P' = \frac{1}{2}\gamma(d_1f_1)^2 K_p \tag{A.79}$$

where K_p is the Rankine passive earth-pressure coefficient. The lines of action of both W and P′ are known. W will act through the centroid of abd_1f_1 if the soil density is uniform. P′ acts at 1/2 of the height d_1f_1 since the Rankine passive pressure distribution for a frictional soil is triangular.

The required passive force can be found for each trial failure plane bd_1c as follows (see Figure A.24):

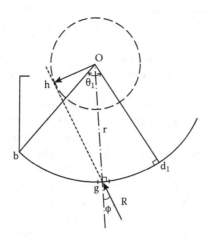

Figure A.25 Development of the friction circle.

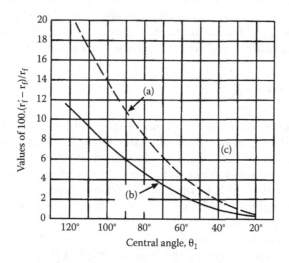

Figure A.26 Taylor's correction chart for the friction circle method (a) for uniform stress distribution on surface of sliding (b) for sinusoidal stress distribution on surface of sliding. $r_f = r \sin \phi'$. r_f' is the correct value.

i. Draw the line of action P′ on the space diagram to intercept the line of action of W (at i).

ii. Construct the force diagram for W and P′, and find the inclination of their resultant (gh).

iii. Draw the line of action of the resultant of W and P′ through i, parallel to gh.

iv. The passive force Q_p' acts at δ′ to the normal to the back of the wall, and for a frictional soil passes through at 1/3 of the wall height. The line of action of Q_p' can therefore be drawn on the space diagram to find the intersection with the resultant of P′ and W (at j).

v. Determine the inclination of the resultant R′. The resultant of the shear and normal forces on bd_1 must pass through j, and is assumed to be tangential to the friction circle. R′ must therefore pass through j and a.

vi. R′ and Q_p' are now known in direction but not magnitude, and can be drawn on the force diagrams to find their intersection at k.

vii. The passive force, Q_p', can now be found by scaling from the force diagram. The failure surface is moved to find the minimum value of Q_p'.

The method described so far determines the effective passive force for a dry frictional cohesionless soil. The force due to any groundwater pressure on ab would have to be added to Q_p' to find total passive force. In addition, the method has not allowed soils with a cohesion intercept, or soils which have an applied surcharge load at their surface.

For this more complex case, it is convenient to separate the elements of passive pressure resulting from the surcharge and the effective cohesion intercept, from the frictional effect already discussed. The Rankine–Bell solution for a cohesive soil with a surcharge, in a passive state, is

$$q_p' = K_p(p' + s) + 2\sqrt{K_p}.c \qquad (A.80)$$

since a uniform surcharge at ground surface will have the same effect as an additional layer of soil giving the same vertical stress increase. On this basis, the friction circle calculations for the effective cohesion and the surcharge are separated as shown in Figure A.27; the left-hand diagram representing the friction circle method is already described.

The additional force, Q_p'', due to the effective cohesion intercept and the friction is obtained as follows (see Figure A.28), on the basis that the soil is weightless and frictionless. Several of the forces can be determined both in magnitude and direction, and their point of application may also be known (e.g. $C_w' = c_w'.ab$, and acts along the wall face). C′ is the resultant of cohesive forces acting along the arc bd_1. The resultant force has magnitude $c'.bd_1$ (i.e. the length of straight line bd_1, not the length of the arc bd_1). The

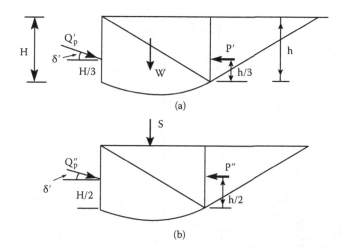

(a)

(b)

Figure A.27 Division of forces for a friction circle calculation for a c′ φ′ soil with surcharge. (a) Q'_p determined for s = 0 and c′ = 0, with φ′ γ_{soil} (i.e. component P′ is depth dependent, P′ = 1/2 γh²K_p) the normal method for frictional soil. (b) Q''_p determined for γ = 0, with c′ soil and surcharge s (i.e. component P″ does not vary with depth, P″ = 2c′hK_p^{0.5} + shK_p).

point of application of C is determined by taking moments of elements of the arc about O.

$$\text{Distance of resultant from O} = r\,\frac{\text{length of arc bd}_1}{\text{length of straight line bd}_1}$$

P″ is the Rankine–Bell passive resistance on d_1f_1, due to s and c′, with γ = 0.

$$P = d_1 f_1 \left[2c'\sqrt{K_p} + s\,K_p \right] \tag{A.81}$$

P acts at $(d_1f_1)/2$ below the ground surface.

S is the resultant of surcharge on af_1, i.e. $s.af_1$

Q''_p is not known, except in direction, and point of application.

The construction of the force polygon for abd_1f_1 is carried out as follows:

a. Draw C'_w and C′ on the polygon and space diagrams, and determine the inclination of their resultant on the force polygon (12). Determine the point of intersection of C'_w and C′ on the space diagram.

b. Draw the line of action of the resultant (12) through 1 on the space diagram, parallel to 12 on the force polygon. Determine the position

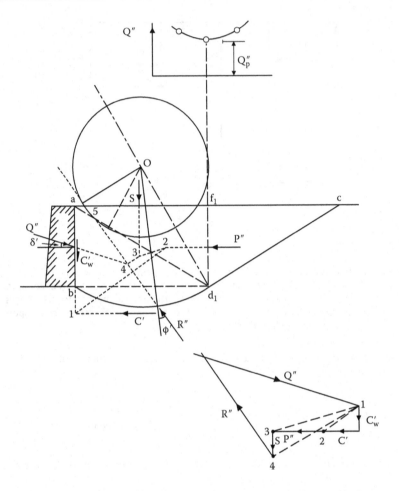

Figure A.28 Friction circle method for passive state—determination of cohesive and surcharge elements of passive force.

of the intercept of the line of action of the resultant and the line of action P″ (at 2).

c. Draw P″ on the force polygon, and determine the direction of the resultant of C'_w, C' and P″ (13), draw a line parallel to this, passing through point 2, on the space diagram. Find the point of intersection of this resultant with S on the space diagram (3).

d. Draw S on the force polygon, and determine the direction of the resultant of C'_w, C', P″ and S (14). Draw a line parallel to this, passing through point 3 on the space diagram. Find the point of intersection of the resultant with the line of action of Q''_p (4).

e. R″ must be tangential to the friction circle, and passes through point 4 on the space diagram. Thus, the direction of R″ (45) can be found.

f. Draw the directions of Q_p'' and R″ on the force polygon, and determine Q_p'' by scaling.

It is to be noted above that the directions of Q_p'' and R″ are the same as in the first part of the problem, since these resultant forces are still affected by the internal friction of the soil, and the wall friction. The total passive force acting on the wall must be determined from the resultant of Q_p' acting at one-third of the wall height, and Q_p'' acting at one-half of the wall height, and Q_p'' acting at one-half of the wall height.

Earth pressure coefficients derived from this method, to be used in the Rankine–Bell equation,

$$q_p' = K_p \cdot p' + K_{pe} \cdot c' \tag{A.82}$$

have been given by Packshaw (1946) for cases with one horizontal ground surface, and parameters c, c_w ϕ, δ (or c′, c_w', ϕ' and δ′). Therefore this method is only very rarely required, perhaps for more complete conditions of groundwater or soil or wall geometry.

A.5 LATERAL PRESSURES DUE TO EXTERNAL LOADS

The soil supported by many types of retaining structures may be subjected to external loads (i.e. loads not derived from the self-weight of the soil itself). For example, a quay wall in a dock will obviously have traffic driven over it, and freight placed upon it. A bridge abutment will be subjected to both the vertical loading of passing vehicles, and also to horizontal breaking forces. Some temporary and permanent retaining structures are built specifically to provide support for preexisting permanent structures, for example, adjacent buildings or power pylons, while temporary excavations for new foundations are made.

External loads normally act to increase the horizontal stresses on a retaining wall. A number of methods exist to predict their effect, but there is little reliable data against which to try these methods or to check that they are of sufficient reliability for design purposes.

The simplest case of an external load is where a uniformly distributed load is placed over the entire ground surface behind or in front of a retaining wall. It can be used to demonstrate the problem which faces the designer, in trying to use hand calculation methods to predict the increased horizontal total stress to be supported by a wall. Three approaches are possible:

- Simple 'elastic' solutions
  —With implied horizontal wall displacements
  —Rigid wall
- Simple 'plastic' solutions
  —Active
  —At-rest
  —Passive
- Numerical modelling of wall soil and construction process

Elastic stress distributions, readily available in texts such as Poulos and Davis (1974), can be used to obtain both the vertical and horizontal stress increases resulting from a wide variety of load geometries. Simple elastic solutions assume that the soil is a linear elastic material, and will normally (in order to make use of existing solutions) also assume that the soil is a semi-infinite half-space (implying that the wall and excavation do not exist) and is homogeneous and isotropic. The horizontal stress is calculated directly from the elastic equations, and the result will be a function of Poisson's ratio, which is one of the least well-known parameters in soil mechanics.

In the simplest case—of a uniformly distributed load of great lateral extent behind a rigid retaining wall—there will be no horizontal strains in the soil. If $\epsilon_h = 0$, from Hooke's law

$$\Delta\sigma_h = \frac{\Delta\sigma_v}{(1-v)} \qquad\qquad (A.83)$$

where v is Poisson's ratio (see below for a discussion on its range of values). Equations to predict the lateral stress increase created by loads of more complex geometry, and where the wall is allowed to deflect, will only be available for certain cases (see Section A.5.1). If the wall is assumed rigid, and the loading geometry is simple, then the horizontal stress distribution calculated from elastic solutions may need to be doubled. If the implied horizontal wall displacements are thought to be realistic, then the value given by elastic theory may be used directly. But for other cases, no simple hand solution is available.

Simple 'plastic' approaches in fact use a combination of elastic and plastic methods, implying that the soil is simultaneously both far from failure and at failure. The vertical stress at different elevations down the back of the wall is calculated using elastic stress distributions (see Section A.5.2), and the horizontal total stress increase down the back of the wall is then obtained by multiplying these values by a suitable earth pressure coefficient (i.e. K_a, or K_p, or K_o). Active and passive coefficients are often used, depending on whether the surcharge is on the retained or excavated side of the wall, and in principal K_o would seem appropriate in the case of a rigid wall. In the simplest case, for a uniform surface surcharge, elastic solutions predict that the vertical stress increase at any depth will be equal to the applied surcharge pressure. In other words, the pressure on the retaining structure will be modified in the same way as if there were an extra layer of soil placed on the ground surface. Therefore, the horizontal total stress at any depth will be given by

$$\sigma_h = (\text{surcharge pressure} \times K) + (\text{effective vertical stress} \times K) + \text{water pressure}. \tag{A.84}$$

The horizontal force due to an external load may also be taken into account for failure conditions by using the graphical techniques described in Section A.4. This will often be done when ground or groundwater conditions are too complex to allow earth pressure coefficients to be used. As a result of this approach, a number of methods of calculation have been proposed which are based loosely upon graphical techniques, but aim only to estimate the influence of the external force, rather than both the soil and any external forces. These techniques are described in Section A.5.3 They can lead to rather irrational results. For example, if a Coulomb wedge analysis is carried out for the general case of a line load behind a wall, it will be found that the force on the wall is increased by a uniform amount for all positions of the line load away from the wall, up to a certain point. Beyond that point, the line load has no effect.

It will be evident, from the discussion above, that for a given loading geometry there may be several possible ways for calculating the horizontal forces and stresses on a wall. Each method will yield a different distribution of horizontal stress, and a different increase in horizontal force, on the wall. If hand calculation methods must be used then it is recommended that the calculations be carried out using all the available techniques, so that the full range of stress and force increase can be appreciated.

However, if the effect of external loading is critical to the design, it is recommended that numerical methods are used. The various options available are discussed in Chapter 8 (Introduction to analysis), where the methods of modelling the soil, the wall, and its construction are also considered. Ideally, a constitutive model invoking both elasticity and plasticity should be used to estimate the effects of external loading, and wall installation and time effects (i.e. due to dissipation of excess pore pressures) should be included. Unfortunately, it will not normally be economically feasible to carry out such an analysis, especially when three-dimensional modelling is required in order to simulate the loading and excavation geometry. For almost all purposes, it will be possible only to model plane strain conditions, so that considerable simplification of the design problem will probably be required before numerical analysis can begin.

A.5.1 Elastic solutions for horizontal stress increase

The classical solutions for loads applied to the surface of a semi-infinite elastic half-space can readily be used to predict the horizontal stress increase on a wall at any depth, due to a load at the soil surface. Table A.1 gives some examples. These solutions required an estimate of Poisson's ratio (v), which is often assumed to be as below

Soil type	Assumed value of Poisson's ratio	
	Short term	Long term
Cohesive	0.50	0.25
Granular	0.25	0.25

In cases 2 and 4 in Table A.1, the loads extend equally on both sides of the wall position, and the implied horizontal deformations at the wall are zero. For the other cases, the elastic equations imply wall yield, since in reality they do not consider a wall at all but simply a load at the surface of an elastic material which extends downward, and laterally in all directions to infinity. For a rigid wall Mindlin (1936) pointed out that as the horizontal displacement at the wall will be zero the 'method of images' (Figure A.29) may be invoked to predict horizontal stresses. The method of images is based upon the principle of superposition. The stresses in an elastic solid due to load P imply horizontal deformations in the x direction at the position of the wall (Figure A.29a) which may be brought back to zero by the application of load P', magnitude equal to P. However, by the principle of superposition, the stresses on the wall will be doubled when P' is applied.

On this basis, Spangler and Mickle (1956) proposed a number of simplified equations for the horizontal stress on a rigid wall, where the horizontal stress from the elastic equation is doubled.

A.5.2 Calculation based on vertical elastic stress distributions

The use of elasticity to predict horizontal stresses has the disadvantage that the implied stresses and strains in the elastic material may not be possible in a soil. For example, elastic solutions can imply tension in some area of the soil mass, and soil cannot sustain tension. Further, the strength of the material is assumed infinite, and therefore any combination of effective principal stress ratio is permissible. In soil, effective principal stress ratios are limited to less than, or equal to, the value at which failure occurs.

It is therefore not unreasonable to calculate the horizontal loads on walls due to external loads by using elastic theory to find the vertical stress change, and then multiplying the vertical stress increase by the relevant earth pressure coefficient, depending on wall yield, etc. The elastic equations for vertical stress

Table A.1 Elastic solutions for horizontal stress increase

Case	Geometry	Solution
1. Point load		$\sigma_h = \dfrac{P}{2\pi R^2}\left[\dfrac{3r^2 z}{R^3} - \dfrac{(1-2v)R}{(R+z)}\right]$ where $r = (x^2 + y^2)^{1/2}$ $R = (x^2 + y^2 + z^2)^{1/2}$
2. Line load, perpendicular to wall		$\sigma_h = 2pvz/\pi R^2$ where $R = (y^2 + z^2)^{1/2}$
3. Line load, parallel to wall		$\sigma_h = 2px^2 z/\pi R^4$ where $R = (x^2 + z^2)^{1/2}$
4. Strip load, perpendicular to wall		$\sigma_h = 2pv\alpha/\pi$
5. Strip load, parallel to wall		$\sigma_h = \dfrac{P}{\pi}[\alpha - \sin\alpha\,\cos(\alpha + 2\delta)]$
6. Loading on a rectangular area		Use principle of super-position. Below the corner of a rectangular area, l by b $\sigma_h = \dfrac{P}{2\pi}\left[\tan^{-1}\dfrac{lb}{zR_3} - \dfrac{lbz}{R_1^2 R_3}\right]$ where $R_1 = (l^2 + z^2)^{1/2}$ $R_3 = (l^2 + b^2 + z^2)^{1/2}$

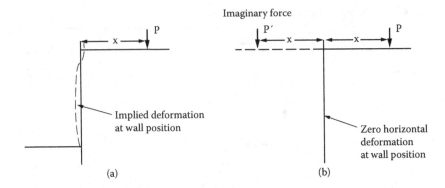

Figure A.29 Principle of Mindlin's method of images. (a) Deformation of wall under a single force, P. (b) Deformation of wall reduced to zero by addition of a second force, P′.

increase are given in Table A.2. Here, once again, the method of images must be used for cases 1, 3, 5 and 6 to predict the maximum stress if the wall is rigid.

A.5.3 Empirical approaches

Because of the lack of data on the effects of external loads, a number of empirical approaches have been used.

For line loads, Terzaghi and Peck (1948) suggest the simple approach shown in Figure A.30. The line d′e′ is drawn from the point of application of the line load W_L at ground surface at an angle of 40° to the horizontal, toward wall. An equivalent line load of $K_a.W_L$ is applied horizontally to the back of the wall at the point (d′) where this line hits the back of the wall. As Terzaghi (1954) points out, this method (which is loosely based on results from the Coulomb wedge graphical approach) gives results which in no way resemble those from experimentation.

The British Code of Practice CP2 (1951) extended this approach to allow for area loads, as follows:

i. The 40° line is constructed from the centre of the loaded area.
ii. If the length of the load (parallel to the wall) is L, and the distance between the back of the wall and the near edge of the loaded area is x, the resultant load on the back of the wall is assumed to be of length (L + x).
iii. If W_L is the total load on the area at ground surface, the resultant horizontal thrust/unit length of wall may be taken as

$$\frac{K_a W_L}{(L + x)}.$$

CP2 described this as a 'tentative approximate method'.

Table A.2 Elastic solutions for vertical stress increase

Case	Geometry	Solution
1. Point load		$\sigma_v = \dfrac{3Pz^3}{2\pi R^5}$ P = point load R = $(x^2 + y^2 + z^2)^{1/2}$
2. Line load, perpendicular to wall		$\sigma_v = 2pz^3/\pi R^4$ p = load/unit length R = $(y^2 + z^2)^{1/2}$
3. Line load, parallel to wall		$\sigma_v = 2pz^3/\pi R^4$ p = load/unit length R = $(x^2 + z^2)^{1/2}$
4. Strip load, perpendicular to wall		$\sigma_v = \dfrac{P}{\pi}[\alpha + \sin \alpha \cos(\alpha + 2\delta)]$ p = load/unit length
5. Strip load, parallel to wall		As (4) above
6. Loading on a rectangular area		Use principle of super-position. Below the corner of a rectangular area l × b $\sigma_v = \dfrac{P}{2\pi}\left[\tan^{-1} \dfrac{lb}{zR_3} \right.$ $\left. + \dfrac{lbz}{R_3}\left(\dfrac{1}{R_1^2} + \dfrac{1}{R_2^2} \right) \right]$ where $R_1 = (l^2 + z^2)^{1/2}$ $R_2 = (b^2 + z^2)^{1/2}$ $R_3 = (l^2 + b^2 + z^2)^{1/2}$

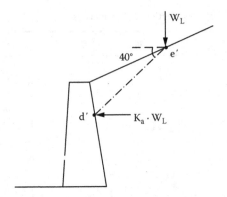

Figure A.30 Method of estimating the effect of a line load. N.B. If point d′ falls below the bottom of the wall, the effects of the line load are ignored. (From Terzaghi, K. and Peck, R.B., *Soil Mechanics in Engineering Practice*. First Edition, John Wiley, New York, 1948.)

Appendix B: Earth pressure coefficients

This appendix contains tables of earth pressure coefficients, and notes on their derivation and use.

It is assumed that the reader will wish to use the simplest method of computation possible, and therefore the use of earth pressure coefficients is recommended unless the complexity of the problem demands a more time-consuming graphical solution. In the writers' experience, this is rare because the amount of subsoil data available is normally inadequate to allow complex geometry to be assumed for either the subsoil or the groundwater.

On this basis, the earth pressure calculation should use the methods given in the table on the following page (see Figure B.1 for definition of terms).

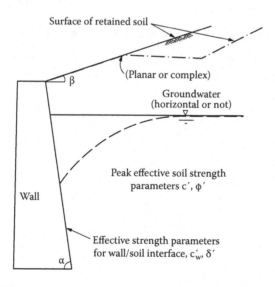

Figure B.1 Definition of terms for selection of earth pressure coefficients.

(See also notes on derivation of tables at the end of this Appendix.)

β	α	c'	ϕ'	c'_w	δ'	Soil surface and groundwater surface geometry	Active/ passive	Method of finding K
0°	90°	0	ϕ'	0	0	Planar/horizontal	A or P	Table B.1
0°	90°	0	ϕ'	0	δ'	"	A	Table B.2
0°	90°	c'	ϕ'	c'_w	δ'	"	A	Table B.2. For K_{ac} see note.
β	90°	0	ϕ'	0	δ'	"	A	Table B.3 (i)
β	90°	c'	ϕ'	c'_w	δ'	"	A	Table B.3 (i). For K_{ac} see note.
0°	α	0	ϕ'	0	δ'	"	A	Table B.3 (ii)
0°	α	c'	ϕ'	c'_w	δ'	"	A	Table B.3 (ii). For K_{ac} see note.
0°	90°	c'	ϕ'	0	0	"	A or P	Table B.4
0°	90°	c_u	0	0	0	"	A or P	Table B.4
0°	90°	c'	ϕ'	c'_w	δ'	"	A	Table B.6
0°	90°	c'	ϕ'	c'_w	δ'	"	P	Table B.7
0°	90°	0	ϕ'	0	δ'	"	P	Table B.8
β	α	0	ϕ'	0	δ'	"	A	Use Müller-Breslau equation. For K_c see note to Table B.3
β	α	0	ϕ'	0	$\delta' < 15°$	"	P	Table B.5
β	α	c'	ϕ'	c'_w	δ'	"	P	Table B.5. For K_{pc} see note.
						Non-planar/not horizontal	A or P	Use graphical method

B.I   TABLES OF EARTH PRESSURE COEFFICIENTS

Table B.1 Coefficients of active and passive earth pressure for cohesionless soils with a vertical back of wall ($\alpha = 90°$) and a horizontal surface ($\beta = 0$) in the Rankine state ($\delta' = 0$)

Table B.2 Coefficients of active earth pressure for cohesionless soils with a horizontal surface ($\beta = 0$), vertical back of wall ($\alpha = 90°$) and wall friction (δ')

Table B.3 (i) Coefficients of active earth pressure for cohesionless soils with wall friction (δ') and a vertical back of wall ($\alpha = 90°$)

(ii) Coefficients of active earth pressure for cohesionless soils with wall friction (δ') and a horizontal soil surface ($\beta = 0$)

Table B.4 Coefficients of active and passive earth pressure for cohesive soils with a horizontal surface ($\beta = 0$) and no wall adhesion ($c'_w = 0$) in the Rankine state ($\delta' = 0$)

Table B.5 (a) Values for the coefficient of passive earth pressure for cohesionless soils, retained by a wall with angle of wall friction (δ') = ϕ', using a log spiral failure surface

(b) Reduction factors to be applied to K_p values in Table B.5 (a), to allow for $\delta' < \phi'$

Table B.6 Coefficients of active earth pressure for cohesive soils with a horizontal surface ($\beta = 0$) and vertical back of wall ($\alpha = 90°$), with wall friction (δ') and adhesion (c'_w), derived with no allowance for tension cracks using an extended Coulomb wedge analysis by Packshaw (1946), as quoted in CP2:1951

Table B.7 Coefficients of passive earth pressure for cohesive soils with a horizontal surface ($\beta = 0$) and vertical back of wall ($\alpha = 90°$), with wall friction (δ' and adhesion (c'_w), derived using the friction circle graphical method by Packshaw (1946), as quoted in CP2:1951

Table B.8 Coefficients of passive earth pressure for cohesionless soils with a horizontal surface ($\beta = 0$), vertical back of wall ($\alpha = 90°$), and wall friction (δ', derived using the 'friction circle' graphical method and given in CP2:1951

Table B.1 Coefficients of active and passive earth pressure for cohesionless soils with a vertical back of wall ($\alpha = 90°$) and a horizontal surface ($\beta = 0$) in the Rankine state ($\delta' = 0$)

ϕ' (degrees)	K_a	K_p
10	0.704	1.420
12	0.656	1.525
14	0.610	1.638
16	0.568	1.761
18	0.528	1.894
20	0.490	2.040
22	0.455	2.198
24	0.422	2.371
26	0.390	2.561
28	0.361	2.770
30	0.333	3.000
32	0.307	3.255
34	0.283	3.537
36	0.260	3.852
38	0.238	4.204
40	0.217	4.599
42	0.198	5.045
44	0.180	5.550
46	0.163	6.126

After: Rankine, W.J.M., *Phil. Trans. Roy. Soc. Lond.* 147, 9–27, 1857.

Table B.2 Coefficients of active earth pressure for cohesionless soils with a horizontal surface ($\beta = 0$), vertical back of wall ($\alpha = 90°$) and wall friction (δ')

ϕ'	δ' (degrees)																
	0	3	6	9	12	15	18	21	24	27	30	33	36	39	42	45	48
9	.729	.701	.679	.662													
12	.655	.631	.611	.596	.584												
15	.588	.567	.55	.536	.526	.517											
18	.527	.509	.494	.483	.473	.466	.46										
21	.472	.456	.444	.433	.425	.419	.414	.411									
24	.421	.408	.397	.389	.382	.376	.372	.369	.368								
27	.375	.364	.355	.347	.342	.337	.333	.331	.33	.33							
30	.333	.324	.316	.31	.305	.301	.298	.296	.295	.296	.297						
33	.294	.287	.28	.275	.271	.268	.266	.264	.264	.264	.265	.267					
36	.259	.253	.248	.243	.24	.237	.236	.235	.234	.235	.236	.238	.241				
39	.227	.222	.218	.214	.211	.209	.208	.207	.207	.208	.209	.211	.214	.217			
42	.198	.194	.19	.187	.185	.184	.183	.182	.182	.183	.184	.186	.189	.192	.196		
45	.171	.168	.165	.163	.161	.16	.159	.159	.159	.16	.161	.163	.166	.168	.172	.176	
48	.147	.144	.142	.14	.139	.138	.138	.138	.138	.139	.14		.142	.144	.147	.15	.154	.159

After: Mayniel, K. *Traite experimental, analytique et practique de la poussee des terres et des murs de revetement*, Paris, 1808; Rankine, W.J.M., *Phil. Trans. Roy. Soc. Lond.* 147, 9–27, 1857, etc.

Table B.3 (i) Coefficients of active earth pressure for cohesionless soils with wall friction (δ') and a vertical back of wall (α = 90°)

φ' (deg.)	δ' (degrees)																
	0	3	6	9	12	15	18	21	24	27	30	33	36	39	42	45	48
Part I: Soil surface (β) slopes at −30°																	
9	.545	.506	.476	.452													
12	.487	.457	.433	.414	.398												
15	.437	.413	.394	.378	.365	.354											
18	.394	.374	.358	.345	.334	.325	.318										
21	.354	.338	.325	.314	.305	.298	.292	.287									
24	.319	.306	.295	.286	.278	.272	.267	.263	.26								
27	.287	.276	.267	.259	.253	.248	.244	.241	.238	.237							
30	.257	.248	.241	.234	.229	.225	.222	.219	.217	.216	.216						
33	.23	.222	.216	.211	.207	.203	.201	.199	.198	.197	.197	.197					
36	.204	.198	.193	.189	.186	.183	.181	.18	.179	.178	.178	.179	.181				
39	.181	.176	.172	.169	.166	.164	.162	.161	.161	.161	.161	.162	.163	.165			
42	.16	.156	.152	.15	.148	.146	.145	.144	.144	.144	.144	.145	.147	.149	.151		
45	.14	.137	.134	.132	.13	.129	.128	.128	.128	.128	.129	.13	.131	.133	.135	.138	
48	.121	.119	.117	.115	.114	.113	.112	.112	.112	.113	.114	.115	.116	.118	.12	.123	.126

(continued)

Table B.3 (i) (Continued) Coefficients of active earth pressure for cohesionless soils with wall friction (δ') and a vertical back of wall (α = 90°)

Part 2: Soil surface (β) slopes at −25°

ϕ' (deg.)	δ' (degrees)																
	0	3	6	9	12	15	18	21	24	27	30	33	36	39	42	45	48
9	.567	.529	.499	.476													
12	.508	.478	.454	.435	.42												
15	.457	.433	.413	.398	.385	.374											
18	.411	.392	.376	.363	.352	.343	.336										
21	.37	.354	.341	.33	.321	.314	.308	.303									
24	.333	.32	.309	.3	.292	.286	.281	.277	.274								
27	.299	.288	.279	.271	.265	.26	.256	.253	.251	.249							
30	.268	.259	.251	.245	.24	.236	.232	.23	.228	.227	.227						
33	.239	.232	.225	.22	.216	.213	.21	.208	.207	.206	.206	.207					
36	.212	.206	.201	.197	.194	.191	.189	.188	.187	.186	.187	.188	.189				
39	.188	.183	.179	.176	.173	.171	.169	.168	.168	.168	.168	.169	.171	.173			
42	.165	.161	.158	.155	.153	.152	.15	.15	.149	.15	.15	.151	.153	.155	.158		
45	.145	.141	.139	.137	.135	.134	.133	.132	.132	.133	.134	.135	.136	.138	.141	.144	
48	.125	.123	.121	.119	.118	.117	.116	.116	.116	.117	.118	.119	.12	.122	.125	.128	.131

Part 3: Soil surface (β) slopes at –20°

9	.591	.554	.525	.502													
12	.53	.501	.478	.459	.444												
15	.477	.453	.434	.419	.406	.395											
18	.43	.41	.394	.381	.371	.362	.355										
21	.387	.371	.358	.347	.338	.33	.324	.32									
24	.348	.334	.323	.314	.307	.301	.296	.292	.289								
27	.312	.301	.292	.284	.278	.273	.269	.266	.264	.262							
30	.279	.27	.262	.256	.251	.247	.243	.241	.239	.239	.239						
33	.249	.241	.235	.23	.225	.222	.22	.218	.217	.216	.216	.217					
36	.221	.215	.209	.205	.202	.199	.197	.196	.195	.195	.195	.196	.198				
39	.195	.19	.186	.182	.18	.178	.176	.175	.175	.175	.175	.177	.178	.181			
42	.171	.167	.164	.161	.159	.157	.156	.156	.155	.156	.156	.158	.159	.162	.165		
45	.149	.146	.144	.141	.14	.138	.138	.137	.137	.138	.139	.14	.142	.144	.147	.15	
48	.129	.127	.125	.123	.122	.121	.12	.12	.12	.121	.122	.123	.125	.127	.13	.133	.136

(continued)

Table B.3 (i) (Continued) Coefficients of active earth pressure for cohesionless soils with wall friction (δ') and a vertical back of wall (α = 90°)

Part 4: Soil surface (β) slopes at −15°

φ' (deg.)	δ' (degrees)																
	0	3	6	9	12	15	18	21	24	27	30	33	36	39	42	45	48
9	.617	.581	.553	.531													
12	.555	.526	.503	.485	.47												
15	.499	.476	.457	.442	.429	.419											
18	.45	.43	.415	.402	.391	.382	.376										
21	.404	.389	.375	.364	.355	.348	.343	.338									
24	.363	.35	.339	.33	.322	.316	.312	.308	.305								
27	.325	.314	.305	.298	.291	.286	.282	.28	.278	.277							
30	.291	.282	.274	.268	.262	.258	.255	.253	.251	.251	.251						
33	.259	.251	.245	.24	.235	.232	.23	.228	.227	.227	.227	.228					
36	.229	.223	.218	.214	.21	.208	.206	.204	.204	.204	.204	.206	.207				
39	.202	.197	.193	.19	.187	.185	.183	.182	.182	.182	.183	.184	.186	.189			
42	.177	.173	.17	.167	.165	.163	.162	.162	.161	.162	.163	.164	.166	.169	.172		
45	.154	.151	.148	.146	.145	.143	.143	.142	.142	.143	.144	.145	.147	.149	.152	.156	
48	.133	.131	.129	.127	.126	.125	.124	.124	.125	.125	.126	.127	.129	.132	.134	.138	.141

Part 5: Soil surface (β) slopes at −10°

9	.647	.613	.586	.565	.500	.446	.399	.359	.323	.292	.264	.24	.217	.197	.179	.162	.147
12	.582	.554	.532	.514	.456	.406	.363	.326	.293	.264	.238	.215	.195	.176	.158	.143	
15	.524	.501	.483	.468	.414	.368	.329	.295	.264	.238	.214	.192	.173	.155	.139		
18	.472	.453	.437	.424	.375	.334	.297	.266	.238	.213	.191	.171	.153	.136			
21	.424	.408	.395	.384	.339	.301	.268	.239	.213	.19	.169	.151	.134				
24	.38	.367	.356	.347	.306	.271	.24	.213	.19	.168	.149	.132					
27	.34	.329	.32	.312	.275	.243	.215	.19	.168	.148	.131						
30	.303	.294	.286	.28	.246	.217	.191	.168	.148	.129							
33	.269	.262	.256	.25	.219	.192	.169	.148	.129								
36	.238	.232	.227	.223	.194	.17	.148	.129									
39	.21	.205	.201	.197	.171	.149	.129										
42	.184	.18	.176	.173	.15	.129											
45	.16	.156	.154	.151	.13												
48	.138	.135	.133	.131	.129												

(continued)

Table B.3 (i) (Continued) Coefficients of active earth pressure for cohesionless soils with wall friction (δ') and a vertical back of wall (α = 90°)

Part 6: Soil surface (β) slopes at –5°

φ' (deg.)	δ' (degrees)																
	0	3	6	9	12	15	18	21	24	27	30	33	36	39	42	45	48
9	.683	.651	.626	.606													
12	.615	.588	.567	.55	.537												
15	.553	.531	.513	.498	.487	.477											
18	.497	.478	.463	.451	.441	.433	.427										
21	.446	.43	.417	.407	.398	.391	.386	.382									
24	.399	.386	.375	.366	.359	.353	.349	.346	.344								
27	.356	.345	.336	.329	.323	.318	.314	.312	.31	.31							
30	.317	.308	.3	.294	.289	.285	.282	.28	.279	.279	.279						
33	.281	.274	.267	.262	.258	.255	.252	.251	.25	.25	.251	.253					
36	.248	.242	.237	.232	.229	.226	.225	.223	.223	.223	.224	.226	.228				
39	.218	.213	.209	.205	.202	.2	.199	.198	.198	.199	.2	.201	.204	.207			
42	.19	.186	.183	.18	.178	.176	.175	.175	.175	.175	.177	.178	.18	.183	.187		
45	.165	.162	.159	.157	.155	.154	.153	.153	.153	.154	.155	.157	.159	.162	.165	.169	
48	.142	.139	.137	.136	.134	.133	.133	.133	.133	.134	.135	.137	.139	.141	.144	.148	.152

Part 7: Soil surface (β) slopes at +5°

9	.799	.777	.761	.749													
12	.711	.69	.673	.66	.651												
15	.634	.615	.599	.587	.578	.571											
18	.565	.548	.534	.523	.515	.508	.504										
21	.504	.489	.477	.467	.459	.453	.449	.446									
24	.448	.435	.424	.416	.409	.404	.4	.398	.397								
27	.397	.386	.377	.37	.364	.36	.357	.355	.354	.354							
30	.351	.342	.334	.328	.323	.32	.317	.316	.315	.316	.317						
33	.309	.302	.295	.29	.286	.283	.281	.28	.28	.28	.282	.284					
36	.271	.265	.26	.256	.252	.25	.248	.247	.247	.248	.25	.252	.255				
39	.237	.232	.228	.224	.221	.22	.218	.218	.218	.219	.22	.222	.225	.229			
42	.206	.202	.198	.195	.193	.192	.191	.191	.192	.192	.193	.195	.198	.201	.206		
45	.178	.174	.171	.169	.168	.166	.166	.166	.166	.167	.168	.17	.173	.176	.18	.184	
48	.152	.149	.147	.145	.144	.143	.143	.143	.144	.145	.146	.148	.15	.153	.156	.16	.165

(continued)

Table B.3 (i) (Continued) Coefficients of active earth pressure for cohesionless soils with wall friction (δ') and a vertical back of wall (α = 90°)

φ' (deg.)	δ' (degrees)																
	0	3	6	9	12	15	18	21	24	27	30	33	36	39	42	45	48
Part 8: Soil surface (β) slopes at +10°																	
12	.811	.797	.787	.781	.777												
15	.703	.687	.675	.666	.659	.655											
18	.618	.603	.591	.581	.574	.569	.566										
21	.545	.531	.52	.511	.504	.499	.496	.495									
24	.481	.469	.459	.451	.445	.44	.437	.436	.436								
27	.424	.414	.405	.398	.393	.389	.386	.385	.384	.386							
30	.373	.364	.357	.351	.346	.343	.34	.339	.339	.34	.342						
33	.327	.32	.313	.308	.304	.302	.3	.299	.299	.3	.302	.305					
36	.286	.28	.274	.27	.267	.265	.263	.263	.263	.264	.266	.268	.272				
39	.249	.244	.239	.236	.233	.231	.23	.23	.23	.231	.233	.235	.239	.243			
42	.215	.211	.207	.205	.203	.201	.2	.2	.201	.202	.203	.206	.209	.212	.217		
45	.185	.182	.179	.177	.175	.174	.173	.173	.174	.175	.177	.179	.181	.185	.189	.184	
48	.158	.155	.153	.151	.15	.149	.149	.149	.15	.151	.152	.154	.157	.16	.163	.168	.173

Part 9: Soil surface (β) slopes at +15°

15	.933	.934	.938	.944	.953	.965											
18	.709	.697	.689	.683	.679	.678	.679										
21	.608	.596	.586	.579	.574	.571	.57	.571									
24	.528	.517	.507	.5	.495	.492	.49	.49	.491								
27	.46	.45	.442	.435	.431	.427	.425	.425	.426	.428							
30	.401	.393	.385	.38	.375	.372	.371	.37	.371	.373	.376						
33	.35	.342	.336	.331	.327	.325	.323	.323	.323	.325	.327	.331					
36	.304	.297	.292	.288	.285	.283	.282	.281	.282	.283	.285	.289	.293				
39	.263	.257	.253	.25	.247	.246	.245	.244	.245	.246	.248	.251	.255	.26			
42	.226	.222	.218	.216	.214	.212	.212	.212	.212	.214	.215	.218	.221	.226	.231		
45	.194	.19	.187	.185	.184	.183	.182	.182	.183	.184	.186	.188	.191	.195	.2	.205	
48	.164	.162	.16	.158	.157	.156	.156	.156	.157	.158	.159	.162	.164	.168	.172	.176	.182

(continued)

Table B.3 (i) (Continued) Coefficients of active earth pressure for cohesionless soils with wall friction (δ') and a vertical back of wall ($\alpha = 90°$)

ϕ' (deg.)	δ' (degrees)																
	0	3	6	9	12	15	18	21	24	27	30	33	36	39	42	45	48
Part 10: Soil surface (β) slopes at +20																	
21	.745	.738	.734	.733	.734	.737	.742	.75									
24	.605	.596	.589	.584	.581	.58	.581	.584	.588								
27	.514	.504	.497	.492	.488	.486	.486	.487	.489	.493							
30	.441	.432	.426	.42	.417	.415	.414	.414	.416	.419	.423						
33	.379	.372	.366	.361	.358	.356	.355	.355	.356	.359	.362	.367					
36	.326	.32	.315	.311	.308	.306	.306	.306	.307	.309	.312	.316	.321				
39	.28	.275	.271	.267	.265	.264	.263	.263	.264	.266	.268	.272	.276	.281			
42	.24	.235	.232	.229	.227	.226	.226	.226	.227	.228	.23	.233	.237	.242	.248		
45	.204	.2	.198	.196	.194	.193	.193	.193	.194	.195	.197	.2	.203	.208	.212	.218	
48	.172	.17	.167	.166	.165	.164	.164	.164	.165	.166	.168	.17	.173	.177	.181	.187	.193

Part 11: Soil surface (β) slopes at +25°

27	.619	.612	.608	.606	.605	.607	.61	.615	.623	.632							
30	.504	.497	.491	.487	.485	.484	.485	.488	.491	.497	.504						
33	.423	.416	.411	.407	.404	.403	.403	.404	.406	.41	.415	.422					
36	.358	.352	.347	.343	.341	.34	.339	.34	.342	.345	.349	.354	.36				
39	.303	.298	.294	.291	.289	.288	.287	.288	.289	.292	.295	.299	.304	.311			
42	.257	.253	.249	.247	.245	.244	.244	.244	.245	.247	.25	.254	.258	.264	.27		
45	.217	.213	.211	.209	.207	.206	.206	.207	.208	.21	.212	.215	.219	.223	.229	.236	
48	.182	.179	.177	.175	.174	.174	.174	.174	.175	.177	.179	.182	.185	.189	.194	.199	.206

(continued)

Table B.3 (i) (Continued) Coefficients of active earth pressure for cohesionless soils with wall friction (δ') and a vertical back of wall (α = 90°)

φ' (deg.)	δ' (degrees)																
	0	3	6	9	12	15	18	21	24	27	30	33	36	39	42	45	48
Part 12: Soil surface (β) slopes at +30°																	
30	.749	.751	.754	.759	.766	.776	.788	.803	.82	.841	.866						
33	.503	.498	.494	.492	.491	.492	.495	.498	.504	.511	.52	.531					
36	.408	.402	.398	.395	.394	.393	.394	.396	.4	.404	.41	.417	.426				
39	.337	.332	.329	.326	.324	.324	.324	.325	.327	.331	.335	.341	.348	.356			
42	.281	.277	.274	.271	.27	.269	.269	.27	.272	.274	.278	.282	.288	.295	.302		
45	.234	.231	.228	.226	.225	.224	.224	.225	.227	.229	.232	.235	.24	.245	.252	.259	
48	.195	.191	.19	.188	.187	.187	.187	.187	.189	.191	.193	.196	.2	.204	.21	.216	.224

After: Müller-Breslau, H. Erddruck auf Stutzmauer, 1906, Alfred Kroner, Stuttgart.

Table B.3 (ii) Coefficients of active earth pressure for cohesionless soils with wall friction (δ′) and a horizontal soil surface (β = 0)

φ′ (deg.)	δ′ (degrees)																
	0	3	6	9	12	15	18	21	24	27	30	33	36	39	42	45	48
Part 1: Back of wall (α) slopes at 60°																	
9	.962	.943	.933	.93													
12	.905	.888	.878	.873	.874												
15	.851	.836	.826	.821	.821	.825											
18	.8	.786	.778	.773	.773	.776	.783										
21	.751	.739	.732	.728	.727	.73	.737	.747									
24	.705	.694	.688	.685	.685	.687	.693	.703	.715								
27	.66	.652	.646	.644	.644	.647	.653	.661	.673	.688							
30	.618	.611	.607	.605	.605	.608	.614	.622	.634	.648	.666						
33	.578	.572	.569	.567	.568	.572	.577	.585	.596	.61	.627	.648					
36	.54	.535	.532	.532	.533	.536	.542	.55	.56	.574	.59	.61	.634				
39	.503	.499	.497	.497	.499	.503	.508	.516	.526	.539	.555	.574	.597	.625			
42	.468	.465	.464	.464	.467	.47	.476	.484	.494	.506	.521	.539	.561	.588	.621		
45	.435	.433	.432	.433	.435	.439	.445	.453	.462	.474	.488	.506	.527	.553	.584	.623	
48	.403	.401	.401	.403	.405	.409	.415	.423	.432	.443	.457	.474	.494	.519	.548	.586	.633

(continued)

Table B.3 (ii) (Continued) Coefficients of active earth pressure for cohesionless soils with wall friction (δ') and a horizontal soil surface (β = 0)

φ' (deg.)	δ' (degrees)																
	0	3	6	9	12	15	18	21	24	27	30	33	36	39	42	45	48
Part 2: Back of wall (α) slopes at 70°																	
9	.853	.831	.817	.808													
12	.792	.772	.759	.75	.745												
15	.735	.717	.705	.696	.691	.69											
18	.681	.666	.655	.647	.642	.641	.641										
21	.631	.618	.608	.601	.597	.595	.596	.599									
24	.584	.572	.564	.558	.554	.553	.554	.557	.562								
27	.539	.529	.522	.517	.514	.513	.514	.517	.522	.529							
30	.497	.489	.483	.479	.476	.476	.477	.48	.485	.492	.501						
33	.458	.451	.446	.442	.441	.441	.442	.445	.45	.456	.465	.475					
36	.421	.415	.411	.408	.407	.407	.409	.412	.416	.423	.431	.441	.453				
39	.386	.381	.378	.376	.375	.375	.377	.38	.385	.391	.399	.408	.42	.434			
42	.353	.349	.346	.345	.344	.345	.347	.351	.355	.361	.368	.377	.388	.401	.41		
45	.322	.319	.317	.316	.316	.317	.319	.322	.327	.332	.339	.348	.358	.371	.386	.403	
48	.292	.29	.288	.288	.288	.29	.292	.295	.3	.305	.312	.32	.33	.342	.356	.372	.392

Part 3: Back of wall (α) slopes at 80°

9	.779	.754	.736	.723	.652	.592	.539	.493	.452	.416	.384	.356	.33	.307	.287	.268	.251
12	.712	.69	.674	.661	.597	.542	.493	.451	.414	.38	.351	.324	.3	.279	.259	.242	
15	.651	.632	.617	.606	.547	.496	.452	.412	.378	.347	.319	.295	.272	.252	.234		
18	.594	.578	.565	.554	.501	.454	.413	.377	.344	.316	.29	.267	.246	.227			
21	.542	.527	.516	.507	.458	.414	.377	.343	.313	.287	.263	.241	.221				
24	.493	.481	.471	.463	.418	.378	.343	.312	.284	.259	.237	.216					
27	.448	.438	.429	.423	.381	.344	.311	.282	.257	.233	.212						
30	.406	.397	.39	.385	.346	.312	.282	.255	.231	.209							
33	.367	.36	.354	.349	.313	.282	.254	.229	.207								
36	.331	.325	.32	.316	.283	.254	.228	.205									
39	.298	.292	.288	.285	.255	.228	.204										
42	.266	.262	.259	.256	.228	.203											
45	.238	.234	.231	.229	.204												
48	.211	.208	.206	.204													

(continued)

Table B.3 (ii) (Continued) Coefficients of active earth pressure for cohesionless soils with wall friction (δ') and a horizontal soil surface (β = 0)

Part 4: Back of wall (α) slopes at 100°

ϕ' (deg.)	\multicolumn{17}{c}{δ' (degrees)}																
	0	3	6	9	12	15	18	21	24	27	30	33	36	39	42	45	48
9	.697	.664	.638	.617													
12	.613	.585	.563	.545	.53												
15	.539	.515	.496	.481	.468	.457											
18	.472	.453	.437	.423	.412	.403	.396										
21	.413	.397	.383	.372	.362	.355	.348	.343									
24	.36	.346	.335	.326	.318	.311	.306	.302	.299								
27	.312	.301	.292	.284	.277	.272	.268	.264	.262	.26							
30	.27	.261	.253	.246	.241	.237	.233	.23	.228	.227	.227						
33	.232	.224	.218	.213	.208	.205	.202	.2	.198	.197	.197	.197					
36	.198	.192	.187	.182	.179	.176	.174	.172	.171	.17	.17	.17	.171				
39	.167	.162	.158	.155	.152	.15	.148	.147	.146	.146	.146	.146	.147	.149			
42	.14	.136	.133	.131	.128	.127	.125	.124	.124	.124	.124	.124	.125	.126	.108		
45	.116	.113	.111	.109	.107	.106	.105	.104	.104	.104	.104	.104	.105	.107	.108	.11	
48	.095	.093	.091	.089	.088	.087	.086	.086	.086	.086	.086	.087	.087	.089	.09	.092	.094

Part 5: Back of wall (α) slopes at 110°

9	.677	.639	.609	.584													
12	.581	.549	.524	.503	.485												
15	.497	.471	.45	.432	.417	.405											
18	.423	.402	.385	.37	.358	.348	.339										
21	.359	.342	.328	.316	.306	.297	.29	.284									
24	.303	.29	.278	.268	.26	.253	.247	.242	.238								
27	.254	.243	.234	.226	.22	.214	.209	.205	.202	.2							
30	.212	.203	.196	.189	.184	.18	.176	.173	.17	.168	.167						
33	.174	.168	.162	.157	.153	.149	.146	.144	.142	.141	.14	.139					
36	.142	.137	.132	.129	.125	.123	.12	.119	.117	.116	.115	.115	.115				
39	.114	.11	.107	.104	.101	.099	.098	.096	.095	.094	.094	.094	.094	.094			
42	.09	.087	.085	.082	.081	.079	.078	.077	.076	.076	.075	.075	.075	.076	.076		
45	.07	.067	.066	.064	.063	.062	.061	.06	.059	.059	.059	.059	.059	.06	.06	.061	
48	.052	.051	.049	.048	.047	.047	.046	.045	.045	.045	.045	.045	.045	.045	.046	.046	.047

(continued)

Table B.3 (ii) (Continued) Coefficients of active earth pressure for cohesionless soils with wall friction (δ') and a horizontal soil surface (β = 0)

φ' (deg.)	δ' (degrees)																
	0	3	6	9	12	15	18	21	24	27	30	33	36	39	42	45	48
Part 6: Back of wall (α) slopes at 120°																	
9	.667	.622	.586	.556													
12	.552	.517	.488	.464	.444												
15	.456	.428	.404	.385	.369	.355											
18	.374	.352	.334	.318	.305	.294	.285										
21	.305	.287	.273	.261	.251	.242	.234	.228									
24	.246	.233	.222	.212	.204	.197	.191	.186	.182								
27	.196	.186	.178	.17	.164	.159	.154	.151	.147	.145							
30	.154	.147	.14	.135	.13	.126	.123	.12	.118	.115	.114						
33	.119	.114	.109	.105	.101	.098	.096	.094	.092	.091	.089	.088					
36	.09	.086	.083	.08	.077	.075	.073	.072	.07	.069	.069	.068	.068				
39	.066	.063	.061	.059	.057	.055	.054	.053	.052	.052	.051	.051	.05	.05			
42	.046	.044	.043	.042	.04	.039	.038	.038	.037	.037	.036	.036	.036	.036	.036		
45	.031	.03	.029	.028	.027	.026	.026	.025	.025	.025	.025	.024	.024	.024	.024	.025	
48	.019	.018	.018	.017	.017	.016	.016	.016	.015	.015	.015	.015	.015	.015	.015	.015	.015

After: Müller-Breslau, H. Erddruck auf Stutzmauert, 1906, Alfred Kroner, Stuttgart.

Table B.4 Coefficients of active and passive earth pressure
for cohesive soils with a horizontal surface
($\beta = 0$) and no wall adhesion ($c'_w = 0$) in the
Rankine state ($\delta' = 0$)

ϕ' (degrees)	Active		Passive	
	K_a	K_{ac}	K_p	K_{pc}
0	1.000	2.000	1.000	2.000
10	0.704	1.678	1.420	2.384
12	0.656	1.620	1.525	2.470
14	0.610	1.563	1.638	2.560
16	0.568	1.507	1.761	2.654
18	0.529	1.453	1.894	2.753
20	0.490	1.400	2.040	2.856
22	0.455	1.349	2.198	2.965
24	0.422	1.299	2.371	3.080
26	0.390	1.250	2.561	3.201
28	0.361	1.202	2.770	3.329
30	0.333	1.155	3.000	3.464
32	0.307	1.109	3.255	3.608
34	0.283	1.063	3.537	3.761
36	0.260	1.019	3.852	3.925
38	0.238	0.975	4.204	4.101
40	0.217	0.933	4.599	4.289
42	0.198	0.890	5.045	4.492
44	0.180	0.849	5.550	4.712
46	0.163	0.808	6.126	4.950

After: Français, J.F., *Memoires de l'Offtcier du Genie*, 4, 1820; Bell, A.L., *Min. Proc. ICE*, 199, 233–272, 1915.

Table B.5 (a) Values for the coefficient of passive earth pressure for cohesionless soils, retained by a wall with angle of wall friction $(\delta') = \phi'$, using a log spiral failure surface

ϕ'	10°	15°	20°	25°	30°	35°	40°	45°	50°
α					$\dfrac{\beta}{\phi'} = +1.0$				
45°	1.21	1.55	2.08	2.99	4.67	8.00	15.3	33.5	89.9
50°	1.32	1.72	2.35	3.43	5.47	9.58	18.8	42.6	119
55°	1.43	1.89	2.63	3.91	6.38	11.4	23.1	54.1	158
60°	1.53	2.06	2.92	4.43	7.40	13.6	28.4	69.1	211
65°	1.62	2.22	3.21	4.99	8.53	16.1	34.7	87.8	281
70°	1.70	2.38	3.51	5.57	9.77	19.0	42.2	111	373
75°	1.77	2.53	3.80	6.18	11.1	22.3	51.1	140	494
80°	1.83	2.66	4.09	6.81	12.6	26.0	61.6	176	652
85°	1.88	2.79	4.37	7.47	14.2	30.1	74.0	220	858
90°	1.93	2.91	4.66	8.16	15.9	34.9	88.7	275	1130
95°	1.96	3.00	4.94	8.89	17.8	40.3	106	344	1480
100°	1.98	3.12	5.23	9.67	19.9	46.6	127	431	1950
105°	2.00	3.21	5.53	10.5	22.3	53.9	153	540	2580
110°	2.01	3.30	5.83	11.4	25.0	62.5	185	680	3420
115°	2.01	3.38	6.15	12.4	28.1	72.8	223	858	4540
120°	2.00	3.45	6.47	13.5	31.6	84.8	271	1090	6040
125°	1.98	3.52	6.80	14.6	35.5	99.0	329	1380	8050
130°	1.94	3.56	7.11	15.8	39.8	115	399	1750	10,700
135°	1.89	3.58	7.38	17.0	44.5	134	482	2200	14,100
α					$\dfrac{\beta}{\phi'} = +0.8$				
45°	1.19	1.52	2.02	2.85	4.29	6.96	12.4	25.0	60.4
50°	1.30	1.69	2.28	3.27	5.01	8.32	15.2	31.8	79.8
55°	1.41	1.85	2.56	3.73	5.84	9.92	18.7	40.4	106
60°	1.51	2.02	2.84	4.22	6.76	11.8	22.9	51.3	141
65°	1.60	2.18	3.12	4.74	7.78	13.9	28.0	65.1	187
70°	1.68	2.33	3.40	5.29	8.89	16.4	34.0	82.3	248
75°	1.75	2.47	3.69	5.86	10.1	19.2	41.1	104	328
80°	1.81	2.60	3.97	6.45	11.4	22.3	49.5	130	432
85°	1.86	2.73	4.24	7.07	12.8	25.8	59.3	162	568
90°	1.90	2.84	4.52	7.71	14.4	29.9	71.0	203	747
95°	1.94	2.95	4.79	8.39	16.1	34.5	85.0	253	981
100°	1.96	3.05	5.07	9.11	18.0	39.8	102	316	1290
105°	1.97	3.14	5.36	9.89	20.1	46.0	122	397	1700

(continued)

Table B.5 (Continued) (a) Values for the coefficient of passive earth pressure for cohesionless soils, retained by a wall with angle of wall friction (δ') = ϕ', using a log spiral failure surface

ϕ'	10°	15°	20°	25°	30°	35°	40°	45°	50°
110°	1.98	3.23	5.66	10.7	22.6	53.4	147	499	2250
115°	1.98	3.31	5.96	11.7	25.3	62.1	178	630	2990
120°	1.97	3.38	6.28	12.7	28.5	72.4	216	797	3980
125°	1.95	3.44	6.59	13.8	32.0	84.4	262	1010	5310
130°	1.92	3.48	6.89	14.9	35.9	98.4	318	1280	7050
135°	1.87	3.50	7.15	16.0	40.1	114	384	1610	9320

α

$$\frac{\beta}{\phi'} = +0.6$$

α	10°	15°	20°	25°	30°	35°	40°	45°	50°
45°	1.14	1.42	1.82	2.47	3.52	5.36	8.85	16.3	35.0
50°	1.25	1.58	2.06	2.83	4.12	6.41	10.9	20.7	46.2
55°	1.35	1.73	2.31	3.23	4.80	7.66	13.4	26.3	61.3
60°	1.45	1.89	2.57	3.66	5.56	9.08	16.4	33.4	81.6
65°	1.54	2.04	2.83	4.12	6.41	10.8	20.0	42.4	108
70°	1.62	2.18	3.09	4.60	7.33	12.7	24.3	53.6	144
75°	1.69	2.32	3.35	5.10	8.33	14.8	29.4	67.5	190
80°	1.75	2.45	3.60	5.62	9.42	17.3	35.4	84.7	251
85°	1.80	2.57	3.86	6.16	10.6	20.0	42.5	106	330
90°	1.84	2.68	4.11	6.72	11.9	23.1	50.9	132	433
95°	1.88	2.78	4.36	7.32	13.3	26.7	61.0	165	569
100°	1.90	2.88	4.62	7.95	14.9	30.9	73.0	206	748
105°	1.92	2.97	4.88	8.64	16.7	35.7	87.7	259	987
110°	1.92	3.05	5.16	9.39	18.7	41.4	106	326	1310
115°	1.92	3.13	5.44	10.2	21.0	48.2	128	411	1740
120°	1.92	3.20	5.72	11.1	23.6	56.1	155	520	2310
125°	1.90	3.25	6.01	12.0	26.5	65.5	188	660	3080
130°	1.86	3.29	6.28	13.0	29.7	76.3	228	836	4090
135°	1.81	3.31	6.52	14.0	33.2	88.7	276	4050	5400

α

$$\frac{\beta}{\phi'} = +0.4$$

α	10°	15°	20°	25°	30°	35°	40°	45°	50°
45°	1.06	1.29	1.60	2.07	2.81	4.03	6.17	10.3	19.7
50°	1.17	1.44	1.81	2.39	2.30	4.83	7.60	13.2	26.1
55°	1.27	1.59	2.04	2.74	3.86	5.79	9.36	16.8	34.7
60°	1.37	1.74	2.28	3.11	4.49	6.90	11.5	21.4	46.3
65°	1.46	1.89	2.52	3.52	5.19	8.19	14.1	27.2	61.6
70°	1.54	2.03	2.76	3.94	5.95	9.66	17.2	34.4	81.9
75°	1.62	2.16	3.00	4.38	6.78	11.3	20.8	43.4	108

(continued)

Table B.5 (Continued) (a) Values for the coefficient of passive earth pressure for cohesionless soils, retained by a wall with angle of wall friction (δ') = ϕ', using a log spiral failure surface

ϕ'	10°	15°	20°	25°	30°	35°	40°	45°	50°
80°	1.68	2.29	3.24	4.84	7.69	13.2	25.1	54.5	143
85°	1.73	2.40	3.48	5.31	8.67	15.3	30.2	66.7	188
90°	1.78	2.51	3.71	5.81	9.74	17.8	36.2	85.4	248
95°	1.81	2.61	3.95	6.34	10.9	20.6	43.4	107	326
100°	1.84	2.70	4.19	6.90	12.2	23.8	52.0	134	429
105°	1.86	2.79	4.44	7.51	13.7	27.5	62.5	168	566
110°	1.87	2.87	4.69	8.17	15.4	32.0	75.4	211	750
115°	1.87	2.95	4.95	8.88	17.3	37.2	91.2	266	997
120°	1.86	3.01	5.21	9.65	19.4	43.4	111	337	1330
125°	1.84	3.07	5.47	10.5	21.8	50.6	134	428	1770
130°	1.81	3.10	5.72	11.3	24.5	58.9	163	542	2350
135°	1.76	3.11	5.93	12.2	27.4	66.4	197	683	3100

$$\frac{\beta}{\phi'} = +0.2$$

α									
45°	0.968	1.14	1.37	1.71	2.20	2.96	4.20	6.39	10.8
50°	1.07	1.28	1.56	1.98	2.60	3.57	5.20	8.16	14.3
55°	1.18	1.42	1.77	2.28	3.06	4.30	6.43	10.4	19.1
60°	1.28	1.57	1.99	2.61	3.57	5.15	7.93	13.4	25.5
65°	1.37	1.72	2.21	2.96	4.15	6.14	9.75	17.0	34.0
70°	1.46	1.86	2.44	3.34	4.78	7.27	11.9	21.6	45.3
75°	1.53	1.99	2.67	3.73	5.48	8.76	14.5	27.4	60.2
80°	1.60	2.12	2.90	4.14	6.23	10.0	17.6	34.5	79.6
85°	1.66	2.24	3.13	4.57	7.06	11.7	21.2	43.3	105
90°	1.71	2.35	3.35	5.02	7.95	13.6	25.5	54.3	138
95°	1.75	2.45	3.58	5.49	8.94	15.7	30.6	68.0	182
100°	1.78	2.55	3.81	5.99	10.0	18.2	36.8	85.2	240
105°	1.81	2.64	4.04	6.53	11.3	21.2	44.3	107	318
110°	1.82	2.72	4.28	7.12	12.7	24.6	53.4	135	421
115°	1.82	2.80	4.52	7.75	14.3	28.7	64.6	170	560
120°	1.82	2.86	4.76	8.43	16.0	33.4	78.4	216	745
125°	1.80	2.91	5.00	9.15	18.0	39.0	95.2	274	993
130°	1.76	2.95	5.22	9.89	20.2	45.4	115	346	1320
135°	1.71	2.95	5.41	10.6	22.5	52.6	139	436	1740

$$\frac{\beta}{\phi'} = 0$$

α									
45°	0.859	0.977	1.14	1.36	1.67	2.11	2.76	3.97	5.59
50°	0.963	1.11	1.32	1.60	1.99	2.56	3.44	4.87	7.46

(continued)

Table B.5 (Continued) (a) Values for the coefficient of passive earth pressure for cohesionless soils, retained by a wall with angle of wall friction $(\delta') = \phi'$, using a log spiral failure surface

ϕ'	10°	15°	20°	25°	30°	35°	40°	45°	50°
55°	1.07	1.25	1.51	1.86	2.36	3.11	4.28	6.27	10.0
60°	1.17	1.39	1.71	2.14	2.78	3.75	5.31	8.05	13.4
65°	1.27	1.54	1.92	2.45	3.25	4.50	6.57	10.3	18.0
70°	1.36	1.68	2.13	2.78	3.78	5.36	8.07	13.2	24.1
75°	1.44	1.82	2.35	3.14	4.35	6.35	9.87	16.7	32.0
80°	1.52	1.95	2.57	3.50	4.98	7.47	12.0	21.2	42.5
85°	1.59	2.07	2.79	3.89	5.67	8.75	14.5	26.7	56.3
90°	1.64	2.19	3.01	4.29	6.42	10.2	17.5	33.5	74.3
95°	1.69	2.30	3.23	4.72	7.25	11.9	21.1	42.1	98.2
100°	1.73	2.40	3.45	5.17	8.17	13.8	25.5	52.9	130
105°	1.76	2.50	3.67	5.66	9.20	16.1	30.7	66.6	172
110°	1.78	2.58	3.90	6.18	10.4	18.7	37.2	84.0	228
115°	1.78	2.66	4.13	6.74	11.7	21.8	45.0	106	303
120°	1.78	2.72	4.35	7.33	13.1	25.5	54.6	135	404
125°	1.76	2.77	4.57	7.95	14.8	29.7	66.3	171	538
130°	1.73	2.80	4.77	8.59	16.5	34.6	80.3	216	714
135	1.67	2.79	4.93	9.21	18.4	40.0	98.6	272	941

α

$$\frac{\beta}{\phi'} = -0.2$$

	10°	15°	20°	25°	30°	35°	40°	45°	50°
45°		0.810	0.913	1.05	1.22	1.44	1.74	2.16	2.77
50°	0.838	0.936	1.07	1.24	1.47	1.77	2.19	2.79	3.72
55°	0.941	1.07	1.24	1.46	1.76	2.17	2.74	3.61	5.02
60°	1.04	1.21	1.42	1.71	2.10	2.64	3.43	4.68	6.78
65°	1.15	1.35	1.62	1.98	2.48	3.19	4.28	6.03	9.13
70°	1.24	1.49	1.82	2.27	2.90	3.84	5.30	7.75	12.3
75°	1.34	1.63	2.03	2.58	3.37	4.58	6.52	9.90	16.4
80°	1.42	1.77	2.24	2.91	3.89	5.43	7.97	12.6	21.9
85°	1.50	1.90	2.45	3.25	4.46	6.39	9.71	15.9	29.1
90°	1.57	2.02	2.67	3.62	5.09	7.50	11.8	20.1	38.6
95°	1.63	2.14	2.88	4.00	5.78	8.78	14.3	25.3	51.1
100°	1.68	2.25	3.10	4.41	6.54	10.2	17.2	31.9	67.7
105°	1.71	2.35	3.31	4.84	7.40	12.0	20.9	40.3	89.8
110°	1.74	2.44	3.53	5.30	8.35	14.0	25.3	51.0	119
115°	1.75	2.52	3.74	5.80	9.42	16.3	30.7	64.5	159
120°	1.74	2.58	3.95	6.31	10.6	19.0	37.2	81.8	212
125°	1.73	2.62	4.15	6.85	11.9	22.2	45.2	104	282
130°	1.69	2.65	4.32	7.39	13.3	25.8	54.7	131	374

(continued)

Table B.5 (Continued) (a) Values for the coefficient of passive earth pressure for cohesionless soils, retained by a wall with angle of wall friction (δ') = ϕ', using a log spiral failure surface

ϕ'	10°	15°	20°	25°	30°	35°	40°	45°	50°
135°	1.63	2.64	4.46	7.90	14.8	29.8	65.8	165	493

α $\dfrac{\beta}{\phi'} = -0.4$

45°									1.16
50°					0.924	1.03	1.17	1.32	1.52
55°		0.804	0.879	0.979	1.10	1.26	1.44	1.68	1.98
60°		0.908	1.01	1.14	1.31	1.51	1.78	2.12	2.59
65°		1.01	1.15	1.32	1.53	1.81	2.17	2.66	3.36
70°		1.12	1.29	1.50	1.78	2.15	2.64	3.33	4.35
75°		1.22	1.43	1.70	2.05	2.52	3.17	4.13	5.60
80°		1.31	1.57	1.90	2.34	2.94	3.79	5.09	7.16
85°		1.40	1.70	2.10	2.64	3.40	4.51	6.23	9.12
90°		1.48	1.83	2.31	2.96	3.90	5.32	7.61	11.6
95°		1.55	1.96	2.52	3.30	4.46	6.27	9.25	14.6
100°		1.61	2.07	2.72	3.66	5.08	7.35	11.2	18.5
105°		1.65	2.18	2.93	4.05	5.77	8.62	13.7	23.4
110°		1.68	2.27	3.14	4.45	6.54	10.1	16.6	29.7
115°		1.70	2.35	3.34	4.87	7.40	11.8	20.2	37.7
120°		1.70	2.42	3.53	5.32	8.34	13.8	24.5	47.8
125°		1.68	2.46	3.70	5.77	9.37	16.1	29.7	60.6
130°		1.64	2.47	3.85	6.21	10.5	18.7	35.9	76.5
135°		1.58	2.46	3.95	6.62	11.6	21.6	43.2	96.0

α $\dfrac{\beta}{\phi'} = -0.6$

45°									
50°									
55°									
60°			0.880	0.953	1.03	1.12	1.22	1.32	1.44
65°	0.874	0.943	1.03	1.13	1.25	1.38	1.54	1.73	1.97
70°	0.970	1.08	1.19	1.33	1.50	1.70	1.94	2.25	2.67
75°	1.08	1.21	1.37	1.55	1.78	2.06	2.42	2.91	3.61
80°	1.18	1.35	1.55	1.79	2.09	2.48	3.00	3.74	4.85
85°	1.28	1.48	1.73	2.04	2.44	2.97	3.70	4.79	6.50
90°	1.37	1.61	1.92	2.31	2.82	3.53	4.54	6.10	8.67
95°	1.45	1.74	2.11	2.60	3.25	4.18	5.55	7.75	11.6
100°	1.51	1.86	2.30	2.90	3.72	4.92	6.77	9.83	15.4

(continued)

Table B.5 (Continued) (a) Values for the coefficient of passive earth pressure for cohesionless soils, retained by a wall with angle of wall friction $(\delta') = \phi'$, using a log spiral failure surface

ϕ'	10°	15°	20°	25°	30°	35°	40°	45°	50°
105°	1.56	1.96	2.49	3.22	4.25	5.79	8.25	12.5	20.5
110°	1.60	2.06	2.68	3.55	4.83	6.80	10.0	15.8	27.3
115°	1.62	2.14	2.86	3.90	5.47	7.97	12.2	20.1	36.5
120°	1.62	2.20	3.02	4.26	6.18	9.32	14.9	25.5	48.7
125°	1.61	2.24	3.17	4.62	6.94	10.9	18.0	32.3	64.8
130°	1.57	2.24	3.29	4.96	7.74	12.60	21.8	40.8	85.8
135°	1.50	2.22	3.37	5.28	8.56	14.45	26.1	51.1	113

$$\frac{\beta}{\phi'} = -0.8$$

α	10°	15°	20°	25°	30°	35°	40°	45°	50°
45°									
50°									
55°									
60°									
70°			0.925	0.961	0.991	1.01	1.03	1.03	1.02
75°	0.957	1.01	1.07	1.12	1.18	1.23	1.28	1.32	1.37
80°	1.05	1.13	1.21	1.30	1.39	1.48	1.59	1.70	1.83
85°	1.15	1.25	1.37	1.49	1.62	1.78	1.95	2.17	2.45
90°	1.24	1.37	1.52	1.69	1.89	2.11	2.39	2.76	3.25
95°	1.31	1.49	1.68	1.91	2.17	2.50	2.93	3.50	4.33
100°	1.38	1.59	1.84	2.13	2.49	2.95	3.57	4.43	5.75
105°	1.44	1.69	2.00	2.37	2.85	3.47	4.34	5.62	7.64
110°	1.47	1.78	2.15	2.62	3.24	4.08	5.28	7.12	10.2
115°	1.50	1.85	2.30	2.88	3.67	4.78	6.43	9.04	13.6
120°	1.50	1.90	2.43	3.15	4.15	5.60	7.82	11.5	18.1
125°	1.48	1.94	2.55	3.41	4.66	6.53	9.48	14.5	24.1
130°	1.44	1.94	2.65	3.67	5.20	7.57	11.5	18.4	31.9
135°	1.38	1.92	2.71	3.90	5.74	8.71	13.7	23.0	42.0

$$\frac{\beta}{\phi'} = -1.0$$

α									
45°									
50°									
55°									
60°									
65°									
70°									
75°									

(continued)

Table B.5 (Continued) (a) Values for the coefficient of passive earth pressure for cohesionless soils, retained by a wall with angle of wall friction (δ') = ϕ', using a log spiral failure surface

ϕ'	10°	15°	20°	25°	30°	35°	40°	45°	50°
75°									
80°									
85°									
90°	0.982	0.966	0.940	0.906	0.866	0.819	0.766	0.707	0.643
95°	1.04	1.04	1.03	1.01	0.985	0.956	0.922	0.883	0.840
100°	1.08	1.10	1.11	1.12	1.12	1.11	1.11	1.10	1.10
115°	1.12	1.16	1.20	1.23	1.26	1.30	1.34	1.38	1.45
120°	1.15	1.22	1.28	1.35	1.43	1.51	1.61	1.74	1.91
125°	1.16	1.26	1.37	1.48	1.61	1.77	1.95	2.20	2.53
130°	1.16	1.30	1.45	1.62	1.82	2.06	2.37	2.78	3.37
135°	1.15	1.32	1.52	1.75	2.04	2.41	2.88	3.53	4.49
140°	1.12	1.33	1.58	1.89	2.29	2.80	3.49	4.47	5.97
145°	1.08	1.31	1.62	2.02	2.54	3.24	4.21	5.63	7.90

After: Caquot, A. and Kerisel, J., *Tables for the Calculation of Passive Pressure, Active Pressure and Bearing Capacity of Foundations*, Gauthier-Villars, Paris, 1948.

Table B.5 (b) Reduction factors to be applied to K_p values in Table B.5 (a), to allow for $\delta' < \phi'$

$\dfrac{\delta'}{\phi'}$	1.0	0.9	0.8	0.7	0.6	0.5	0.4	0.3	0.2	0.1	0.0
ϕ'											
10	1.00	0.991	0.989	0.978	0.962	0.946	0.929	0.912	0.898	0.881	0.864
15	1.00	0.986	0.979	0.961	0.934	0.907	0.881	0.854	0.830	0.803	0.775
20	1.00	0.983	0.968	0.939	0.901	0.862	0.824	0.787	0.752	0.716	0.678
25	1.00	0.980	0.954	0.912	0.860	0.808	0.759	0.711	0.666	0.620	0.574
30	1.00	0.980	0.937	0.878	0.811	0.746	0.686	0.627	0.574	0.520	0.467
35	1.00	0.980	0.916	0.836	0.752	0.674	0.603	0.536	0.475	0.417	0.362
40	1.00	0.980	0.886	0.783	0.682	0.592	0.512	0.439	0.375	0.316	0.262
45	1.00	0.979	0.848	0.718	0.600	0.500	0.414	0.339	0.276	0.221	0.174
50	1.00	0.975	0.797	0.638	0.506	0.399	0.313	0.242	0.185	0.138	0.102
55	1.00	0.966	0.731	0.543	0.401	0.295	0.215	0.153	0.108	0.0737	0.0492
60	1.00	0.948	0.647	0.434	0.290	0.193	0.127	0.0809	0.0505	0.0301	0.0178

After: Caquot, A. and Kerisel, J., *Tables for the Calculation of Passive Pressure, Active Pressure and Bearing Capacity of Foundations*, Gauthier-Villars, Paris, 1948.

Table B.6 Coefficients of active earth pressure for cohesive soils with a horizontal surface ($\beta = 0$) and vertical back of wall ($\alpha = 90°$), with wall friction (δ') and adhesion (c'_w), derived with no allowance for tension cracks using an extended Coulomb wedge analysis by Packshaw (1946), as quoted in CP2:1951

			ϕ'					
Coeff.	δ'	c'_w/c'	0°	5°	10°	15°	20°	25°
K_a	0	All	1.00	0.85	0.70	0.59	0.48	0.40
K_{ac}	ϕ'	values	1.00	0.78	0.64	0.50	0.40	0.32
	0	0	2.00	1.83	1.68	1.54	1.40	1.29
	0	1.0	2.83	2.60	2.38	2.16	1.96	1.76
	ϕ'	0.5	2.45	2.10	1.82	1.55	1.32	1.15
	ϕ'	1.0	2.83	2.47	2.13	1.85	1.59	1.41

Table B.7 Coefficients of passive earth pressure for cohesive soils with a horizontal surface ($\beta = 0$) and vertical back of wall ($\alpha = 90°$), with wall friction (δ') and adhesion (c'_w), derived using the friction circle graphical method by Packshaw (1946), as quoted in CP2:1951

			ϕ'					
Coeff.	δ'	c'_w/c'	0°	5°	10°	15°	20°	25°
	0	All	1.0	1.2	1.4	1.7	2.1	2.5
K_a	ϕ'	values	1.0	1.3	1.6	2.2	2.9	3.9
K_{pc}	0	0	2.0	2.2	2.4	2.6	2.8	3.1
	0	0.5	2.4	2.6	2.9	3.2	3.5	3.8
	0	1.0	2.6	2.9	3.2	3.6	4.0	4.4
	ϕ'	0.5	2.4	2.8	3.3	3.8	4.5	5.5
	ϕ'	1.0	2.6	2.9	3.4	3.9	4.7	5.7

After: Packshaw, S., J. Instn. Civ. Engrs., 25, 233–256, 1946.

Table B.8 Coefficients of passive earth pressure for cohesionless soils with a horizontal surface ($\beta = 0$), vertical back of wall ($\alpha = 90°$), and wall friction (δ', derived using the 'friction circle' graphical method and given in CP2:1951

δ' (degrees)	ϕ' (degrees)			
	25	30	35	40
0	2.5	3.0	3.7	4.6
10	3.1	4.0	4.8	6.5
20	3.7	4.9	6.0	8.8
30		5.8	7.3	11.4

After: Graphical method as given in BS Code of Practice No. 2 (1951) *Earth Retaining Structures.* Institution of Structural Engineers, London.

These values have been criticized by Rowe and Peaker (1965) on the basis that progressive failure in dense sands, and the need to limit wall deflections in loose sands, lead to much lower values in practice. A 'factor of safety' on the coefficient of passive earth pressure of about 1.5 is required to allow for these phenomena, if the peak triaxial effective angle of friction is used for ϕ'.

B.2 NOTES ON DERIVATION OF TABLES

Notes on derivation of tables

q'_z = effective stress normal to back of wall
γ = bulk unit weight of soil

Comments on derivation	Use
Table B.I	
Theoretical values from Rankine's (1857) analysis, for granular soil with a horizontal ground surface. The analysis assumes δ' (= the angle of wall friction) = 0. These values are only relevant for	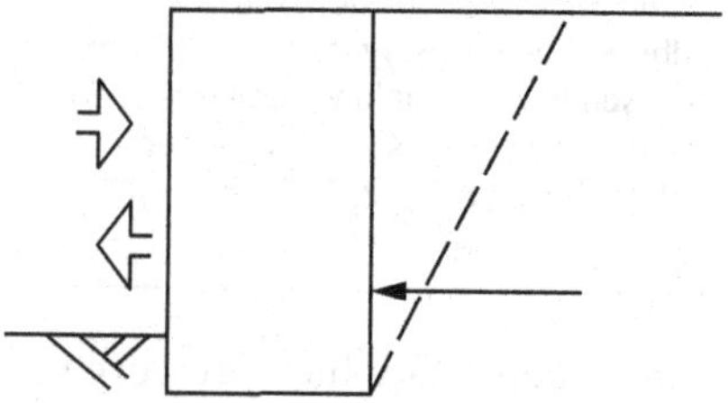

(continued)

Comments on derivation	Use
(a) smooth walls	
(b) walls subjected to vibration	
(c) L or T cantilever walls with a long heel	
(d) walls which cannot sustain vertical loads	
There is no shear stress on the back of the vertical wall.	$q'_z = (\gamma \cdot z - u)K_a$ $\tau = 0$

Table B.2

Theoretical values from Mayniel's (1808) analysis, based upon a planar active failure surface, and friction between a granular soil and the back of the wall. Ground surface is horizontal and the back of the wall is vertical. The resultant force is inclined at δ' to the normal to the back of the wall.	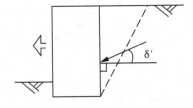
Potts and Burland (1983) suggest that when effective cohesion and wall adhesion are present, these values may be used in the Rankine–Bell equation (i.e. $q'_z = (\gamma \cdot z - u).K_a - c'.K_{ac}$ – see below), using $K_{ac} = 2(K_a(1 + c'_w/c'))^{1/2}$	$q'_z = (\gamma.z - u)K_a \cos\delta'$ $\tau = q'_z \sin\delta'$

Table B.3

Theoretical values from Müller-Breslau's (1906) analysis. Based on a planar active failure surface, but with inclined ground surface and back of wall. The soil is purely frictional, and the resultant force acts at δ' to the normal to the back of the wall. Note that β cannot exceed ϕ', because of slope instability.	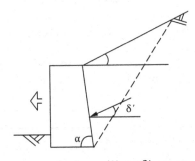
Potts and Burland (1983) suggest that when effective cohesion and wall adhesion are present, these; values may be used in (the Rankine–Bell equation (i.e. $q'_z = (\gamma \cdot z - u).K_a - c'.K_{ac}$ – see below), using $K_{ac} = 2(K_a(1 + c'_w/c'))^{1/2}$	$q'_z = (\gamma.z - u)K_a \cos\delta'$ $\tau = q'_z \sin\delta'$

(continued)

z = depth at which pressure is calculated
u = pore water pressure at depth z
K = earth pressure coefficient (K_a-active, K_p-passive)

Comments on derivation	Use

Table B.4

Theoretical values from Bell's (1915) analysis. Soil is both cohesive and frictional, but in the Rankine state where the ground surface is horizontal and therefore the effective angle of wall friction, δ', is zero.

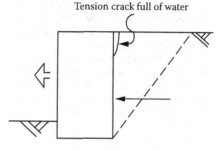

Tension crack full of water

The back of the wall is vertical, and there is no adhesion between the soil and the back of the wall.

$$q'_z = (\gamma.z - u)K_a - c'.K_a$$

for active case.

Calculations will produce tension next to the crest of the wall. The effective stress in this zone should be taken as equal to zero, as tension cracks will form. Tension cracks should be assumed to be full of water, for the worst case.

$$q'_z = (\gamma \cdot z - u)K_p + c'.K_{pc}$$

For passive case

$$\tau = 0$$

Table B.5

Values of passive earth pressure coefficient derived using a log spiral by Caquot and Kensel (1948).

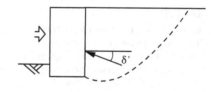

The soil is purely frictional, but Potts and Burland (1983) suggest that values of K_{pc} for use in the Rankine–Bell equation, $q'_z = (\gamma.z - u)K_p + c'.K_{pc}$, can be taken as $K_{pc} = 2(K_p(1 + c'_w/c'))^{1/2}$ when cohesion and wall friction are present. The wall is rough, the failure surface is non-planar, and the coefficients give the pressure inclined at to δ' the normal to the back of the wall.

$$q'_z = (\gamma.z - u)K_a \cos \delta'$$

$$\tau = (\gamma.z - u)K_p \sin \delta'$$

(continued)

τ = shear stress on back of wall at depth z.

Comments on derivation	*Use*

Tables B.6, B.7[a]

Values of active and passive earth pressure
as presented by Packshaw (1946). Active
values are derived from graphical analysis,
using a planar failure surface.

The soil is frictional or cohesive, has a
horizontal upper surface, and is retained
by a rough vertical wall. Wall adhesion is
allowed for. In the active case there is no
allowance for tension cracks.

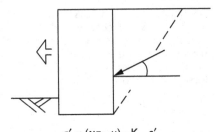

$$q'_z = (\gamma z - u) - K_{ac}.c'$$

$$\tau = q'_z . \tan\delta' + c'_w$$

Passive values are obtained from the
'ϕ-circle' method. The results allow
analysis of a general case for a frictional
or cohesive soil with a horizontal ground
surface, retained by a rough, adhesive,
vertical wall. The failure surface is
non-planar.

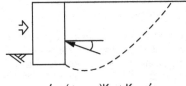

$$q'_z = (\gamma.z - u)K_p + K_{pc}.c'$$

$$\tau = q'_z . \tan\delta' + c'_w$$

Table B.8[b]

Passive pressure coefficients given in
CP2:1951, derived by the 'ϕ-circle'
method for a purely frictional soil, with a
horizontal ground surface and rough
vertical wall. The failure surface is
non-planar.

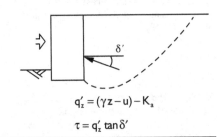

$$q'_z = (\gamma z - u) - K_a$$

$$\tau = q'_z \tan\delta'$$

[a] In both tables, the coefficients quoted will give the horizontal component of the lateral thrust.
[b] Tabulated coefficients will give the horizontal component of thrust.

References

American Concrete Institute. 1992. Prediction of creep, shrinkage and temperature effects in concrete structures, ACI209R.

Anon. 2006. *A Designers Simple Guide to BS EN 1997*. UK Department for Communities and Local Government. http://webarchive.nationalarchives.gov.uk/20120919132719; http://www.communities.gov.uk/documents/planningand building/pdf/153986.pdf.

ArcelorMittal. 2008. *ArcelorMittal Piling Handbook*. 8th edition. Formerly the British Steel Piling Handbook. http://www.arcelormittal.com/sheetpiling/page/index/name/arcelor-piling-handbook.

Atkinson, J.H. 1981. *Foundations and Slopes*. McGraw-Hill, Maidenhead.

Banerjee, P.K. and Butterfield, R. 1981. *Boundary Element Methods in Engineering Science*. McGraw-Hill, Maidenhead.

Barker, K.J. and Carder, D.R. 2006. The long term monitoring of stresses behind three integral bridge abutments, Concrete Bridge Development Group. Technical Paper 10, January, 31 pp. Unpublished Project Report UPS IS/045/05. http://www.cbdg.org.uk/pdfs/tp10.pdf. Accessed June 18, 2011.

Barley, A.D. 1997. The single bore multiple anchor system. In *Ground Anchorages and Anchored Structures* (Ed. Littlejohn, G.S.), 65–75. Thomas Telford, London.

Batten, M. and Powrie, W. 2000. Measurement and analysis of temporary prop loads at Canary Wharf underground station, east London. *Proc. Instn. Civ. Engrs. Geotech. Engg.* 143, 151–163.

Bell, A.L. 1915. The lateral pressure and resistance of clay and the supporting power of clay foundations. *Min. Proc. ICE* 199 (1), 233–272.

Bica, A.V.D. and Clayton, C.R.I. 1992. The preliminary design of free embedded cantilever walls in granular soil. *Proc. Int. Conf. on Retaining Structures*, Cambridge, 731–740.

Bishop, A.W. 1955. The use of the slip circle in the stability analysis of slopes. *Géotechnique* 5, 7–17.

Bishop, A.W. and Morgenstern, N. 1960. Stability coefficients for earth slopes. *Geotechnique* 10, 129–150.

Bishop, A.W., Webb, D.C. and Lewin, P.I. 1965. Undisturbed samples of London Clay from Ashford Common Shaft: Strength/effective stress relationships. *Géotechnique* 15, 1–31.

Bjerrum, L. and Eide, O. 1956. Stability of strutted excavations in clay. *Géotechnique* 6, 32–47.

Black, W.P.M. 1962. A method of estimating the California bearing ratio of cohesive soil fromplasticity data. *Géotechnique* 12 (4), 271–282.

Black, W.P.M. and Lister, N.W. 1979. The strength of clay fill subgrades: Its prediction and relation to road performance. *Proc. ICE Conf. on Clay Fills*, London, 37–48.

Bloodworth, A.G., Xu, M., Banks, J.R. and Clayton, C.R.I. 2011. Predicting the earth pressure on integral bridge abutments. *ASCE Journal of Bridge Engineering.* doi: 10.1061/(ASCE)BE.1943-5592.0000263.

Blum, H. 1931. Einspannungsverhdltnisse bei Bohlwerken. Wilh. Ernst und Sohn, Berlin.

Blum, H. 1950. Beitrag zur Berechnung von Bohlwerken. Die Bautechnik 27 (2), 45.

Blum, H. 1951. Beitrag zur Berechnung von Bohlwerken unter Beriicksichtigung der Wandverforming, insbesondere bei mil der Tiefe linear zunehmender Widerstandsziffer. Wilh. Ernst und Sohn, Berlin.

Bolton, M.D. 1986. The strength and dilatancy of sands. *Géotechnique* 36, 1, 65–78.

Bolton, M.D. and Powrie, W. 1987. Collapse of diaphragm walls retaining clay. *Géotechnique* 37 (3), 335–353.

Bond, A. and Harris, A. 2008. *Decoding Eurocode 7.* Taylor and Francis, 598 pp.

Borin, D.L. 1988. WALLAP Anchored and Cantilevered Retaining Wall Analysis Program: User's Manual (Version 3). Geosolve, London.

Boussinesq, J. 1885. Application des potentials a l'etude de l'equilibre et du mouvement des solides elastiques. Gauthier-Villars, Paris.

Bowles, J.E. 1968. *Foundation Analysis and Design.* McGraw-Hill, New York.

Bowles, J.E. 1974. *Analytical and Computer Methods in Foundation Engineering.* McGraw-Hill, New York.

Briaud, J.L. and Lim, Y. 1997. Soil nailed wall under piled bridge abutment: Simulation and guidelines. *J. Geotech. Geoenviron. Eng.* 123 (11), 1043–1050.

Brinch Hansen, J. 1953. Earth pressure calculation; Application of a new theory of rupture to the calculation and design of retaining walls, anchor slabs, free sheet walls, anchored sheet walls, fixed sheet walls, braced walls, double sheet walls and cellular cofferdams. Thesis., Danish Technical Press (Copenhagen) (for the Institution of Danish Civil Engineers).

Brinch Hansen, J. 1958. *Earth Pressure Calculation.* Danish Technical Press, Copenhagen.

Brinch Hansen, J. 1961. The ultimate resistance of rigid piles against transverse forces. *Dansk Geoteknisk Inst. Bull* 12, 5–9.

Bromhead, E.N. 1979. A simple ring shear apparatus. *Ground Engineering* 12 (5), 40, 42, 44.

Bromhead, E.N. 1998. *The Stability of Slopes.* Spon, London.

Broms, B.B. 1965. Lateral resistance of piles in cohesionless soils. *Proc. ASCE, J. Soil Mech. Found Div.* 90 (SM3), 123–156.

Broms, B.B. 1968. Swedish tie-back system for sheet pile walls. *Proc. 3rd Budapest Conf. Soil Mech. Found. Engg.*, Budapest, 391–403.

Broms, B. 1971. Lateral pressure due to compaction of cohesionless soils. *In Proc. 4th Int. Conf. Soil Mech. Found. Engg.*, Budapest, 373–384.

Bruce, D.A. and Jewell, R.A. 1986. Soil nailing: Application and practice. *Ground Engineering*, Part I, Nov. 86 (10–15); Part 2, Jan. 87 (21–33).

BS 1377. 1990. *Methods of Test for Soils for Civil Engineering Purposes*. British Standards Institution, London.

BS EN 1997-1. 2004. Eurocode 7—Geotechnical design—Part 1: General Rules. British Standards Institution, London.

BS EN 1997-2. 2007. Eurocode 7—Geotechnical Design—Part 2: Ground investigation and testing. British Standards Institution, London.

BS 5930. 1999. *Code of Practice for Site Investigations (formerly CP 2001)*. British Standards Institution, London.

BS 6031. 2009. *Code of Practice for Earthworks*. British Standards Institution, London.

BS 6349. 2000. *Codes of Practice for Maritime Structures, Part 1*. General Criteria. British Standards Institution, London.

BS 6349-2:2010. 2010. *Maritime Works—Part 2: Code of Practice for the Design of Quay Walls, Jetties and Dolphins*. British Standards Institution, London.

BS 8002. 1994. *Code of Practice for Earth Retaining Structures*. British Standards Institution, London.

BS 8004. 1986. *Code of Practice for Foundations*. British Standards Institution, London.

BS 8006-1. 2010. *Code of Practice for Strengthened/Reinforced Soils and Other Fills*. British Standards Institution, London.

BS 8006-2. 2011. *Code of Practice for Strengthened/Reinforced Soils. Soil Nail Design*. British Standards Institution, London.

BS 8081. 1989. *Code of Practice for Ground Anchorages*. British Standards Institution, London.

BS EN 1992-1-1:2004 Eurocode 2: Design of concrete structures. General rules and rules for buildings British Standards Institution, London.

BSC 1997. *Piling Handbook*. British Steel Corporation.

BS Code of Practice No. 2. 1951. *Earth Retaining Structures*. Institution of Structural Engineers, London.

Buisman, A.S.K. 1940. Grondmechanica. Waltman, Delft.

Bullet, P. 1691. L'architecture practique. Paris.

Burland, J.B. and Burbidge, M.C. 1985. Settlement of foundations on sand and gravel. *Proc. I.C.E.*, Part 1, 78, 1325–1371.

Burland, J.B. and Fourie, A.B. 1985. The testing of soils under conditions of passive stress relief. *Géotechnique* 35 (2), 193–198.

Burland, J.B. and Hancock, R.J.R. 1977. Underground car park at the House of Commons: Geotechnical aspects. *The Structural Engineer* 55 (2), 87–100.

Burland, J.B., Potts, D.M. and Walsh, N.M. 1981. The overall stability of free and propped cantilever retaining walls. *Ground Engineering* July, 28–38.

Burland, J.B., Simpson, B. and St John, H.D. 1979. Movements around excavations in London clay. Proc. 7th Eur. Conf Soil Mech., Brighton, 1, 13–29.

Butler, F.G. 1975. Heavily over-consolidated clays. Review paper: Session III. Conf. on Settlement of Structures, Pentech Press, London.

Byrne, R.J., Cotton, D., Porterfield, J., Wolschlag, C. and Ueblacker, G. 1998. Manual for Design and Construction Monitoring of Soil Nail Walls, Report FHWA-SA-96-69R, Federal Highway Administration, Washington, DC.

Canadian Geotechnical Society. 1978. *Canadian Foundation Engineering Manual*. The Foundations Committee, Canadian Geotechnical Society.

Caquot, A. and Kerisel, J. 1948. Tables for the Calculation of Passive Pressure, Active Pressure and Bearing Capacity of Foundations. Gauthier-Villars, Paris.

Caquot, A. and Kerisel, J. 1953. Traite de mecanique des sols. Gauthier-Villars, Paris.

Carder, D.R. 2005. Design guidance on the use of a row of spaced piles to stabilise clay highway slopes. *TRRL Report LR632*. Transport and Road Research Laboratory, Crowthorne, UK.

Carder, D.R., Murray, R.T. and Krawczyk, J.V. 1980. Earth pressure against an experimental retaining wall backfilled with silty clay. *TRRL Report LR946*. Transport and Road Research Laboratory, Crowthorne, UK.

Carder, D.R., Pocock, R.G. and Murray, R.T. 1977. Experimental Retaining Wall Facility—Lateral Stress Measurements with Sand Backfill. *TRRL Report LR766*. Transport and Road Research Laboratory, Crowthorne, UK.

Carder, D.R. and Symons, I.F. 1989. Long-term performance of an embedded cantilever retaining wall in stiff clay. *Géotechnique* 39, 55–76.

CD 209. 1980. Crib walling and notes. New Zealand Specification MWD.

Cedergren, H.R. 1989. *Seepage, Drainage and Flow Nets*. 3rd edn. John Wiley, New York.

Cernica, J.H. 1995. *Foundation Design*, John Wiley & Sons Ltd., New York, USA.

Chandler, R.J. 1984. Recent European experience of landslides in overconsolidated clays and soft rocks. State-of-the-Art Lecture. In Proc. IVth Int. Symp. on Landslides, 2, 61–81.

Chandler, R.J. and Skempton, A.W. 1974. The design of permanent cutting slopes in stiff fissured clays. *Géotechnique* 24 (4), 457–466.

Chettoe, C.S. and Adams, H.C. 1938. *Reinforced Concrete Bridge Design*. 2nd edn. Chapman and Hall, London.

CIRIA. 1974. *A Comparison of Quay Wall Design Methods*. Report No. 54, Construction Industry Research and Information Association, London.

CIRIA. 1984. Report No. 104. See Padfield and Mair (1984).

Clayton, C. et al. 1993. *Earth Pressure and Earth-Retaining Structures*, 2nd edn. Taylor & Francis, New York.

Clayton, C.R.I. 1995. *The Standard penetration Test (SPT): Methods and Use*. Construction Industry Research and Information Association (CIRIA, London), Report R143, 144 pp.

Clayton, C.R.I. 2001. *Managing Geotechnical Risk. Improving productivity in UK building and construction*. Thomas Telford, London.

Clayton, C.R.I. 2011. Stiffness at small strain: Research and practice. *Géotechnique* 61 (1), 5–37.

Clayton, C.R.I. and Heymann, G. 2001. The stiffness of geomaterials at very small strains. *Géotechnique* 51 (3), 245–256.

Clayton, C.R.I., Matthews, M.C. and Simons, N.E. 1995. *Site Investigation. Blackwell Science*, 2nd ed., 584 pp.

Clayton, C.R.I. and Militisky, J. 1983. Installation effects and the performance of bored piles in stiff clay. *Ground Engineering* 16 (2), 17–22.

Clayton, C.R.I. and Siddique, A. 1999. Tube sampling disturbance—Forgotten truths and new perspectives. *Proc. Inst. Civil Engrs, Geotechnical Engineering*, 137, July, 127–135.

Clayton, C.R.I., Siddique, A. and Hopper, R.J. 1998. Effects of sampler design on tube sampling disturbance—Numerical and analytical investigations. *Géotechnique* 48 (6), 847–867.

Clayton, C.R.I., Simons, N.E. and Matthews, M.C. 1995. Site Investigation. 2nd edn. Blackwell Scientific, Oxford. Downloadable from http://www.geotechnique.info.

Clayton, C.R.I., Symons, I.F. and Hiedra Cobo, J.C. 1991. The pressure of clay backfill against retaining structures. Canadian Geotech. Journal 28 April, 282–297.

Clayton, C.R.I., Xu, M. and Bloodworth, A. 2006. A laboratory study of the development of earth pressure behind integral bridge abutments. Géotechnique 56 (8), 561–571.

Clough, G.W. and Denby, G.M. 1977. Stabilizing berm design for temporary walls in clay. J. Geot. Engg Div. ASCE 103 (GT2), 75–90.

Clough, G.W. and Hansen, L.A. 1977. A finite element study of the behavior of Willow Island Cofferdam. J. Geotech. Engg Div, ASCE 3 (4), 521–544.

Clough, G.W. and O'Rourke, T.D. 1990. Construction induced movements of in situ walls. Proc. of Design and Construction of Earth-Retaining Structures, Ithaca, NY ASCE GSP 25, 430–470.

CMP. 1987. BEASY User Manual. Computational Mechanics, Southampton.

Coduto, D.P. 2001. Foundation Design Principles and Practices. 2nd edn. Prentice-Hall, USA.

Committee for Waterfront Structures (EAU). 1978. Recommendations of the Committee for Waterfront Structures, EAU (1975), translation of Empfehlungen des Arbeitsausschusses 'Ufer Einfassungen' (EAU 1975). Wilh. Ernst und Sohn, Berlin.

Coulomb, C.A. 1776. Essai sur une application des regles de maximis et minimis a quelques problemes de statique, relatifs a l'architecture. Memoires de Mathematique et de Physique présentés a l'Academic Royale des Sciences, Paris, 1773, 1, 343–382.

Couplet, P. 1726–1728. De la poussee des terres contre leurs revestements, et de la force desrevestments qu'on leur doit opposer. Histoire de l'Academie Royale des Sciences, (1726, 106) (1727, 139) (1728, 113), Paris.

Cummings, E.M. 1957. Cellular cofferdams and docks. Trans. ASCE 125, 13–34.

Cundall, P. 1976. Explicit finite-difference methods in geomechanics. Proc. 2nd Int. Conf. Num. Meth. in Geomechanics, Blackshurg, 1, 132–150.

Daly, M.P. and Powrie, W. 2001. Undrained analysis of earth berms as temporary supports for embedded retaining walls. Proceedings of the Institution of Civil Engineers (Geotechnical Engineering) 149 (4), 237–248.

D'Appolonia, D.J. 1971. Effects of foundation construction on nearby structures. In Proc. 4th Pan-Am. Conf. Soil. Mech. Found. Engin, State-of-the-Art, 1, 189–236.

Darbin, M., Jailloux, J.M. and Montuelle, J. 1988. Durability of reinforced earth structures: The results of a long-term study conducted on galvanised steel. Proc. Instn Civ. Engrs, Part 1, 84, 1029–1057.

de Beer, E.E. 1949. Groundmechanica. Deel II, Funderingen. N.V. Standard Boekhandel, Antwerpen.

de Belidor, B.F. 1729. La science des ingenieurs dans la conduite des travaux de fortification etd'architecture civile. Paris.

Diakoumi, M. and Powrie, W. 2009. Relative soil/wall stiffness effects on retaining walls propped at the crest. 2nd Int. Conf. on New Developments in Soil Mechanics and Geotechnical Engineering, Near East University, Nicosia, North Cyprus, 488–495.

Di Biagio, E. and Roti, J.A. 1972. Earth pressure measurements on a braced slurry trench wall in soft clay. *In Proc. 5th Eur. Conf. Soil Mech. Found. Engg,* Madrid, 1, 473–484.

DIN 4094. 1974. Dynamic and Static Penetrometers: Dimension of Apparatus and Method of Operation, Deutsches Institut fur Normung, Berlin.

Driscoll, R., Scott, P. and Powell, J. 2008. EC7—Implications for UK practice, CIRIA C641, London, 126 pp.

DTp. 1978. Reinforced Earth Retaining Walls and Bridge Abutments for Embankments. Tech. Memo. (Bridges) BE 3/78 (see also amendments). Department of Transport.

DTp. 1987. Reinforced Earth Retaining Walls and Bridge Abutments. Technical Memorandum (Bridges) BE 3/78 (revised 1987). Department of Transport, London: HMSO.

Duncan, J.M. and Chang, C.Y. 1970. Nonlinear analysis of stress and strain in soil. *J. Soil Mech. Found. Div. ASCE* 96, 1629–1653.

Duncan, J.M. and Wright, S.G. 1980. The accuracy of equilibrium methods for slope stability analysis. In *Proc. Int. Symp. on Landslides,* New Delhi, 1, 247–254.

Durgunoglu, H.T. and Mitchell, J.K. 1975. Static penetration resistance of soils: I Analysis, II Evaluation of theory and implications for practice. *Proc. A.S.C.E. Symp. on In-situ Measurement of Soil Properties,* Rayleigh, 1, 151–171.

Edelman, T., Joustra, K., Koppejan, A.W., van der Veen, C. and van Weele, A.E. 1958. Comparative sheet piling calculations. In *Proc. Brussels Conf. on Earth Pressure Problems,* 2, 71–81.

Elias, V. and Juran, I. 1991. Soil nailing for stabilization of highway slopes and excavations. *Report FHWA-RD-89-198,* FHWA, Washington, DC.

Emerson, M. 1976. Bridge temperatures estimated from the shade temperature, *TRRL Report LR696.* Transport and Road Research Laboratory, Crowthorne, UK.

Fanourakis, G.C. and Ballim, Y. 2006. An assessment of the accuracy of nine design models for predicting creep in concrete. *J. South African Institution of Civil Engineering* 48, 4, 2–8.

Fellenius, W. 1936. Calculation of the stability of earth dams. In *Proc. 2nd Congr. on Large Dams,* 4, 455–463.

Fernie, R. and Suckling, T.P. 1996. Simplified approach for estimating lateral wall movement of embedded walls in UK ground. *Proc. Int. Conf. on Geotechnical Aspects of Underground Construction in Soft Ground,* London.

FHWA. 1991. Recommendations Clouterre, French national project report, translated into English by Federal Highways Authority, Report FHWA-SA-93–026, Washington, DC, USA.

FHWA. 1999. Ground Anchors and Anchored Systems. Geotechnical Engineering Circular No. 4, Publication No. FHWA-IF-99-015. Washington, DC. http://www.fhwa.dot.gov/engineering/geotech/library_sub.cfm?keyword=020.

FHWA. 2003. Soil Nail Walls. Report No. FHWA0-IF-03-017, FHWA, Washington, DC, USA. http://www.fhwa.dot.gov/engineering/geotech/library_sub.cfm?keyword=020.

FHWA. 2010. Design and Construction of Mechanically Stabilized Earth Walls and Reinforced Soil Slopes. Federal Highway Administration, Washington, DC, report FHWA-NHI-10-024.

Fitzgerald, M.F. 1915. Appendix II to Bell (1915). *Min. Proc. ICE* 199 (1), 269–272.

Flamant, A.A. 1892. Sur la repartition des pressions dans une solide rectangulaire chargé transversalement. *Compte Rendu à l'Académie des Sciences*, 114, 1465–1468.

Fleming, E.G.K., Weltman, A.J., Randolph, M.F. and Elson, W.K. 2008. Piling Engineering. Surrey University Press, Glasgow and London.

Fleming, W.K. and Sliwinski, Z.J. 1977. The use and influence of bentonite in bored pile construction. CIRIA Report PO-3. *Construction Industry Research and Information Association*, London.

Fleming, W.K. and Sliwinski, Z.J. 1986. The use and influence of bentonite in bored pile construction. CIRIA Report PO-3. *Construction Industry Research and Information Association*, London.

Fleming, E.G.K., Weltman, A.J., Randolph, M.F. and Elson, W.K. 2008. Piling Engineering. Third Edition, Taylor & Francis, Abingdon, U.K.

Forrest, W.S. and Orr, T.L.L. 2011. The Effect of Model Uncertainty on the Reliability of Spread Foundations. ISGSR 2011 (ed. Vogt, Schuppener, Straub and Bräu), Bundesanstalt für Wasserbau. 401–408, ISBN 978-3-939230-01-4.

Fourie, A.B. and Potts, D.M. 1989. Comparison of finite element and limiting equilibrium analyses for an embedded cantilever retaining wall. *Géotechnique* 39, 175–188.

Fookes, P.G. 1997. Geology for engineers: The geological model, prediction and performance. *Q.J. Engineering Geology and Hydrogeology* 30, 293–424.

Français. 1820. Recherches sur la poussee des terres. Memoires de l'Offtcier du Genie, No. 4.

Franzius, O. 1924. Versuche mil passivem Druck. Bauingenieur (Berlin), 314–320.

Fredlund, D.G. and Krahn, J. 1977. Comparison of slope stability methods of analysis. *Can. Geotech. J.* 14 (3), 429–440.

Fredlund, D.G. and Rahardjo, H. 1993. *Soil mechanics for unsaturated soils*. Wiley, New York, 517 pp.

Gaba, A.R., Simpson, B., Powrie, W. and Beadman, D.R. 2003. Embedded Retaining. Walls—Guidelines for Economic Design. CIRIA Report C580. *Construction Industry Research and Information Association*, London.

Gassler, G. 1991. In situ techniques of reinforced soil. Proc Int Reinforced Soil Conf. organised by the British Geotech. Society, Glasgow, (A. McGown, K. Yeo and K.Z. Andrawes, eds.), September 10–12, 185–197.

Gautier, H. 1717. Dissertation sur l'epaisseur des Culees des fonts. Paris.

GCO 1982. Guide to Retaining Wall Design. Geoguide 1, Geotechnical Control Office, Eng. Development Dept., Hong Kong.

Gere, J.M. and Timoshenko, S.P. 1991. *Mechanics of materials: Solutions Manual*. Chapman and Hall.

Goldberg, D.T., Jaworski, W.E. and Gordon, M.D. 1976. Lateral Support Systems and Underpinning: Vol. 1 Design and Construction. Fed. Highway Admin. Report No. FHWA RD-75-128.

Griffiths, D.V. and Smith, I.M. 2006. *Numerical Methods for Engineers—A programming approach*. 2nd ed. Chapman and Hall, London.

Grundbau Taschenbuch 1955. Band 1, Section 2.02, Spundwiinde by Lackner, E. (4. Berechnungsverfahren fur Spundwande). Wilh. Ernst und Sohn, Berlin.

Gunn, M.J. and Clayton, C.R.I. 1992. Installation effects and their importance in the design of earth-retaining structures. *Géotechnique* 42, 137–141.

Gunn, M.J., Satkunananthan, A. and Clayton, C.R.I. 1992. Finite element modelling of installation effects. *Proc. I.C.E. Conf. on Retaining Structures*, Cambridge, 46–55.

Haefeli, R. 1948. The stability of slopes acted upon by parallel seepage. *Proc. 2nd Int. Conf. on Soil Mech. Found. Engg.*, Rotterdam, 1, 57–62.

Hagerty, D.J. and Nofal, M.M. 1992. Normalisation of analytical results for anchored bulkhead design. *Proc. Int. Cont. on Retaining Structures*, Cambridge, 741–749.

Haliburton, T.A. 1968. Numerical analysis of flexible retaining structures. *J. Soil Mech. Fndn. Engg. Divn. ASCE* 94 (SM6), 1233–1251.

Hambly, E.C. 1979. Bridge Foundations and Sub-Structures. Building Research Establishment Report, HMSO, London.

Harr, M.E. 1962. *Groundwater and Seepage*. McGraw-Hill, New York.

Haseltine, B.A. and Tutt, J.N. 1977. *Brickwork Retaining Watts*. The Brick Development Association, Windsor, 32 pp. Available for download from http://www.brick.org.uk.

Hausmann, M.R. 1990. *Engineering Principles of Ground Modification*. McGraw-Hill, New York.

Head, J.M. and Wynne, C.P. 1985. Designing retaining walls embedded in stiff clay. *Ground Engineering* 18 (3), 30, 32, 33.

Health and Safety Executive. 2007. Managing health and safety in construction, Construction (Design & Management) Regulations 2007. Approved Code of Practice L144 HSB Books 2007. ISBN 978 0 7176 6223 4.

Hendron, A.J. 1963. *The Behaviour of Sand in One-Dimensional Compression*. Ph.D. Thesis, University of Illinois.

Henkel, D.J. 1971. The Calculation of Earth Pressures in Open Cuts in Soft Clays. *The Arup Journal* 6 (4), 14–15.

Hetenyi, M. 1946. *Beams on Elastic Foundations*. Ann Arbour, University of Michigan Press.

Heyman, J. 1972. Coulomb's Memoir on Statics. An Essay in the History of Civil Engineering. Cambridge University Press, Cambridge.

Higgins, K.G., Potts, D.M. and Symons, I.F. 1989. Comparison of predicted and measured performance of the retaining walls of the Bell Common tunnel. Transport and Road Research Laboratory, Contractor Report 124.

Hight, D.W. 1986. Laboratory testing: Assessing BS5930. In Site investigation practice: Assessing BS5930. *Eng. Geol. Special Pub. No. 2*, Geological Society, 43–51.

Hill, R. 1950. *The Mathematical Theory of Plasticity*. Oxford University Press.

Holl, D.L. 1941. Plane strain distribution of stress in elastic media. *Iowa Engin. Exp. Station Bulletin* 148–163.

Hooper, J. A. 1973. *Observations on the behaviour of a pile raft foundation on London Clay*. Proc. Inst. Civ. Engrs., Part 2 55, 855–877.

Hubbard, H.W., Potts, D.M., Miller, D. and Burland, J.B. 1984. Design of the retaining walls for the M25 cut and cover tunnel at Bell Common. *Géotechnique* 34 (4), 495–512.

Hvorslev, M.J. 1949. *Subsurface Exploration and Sampling of Soils for Civil Engineering Purposes*. Waterways Experimental Station, Viicksburg, USA.

Ingold, T.S. 1978. Some simplifications of Coulomb's active earth pressure theory. *Ground Engineering* 11 (4), 23, 24, 42.

Ingold, T.S. 1979. The effects of compaction on retaining walls. *Geotechnique* 29 (3), 265–283.

Ingold, T.S. 1980. Author's reply, *Proc. ASCE, J. Geotech. Engg. Div.* 106 (GT9), 1066–1068.

Iqbal, Q. 2009. *The performance of diaphragm type cellular cofferdams.* PhD thesis, University of Southampton.

Jaky, J. 1944. The coefficient of earth pressure at rest. *J. Soc. Hungarian Architects and Engrs.* 78 (22), 355–358. For translation see Fraser, A.M. (1957), Appendix I of The Influence of Stress Ratio on Compressibility and Pore Pressure Coefficients in Compacted Soils, Ph.D. Thesis, University of London.

Janbu, N. 1973. Slope stability computations. In 'Embankment dam engineering. Casagrande Memorial Volume' eds. Hirschfield and Poulos, Wiley, New York, 47–86.

Jardine, R., Potts, D.M., Fourie, A.B. and Burland, J.B. 1986. Studies of the influence of non-linear stress-strain characteristics in soil-structure interaction. *Géotechnique* 36 (3), 377–396.

Jaworski, W.E. 1973. *Evaluation of the Performance of Braced Excavations.* Ph.D. Thesis, Massachusetts Institute of Technology.

Jewell, R.A., Milligan, G.W.E., Sarsby, R.W. and DuBois, D.D. 1984. Interaction between soil and geogrids. *Polymer Grid Reinforcement in Civil Engineering.* Thomas Telford, London, 70–81.

Jewell, R.A. and Wroth, C.P. 1987. Direct shear tests on reinforced sand. *Géotechnique* 37 (1), 53–68.

John, N.W.M. 1987. *Geotextiles.* Blackie, Glasgow.

Jones, C.J.F.P. 1985. *Earth Reinforcement and Soil Structure.* Butterworth, London.

Kenney, T.C. 1959. Discussion. *Proc. ASCE, J. Soil Mech. and Found. Div.* 85 (SM3), 67–79.

Kerisel, J. and Absi, E. 1990. *Active and passive earth pressure tables.* Taylor & Francis, 236 pp.

Kerisel, J. and Persoz, B. 1978. Coulomb and Rheology, Cahiers du Groupe Francaise de Rheologie (in French), 4, 243–247.

Kezdi, A. 1972. Stability of rigid structures. *Proc. 5th Eur. Conf. Soil Mech. Found. Engg.*, 2, 105–130.

Kranz, E. 1953. Uber die Verankeruny von Spundwiinden, 2. Aufl. Berlin.

Krebs Ovesen, N. 1962. *Cellular cofferdams, calculation methods and model tests.* The Danish Geotechnical Institute, Bulletin No. 14, Copenhagen. http://www.geoteknisk.dk/media/4919/geo.dgi.bulletin.no.14.pdf.

Krebs Ovesen, N. 1964. Anchor Slabs: Calculation Methods and Model Tests. Danish Geotech. Inst. Bulletin No. 16, Copenhagen.

Krey, H. 1936. Erddruck, Erdwiderstand und Tragfahigkeit des Baugraundes. 5th edn. Wilh. Ernst und Sohn, Berlin.

Kutmen, G. 1986. The influence of the construction process of bored piles and diaphragm walls: A numerical study. M.Phil thesis, University of Surrey.

Lambe, T.W. 1970. Braced excavations. In *Proc. ASCE Specialty Conf. on Lateral Stresses in the Ground and Design of Earth-Retaining Structures*, 149–218.

Lambe, T.W., Holfskill, L.A. and Hong, I.H. 1970. Measured performance of braced excavations. *Proc. ASCE, J. Soil Mech. Found. Div.* 26, 817–836.

Lambe, T.W. and Whitman, R.V. 1969. *Soil Mechanics.* John Wiley, New York.

Lamboj, L. and Fang, H.Y. 1970. Comparison of maximum moment, tie rod force and embedment depth of anchored sheet pile. Fritz Engin. Lab. Report No. 365.1, Leligh University, Bethlehem.

La Rochelle, P. and Marsal, R.J. 1981. Slope stability, State-of-Art Report. In *Proc. 10th Int. Conf. Soil Mech. Found. Engg.*, 4, 485–507.

Lehane, B. M. 1999. Predicting the restraint provided to integral bridge deck expansion. *Proc. 12th Eur. Conf. Soil Mech. Geotech. Engg.*, Amsterdam 2, 797–802.

Lindsell, P. 1985. Monitoring of Spill-through Bridge Abutments. Final Internal Report, Dept. of Civil Engineering, University of Surrey, UK.

Littlejohn, G.S. 1990. Ground anchorage practice. *Proc. ASCE Conf., Geotech. Special Pub. 25*, (Eds. Lambe, P.C. and Hansen, L.A.) 692–733.

Lohmeyer, E. 1930. Der Grundbau, 4th edn. Ernst & Sohn, Berlin.

Love, A.E.H. 1927. *A Treatise on the Mathematical Theory of Elasticity.* 4th edn. Cambridge University Press.

Lundgren, H. and Brinch Hausen, J. 1958. Geoteknik. Teknisk Forlag, Copenhagen.

Lunne, T., Robertson, P.K. and Powell, J.J.M. 1997. *Cone Penetration Testing in Geotechnical Practice* Spon, London, 312 pp.

Lupini, J.F., Skinner, A.E. and Vaughan, P.R. 1981. The drained residual strength of cohesive soils. *Géotechnique.* 31 (2), 181–213.

Maitland, J.K. and Schroeder, W.L. 1979. Model study of circular sheet pile cells. *J. Geotech. Engg. Div., ASCE.* 105 (GT7), 805–821.

Mana, A.I. and Clough, G.W. 1981. Prediction of movements for braced cuts in clay. *J. Geot. Engg. Divn. ASCE.* 107 (GT6), 759–778.

Marsland, A. 1972. Clays subjected to in situ plate tests. *Ground Engineering 5* (6), 24–31.

Masters-Williams, H., Heap, A., Kitts, H., Greenshaw, L., Davis, S., Fisher, P., Hendry, M. and Owens, D. 2001. Control of water pollution from construction sites. Guidance for consultants and contractors. *Construction Industry Research and Information Association report C532*, 256 pp. CIRIA, London.

Matlock, H. and Reese, L.C. 1960. Generalized solutions for laterally loaded piles. *J. Soil Mech. Fndn. Engg. Divn., ASCE 86* (SM5), 63–91.

Mawditt, J.M. 1989. Discussion on Symons and Murray. *Proc. Instn Civ. Engrs*, Part 1, 86, October, 980–986.

Mayniel, K. 1808. Traite experimental, analytique et practique de la poussee des terres et des murs de revetement. Paris.

Mayne, P.W. and Kulhawy, F.H. 1982. K0-OCR relationships in soil. *J. Geot. Eng. Div. ASCE* 108 (GT6), 851–872.

McClelland, B. and Focht, J.A. 1958. Soil modulus for laterally loaded piles. Transactions ASCE 123, 1049–1086.

MdT. 1979. Les ouvrages en terre armée. Recommendations et règles de l'art. LCPC, Service d'etudes techniques des routes et autoroutes, Ministère des Transportes, France.

Meem, J.C. 1908. The bracing of trenches and tunnels. Trans. ASCE. LX, June, 1.

Meigh, A.C. 1987. Cone penetration testing—Methods and interpretation. CIRIA Report. Butterworth, London, 141 pp.

Meyerhof, G.G. 1953. The bearing capacity of foundations under eccentric and inclined loads. In *Proc. 3rd Int. Conf. Soil Mech. Found. Engg.*, 1, 440–445.

Meyerhof, G.G. 1976. Bearing capacity and settlement of pile foundations. *J. Geot. Engg., ASCE* 102 (GT3), 197–228.

Milititsky, J.M. 1983. *Installation of bored piles in stiff clays: An experimental study of local changes in soil conditions.* Ph.D. thesis, University of Surrey.

Mindlin, R.D. (1936). Discussion: Pressure distribution on retaining walls. In *Proc. 1st Int. Conf. Soil Mech. Found. Engg.,* 3, 155–156.

Ministry of Works and Development 1980. *Crib walling and notes.* MWD New Zealand Specification CD209.

Mitchell, J.K., Guzikowski, F. and Villet, W.C.B. 1978. The measurement of soil properties in situ, present methods—Their applicability and potential. US Dept. of Energy report, Dept. of Civil Engineering, University of California, Berkeley.

Moore, S.R. 1985. Earth Pressures on Spill-Through Abutments. Ph.D. Thesis, University of Surrey.

Morgenstern, N.R. and Price, V.E. 1965. The analysis of the stability of general slip surfaces. *Géotechnique* 15, 79–93.

Morley, A.S. 1912. *Theory of Structures.* Longmans, Green & Co, London.

Moulton, H.G. 1920. Earth and rock pressures. *Trans. Am. Soc. Min. and Metall. Eng.*

Müller-Breslau, H. 1906. Erddruck auf Stutzmauert. Alfred Kroner, Stuttgart.

Murray, R.T. 1980. Discussion on Ingold (1979) *Proc. ASCE J. Geotech. Eng. Div.* 106 (GT9), 1062–1066.

Murray, R.T. 1983. *Studies of the behaviour of reinforced and anchored earth.* Ph.D. Thesis, Heriot-Watt University, Edinburgh.

NAVFAC—DM7. 1982. *Design Manual: Soil Mechanics, Foundations and Earth Structures.* US Dept. of the Navy, Washington, DC.

Naylor, D.J. 1974. Stresses in nearly incompressible materials by finite elements with application to the calculation of excess pore pressures. *Int. J. Num. Meth. Engg.* 8, 443–460.

Newman, M. 1976. *Standard Cantilever Retaining Walls.* McGraw-Hill, New York.

Officine Maccaferri S.p.A. 1987. *Flexible Gabion Structures in Earth Retaining Works.* Officine Maccaferri S.p.A. Bologna, Italy, 22 pp.

Ohde, J. 1938. Zur Theorie des Erddruckes unter besonderer Beriicksichtigung der ErddruckVerteilung. *Die Bautechnik* 16, 150–159, 176–180, 241–245, 331–335, 480–487, 570–571, 753–761.

O'Rourke, T.D. 1987. Lateral stability of compressible walls. *Géotechnique* 37 (2), 145–150.

Osman, A.S. and Bolton, M.D. 2004. A new design method for retaining walls in clay. *Can. Geotechnical Journal* 41, 451–466.

Packshaw, E. 1946. Earth pressure and earth resistance. *J. Instn. Civ. Engrs.* 25, 233–256.

Packshaw, S. 1962. Cofferdams. *Proc. Instn. Civ. Engrs.* 21, 367–398.

Padfield, C.J. and Mair, R.J. 1984. Design of Retaining Walls Embedded in Stiff Clays. CIRIA Report no. 104, *Construction Industry Research and Information Association,* London.

Pappin, J.W., Simpson, B., Felton, P.J. and Raison, C. 1986. Numerical analysis of flexible retaining walls. Proceedings of the International Conference on Numerical Methods in Engineering: Theory and Applications, Swansea. January 1985, 789–802.

Pasternak, P.L. 1954. On a new method of analysis of an elastic foundation by means of two foundation constants (in Russian). Gasudarstvennoe Izdatelstvo Literaturi po Stroitelstvui Arkhitekure, Moscow, USSR.

Peck, R.B. 1969. Deep excavations and tunnelling in soft ground. In *Proc. 7th Int. Conf. Soil Mech. Found. Engg.*, State-of-the-Art volume, 225–250.

Peck, R.B., Hanson, W.E. and Thornburn, T.H. 1974. *Foundation Engineering.* John Wiley, New York.

Pennoyer, R.L. 1934. Gravity bulkheads and cellular cofferdams. *Civil Engineering* 4 (6), 301–305 and 4 (10), 547.

Petterson, K.E. 1955. The early history of circular sliding surfaces. *Géotechnique* 5 (4), 275–296.

Phear, A., Dew, C., Ozsoy, B., Wharmby, N.J., Judge, J. and Barley, A.D. 2005. Soil Nailing—Best practice. CIRIA Report C637, CIRIA, London.

Potts, D.M., Addenbrooke, T.I. and Day, R.A. 1992. The use of soil berms for temporary support of retaining walls. *Proc. Int. Conf. Retaining Structures*, Cambridge, 440–447.

Potts, D.M. and Burland, J.B. 1983. A parametric study of the stability of embedded earth retaining structures. Transport and Road Research Laboratory, Crowthorne, UK. Supplementary Report 813.

Potts, D.M. and Fourie, A.B. 1985. The effect of wall stiffness on the behaviour of a propped retaining wall: Results of a numerical experiment. *Géotechnique* 3 (3), 347–352.

Potts, D.M. and Zdravkovic, L. 2001. Finite element analysis in geotechnical engineering: Application. London, Thomas Telford.

Potts, D.M. and Zdravkovic, L. 1999. Finite element analysis in geotechnical engineering: Theory. London, Thomas Telford.

Poulos, H.G. and Davis, E.H. 1974. *Elastic Solutions for Soil and Rock Mechanics.* Wiley, New York.

Powrie, W. 2004. *Soil mechanics—Concepts and applications.* Spon, Oxford, 675 pp.

Powrie, W. and Batten, M. 2000. Comparison of measured and calculated temporary-prop loads at Canada Water Station. *Géotechnique* 50 (2), 127–140.

Powrie, W. and Daly, M.P. 2002. Centrifuge model tests on embedded retaining walls supported by earth berms. *Géotechnique* 52 (2), 89–106.

Prandtl, L. 1921. Uber die Eindringungsfestigkeit plastischer Baustoffe und die Festigkeit von Schneiden. *Zeitsch. Angew. Mathematik und Mechanik* 1 (1), 15–20.

Pratt, M. 1984. Personal communication.

Prony, R. de. 1802. Recherches sur la poussee des terres, et sur la forme et les dimensions adonner aux murs die revetement. Bull. Societe Philomatique, Paris.

Preene, M., Roberts, T.O.L., Powrie, W. and Dyer, M.R. 2000. *Groundwater control—Design and practice.* Construction Industry Research and Information Association report C515, 204 pp. CIRIA, London.

Puller, M. and Lee, C.K.T. 1996. A comparison between the design methods for earth retaining structures recommended by BS 8002:1994 and previously used methods. *Proc. I.C.E. Geotech. Engg.* 119, 29–34.

Puller, M.B. 2003. *Deep excavations: A practical manual.* Thomas Telford, London.

Rankine, W.J.M. 1857. On the stability of loose earth. *Phil. Trans. Roy. Soc. London* 147 (2), 9–27.

Rondelet, J. 1812. Traite theorique et pratique de l'art de batir. 5th edn, Paris.

Rowe, P.W. 1952. Anchored sheet-pile walls. *Proc. ICE* 1(1), 27–70.

Rowe, P.W. 1957. Sheet pile walls in clay. *Proc. ICE* 1, 629–654.

Rowe, P.W. and Peaker, K. 1965. Passive earth pressure measurements. *Géotechnique* 15, 57–78.

Saint-Venant, A.J.C.B. 1855. Mémoire sur la torsion des prismes. *Mem. Savants Etrangers* 14, 233–560.

Sarma, S.K. 1973. Stability analysis of embankments and slopes. *Géotechnique* 23 (3), 423–433.

Schroeder, W.L. and Maitland, J.K. 1979. Cellular bulkheads and cofferdams. *J. Geot. Engg. Div.* 105 (GT-7), 823–837.

Schuster, R.L., Jones, W.V., Sack, R.L. and Smart, S.M. 1975. Timber Retaining Structures. Transport Research Board, National Research Council. Special Report 160: Low Volume Roads, 116–127.

Seed, R.B. and Duncan, J. 1986. FE analyses: Compaction induced stresses and deformations. *ASCE J. Geotech. Engg.* 112 (1), 23–43.

Shannon and Wilson, Inc. 1982. Final Report Tasks 3.2, 3.3 and 3.4: Finite element models for Lock and Dam No. 26, (Replacement) First Stage Cofferdam. Technical report for US Army Engineer District St. Louis.

Simons, N., Menzies, B. and Matthews, M. 2001. Soil and rock slope engineering. Thomas Telford, London.

Simpson, B. 1984. Finite element design with particular reference to deep basements in London Clay. In *Finite Elements in Geotechnical Engineering* (eds. Naylor et al.). Pineridge Press, Swansea.

Simpson, B., Blower, T., Craig, R.N. and Wilkinson, W.B. 1989. The engineering implications of rising groundwater levels in the deep aquifer beneath London. *Construction Industry Research and Information Association report SP069M*, 116 pp. CIRIA, London.

Simpson, B., O'Riordan, N.J. and Croft, D.D. 1979. A computer model for the analysis of ground movements in London Clay. *Géotechnique* 29 (2), 149–175.

Skempton, A.W. 1951. The bearing capacity of clays. In *Proc. Building Research Congress*, London, 180–189.

Skempton, A.W. 1953. Earth pressure, retaining walls, tunnels and strutted excavations. In Proc. 3rd Int. Conf. Soil Mech. Found. Engg. 2, 353–361.

Skempton, A.W. 1961. Horizontal stresses in overconsolidated Eocene clay. *Proc. 5th Int. Conf. Soil Mech. Found. Engg.* 351–357.

Skempton, A.W. and Delory, F.A. 1957. Stability of natural slopes in London Clay. *Proc. 4th Int. Conf. Soil Mech. Found. Engng.*, London, 2, 378–381.

Smethurst, J.A. and Powrie, W. 2007. Monitoring and analysis of the bending behaviour of discrete piles used to stabilise a railway embankment. *Géotechnique* 57 (8), 663–677.

Smethurst, J.A. and Powrie, W. 2008. Effective-stress analysis of berm-supported retaining walls. *Proc. I.C.E.—Geotechnical Engineering* 161 (GE1), 39–49.

Smith, C. and Gilbert, M. 2010. Ultimate limit state design to Eurocode 7 using numerical methods. *Proc. 7th Eur. Conf. Num. Meth. Geo. Engg.*, Trondheim, 947–952.

Smith, I.M. and Griffiths, D.V. 1990. Programming the Finite Element Method. John Wiley & Sons, Chichester and New York.

Smith, J.E. 1957. *Tests on Concrete Deadman Anchorages in Sand*. ASTM Spec. Tech. Pub. no. 206.

Smith, R.J.H. and Worrall, P.K. 1991. Foundation options for Reinforced Earth on poor ground. In *Performance of Reinforced Soil Structures* (eds. McGown et al.) Thomas Telford, London, 427–432.

Sokolovski, V.V. 1960. *Statics of Soil Media*. Butterworth, London.

Soubra, A.H. and Kastner, R. 1992. Influence of seepage flow on the passive earth pressures. *Proc. ICE Conf. on Retaining Structures*, Cambridge, 67–76.

South African Code of Practice 1989. Lateral Support in Surface Excavations. South African Institution of Civil Engineers, Geotechnical Division.

Sowers, G.F., Robb, A.D., Mullis, C.J. and Glen, A.J. 1957. The residual lateral pressures produced by compacting soils. *Proc. 4th Int. Conf. Soil Mech. and Found. Engg.* 2, 243–247.

Spangler, M.G. and Mickle, J.L. 1956. Lateral pressures on retaining walls due to back fillsurface loads. *Bull. Highway Res. Board* 141, 1–18.

Spencer, E. 1967. A method of analysis for the stability of embankments assuming parallel inter-slice forces. *Géotechnique* 17, 11–26.

Springman, S.M., Norrish, A. and Ng, C.W.W. 1996. Cyclic loading of sand behind integral bridge abutments. *TRRL Report 146*. Transport and Road Research Laboratory, Crowthorne, UK.

Steinbrenner, W. 1934. Tafeln zur Setzungsberechnung. *Die Strasse* 1 (Oct.), 121–124.

Stroud, M.A. 1975. The standard penetration test in insensitive clays and soft rocks. In *Proc. Eur. Symp. on Penetration Testing*, Swedish Geotech. Soc., Stockholm, 2.2, 367–375.

Symons, I.F., Clayton, C.R.I. and Darley, P. 1989. Earth pressure against an experimental retaining wall backfilled with silty clay. *TRRL Report RR192*. Transport and Road Research Laboratory, Crowthorne, UK.

Symons, I.F. and Murray, R.T. 1988. Conventional retaining walls: Pilot and full-scale studies. Proc. Instn. Civ. Engrs., Part 1, 84, June, 519–538.

Tamaro, G.J. 1990. Slurry Wall Design and Construction, Design and Performance of Earth Retaining Structures, ASCE Conference, Cornell University, Ithaca, NY.

Taylor, D.W. 1937. Stability of earth slopes. *J. Boston Soc. Civil Engrs.* 24, 137–246.

te Kamp, W.C. 1977. Sondern en funderingen op palen in zand. Fugro Sounding Symp., Utrecht.

Tedd, P., Chard, B.M., Charles, J.A. and Symons, I.F. 1984. Behaviour of a propped embedded retaining wall in stiff clay at Bell Common Tunnel. *Géotechnique* 34 (4), 513–532.

Teng, W.C. 1962. *Foundation Design*. Prentice-Hall, New Jersey.

Terzaghi, K. 1936. A fundamental fallacy in earth pressure computations. *J. Boston, Soc. Civil Engrs.* 23, 71–88.

Terzaghi, K. 1941. General wedge theory of earth pressure. *Trans. ASCE* 106, 68–97.

Terzaghi, K. 1943. *Theoretical Soil Mechanics*, John Wiley, New York.

Terzaghi, K. 1944. Stability and stiffness and cellular cofferdams. *Proc. ASCE* Sept., 70, 1015–1050, 71 (1945), 980–995. *ASCE Transactions* 110 (1945), 1083–1202. Harvard Soil Mechanics Series, No. 26.

Terzaghi, K. 1954. Anchored bulkheads. *Trans. ASCE* 119, 1243–1280.

Terzaghi, K. 1955. Evaluation of coefficients of subgrade reaction. *Géotechnique* 5 (4), 297–326.

Terzaghi, K. and Peck, R.B. 1948. *Soil Mechanics in Engineering Practice*. 1st edn. John Wiley, New York.

Terzaghi, K. and Peck, R.B. 1967. *Soil Mechanics in Engineering Practice*. 2nd edn. John Wiley, New York.

Thorburn, S. and Smith, I.M. 1985. Major gabion wall failure. In *Failures in Earthworks*. Thomas Telford, London, 279–293.

Timoshenko, S.P. and Goodier, J.N. 1970. *Theory of Elasticity*. 3rd edn. McGraw-Hill, New York.

Timoshenko, S.P. and Woinowsky-Krieger, S. 1959. *Theory of Plates and Shells*. 2nd edn. McGraw-Hill, New York.

Tschebotarioff, G.P. 1962. Retaining structures. Ch. 5 in *Foundation Engineering* (ed. G.A. Leonards). McGraw-Hill, New York.

Tschebotarioff, G.P. 1973. *Foundations, Retaining and Earth Structures*. McGraw-Hill, New York.

Tsinker, G.P. 1983. Anchored sheet pile bulkheads: Design practice. *Proc. ASCE, J. Geotech. Div.* 109 (8), 1021–1038.

Tennessee Valley Authority (TVA) 1957. *Sheet piling cellular cofferdams on rocks*. TVA Technical Monograph No. 75, Vol. 1.

Twine, D. and Roscoe, H. 1997. Prop loads: Guidance on design. Construction Industry Research and Information Association (CIRIA) Report FR/CP/48. CIRIA, London.

Twine, D. and Roscoe, H. 1999. Temporary propping of deep excavations—Guidance on design (CIRIA Report C517). CIRIA, London.

Uriel, S. and Otero, C.S. 1977. Stress and Strain beside a circular trench wall. *Proc. 9th Int. Conf. Soil. Mech. Found. Eng.*, Tokyo, 1, 781–788.

US Army Corps of Engineers. 1989. Engineering and Design: Retaining and Flood Walls. Report EM 1110-2-2502. Dept. of the Army, Washington DC.

US Army Corps of Engineers. 1953. Filter Experiments and Design Criteria. Tech. Memo. 3–360, Waterways Exp. Station, Vicksburg, Mass.

USSI 1984. *USS Steel Sheet Piling Design Manual*. United States Steel International. Updated and reprinted by US Dept. of Transportation (FHWA).

Vauban, Marquis de 1704. Traite de l'attaque des places. Paris.

Vesic, A.S. 1975. Bearing capacity of shallow foundations. In *Foundation Engineering Handbook* (eds. H.F. Winterkorn and H.-Y. Fang), Van Nostrand Reinhold, New York, chap. 3.

Ward, W.H. 1955. Experiences with some sheet-pile cofferdams at Tilbury. Discussion. *Géotechnique* 5 (4), 327–332.

Williams B.P. & Waite D. 1993. The design and construction of sheet-piled cofferdams. CIRIA Special Publication 95, *Construction Industry Research and Information Association*, London.

Whittle, A.J. and Davis, R.V. 2006. Nicoll Highway collapse: Evaluation of geotechnical factors affecting design of excavation support system. *Int. Conf. on Deep Excavations*, Singapore.

Whittle, A.J., Hashash, Y.M.A. and Whitman, R.V. 1993. Analysis of Deep Excavation in Boston, *J. Geot. Engg.* 119 (1), Jan., 69–90.

Winkler, E. 1867. Die Lehre von der Elastizität und Festigkeit. Dominicus, Prague.

Woltmann, R. 1794. Hydraulische Architectur. Vol. 3.

Woods, R.I. 2003. *The Application of Finite Element Analysis to the Design of Embedded Retaining Walls*. PhD Thesis, University of Surrey.

Woods, R.I. and Clayton, C.R.I. 1993. Application of the CRISP finite element program to practical retaining wall design. Proc. Int. Conf. Retaining Structures, Cambridge, 101–111.

Woods, R.I. and Jewell, R.A. 1990. A computer based design method for reinforced slopes and embankments. *Int. J. Geotextiles and Geomembranes* 9 (3), 233–259.

Wrigley, N.E. 1987. Durability and long-term performance of Tensar polymer grids for soil reinforcement. *Materials Science and Technology* 3, March, 161–170.

Wu, T.H. 1975. Retaining walls. Ch. 12 in *Foundation Engineering Handbook* (eds. H.F. Winterkorn and H.-Y. Fang), Van Nostrand Reinhold, New York.

Zienkiewicz, O.C. 1977. *The Finite Element Method*. 3rd edn. McGraw-Hill, London.

Index

Page numbers followed by f and t indicate figures and tables, respectively.

Printed in the United States
by Baker & Taylor Publisher Services